The Geometry of Higher-Order Lagrange Spaces

Fundamental Theories of Physics

An International Book Series on The Fundamental Theories of Physics: Their Clarification, Development and Application

Volume 82

The Geometry of Higher-Order Lagrange Spaces

Applications to Mechanics and Physics

by

Radu Miron

Faculty of Mathematics,
"Al. I. Cuza" University of Iaşi,
Iaşi, Romania

KLUWER ACADEMIC PUBLISHERS

DORDRECHT / BOSTON / LONDON

A C.I.P. Catalogue record for this book is available from the Library of Congress

ISBN 978-90-481-4789-2

Published by Kluwer Academic Publishers,
P.O. Box 17, 3300 AA Dordrecht, The Netherlands.

Kluwer Academic Publishers incorporates
the publishing programmes of
D. Reidel, Martinus Nijhoff, Dr W. Junk and MTP Press.

Sold and distributed in the U.S.A. and Canada
by Kluwer Academic Publishers,
101 Philip Drive, Norwell, MA 02061, U.S.A.

In all other countries, sold and distributed
by Kluwer Academic Publishers Group,
P.O. Box 322, 3300 AH Dordrecht, The Netherlands.

Printed on acid-free paper

Contents

Preface

This monograph is mostly devoted to the problem of the geometrizing of Lagrangians which depend on higher order accelerations. It naturally prolongs the theme of the monograph "The Geometry of Lagrange spaces: Theory and Applications", written together with M. Anastasiei and published by Kluwer Academic Publishers in 1994.

The existence of Lagrangians of order $k > 1$ has been contemplated by mechanicists and physicists for a long time. Einstein had grasped their presence in connection with the Brownian motion. They are also present in relativistic theories based on metrics which depend on speeds and accelerations of particles or in the Hamiltonian formulation of nonlinear systems given by Korteweg–de Vries equations. There resulted from here the methods to be adopted in their theoretical treatment. One is based on the variational problem involving the integral action of the Lagrangian. A second one is derived from the axioms of Analytical Mechanics involving the Poincaré–Cartan forms.

The geometrical methods based on the study of the geometries of higher order could invigorate the whole theory. This is the way adopted by us in defining and studying the Lagrange spaces of higher order.

The problems raised by the geometrization of Lagrangians of order $k > 1$ investigated by many scholars: Ch. Ehresmann, P. Libermann, J. Pommaret; J.T. Synge, M. Crampin, P. Saunders; G.S. Asanov, P.Aringazin; I. Kolar, D. Krupka; M. de Léon, W. Sarlet, P. Cantrjin, H. Rund, W.M. Tulczyjew, A. Kawaguchi, K. Yano, K. Kondo, D. Grigore, R. Miron et al. (see References).

The geometries of higher order, defined here as the study of the category of bundles of jets $(J_0^k M, \pi, M)$, can be developed in at least two ways. One is that due to Ch. Ehresmann, based on a direct approach, without local coordinates, of the properties of objects and morphisms in the above mentioned category. Another one is that suggested by the geometries of order 1 as, for instance, those of Finsler and Lagrange spaces. In this monograph the second way was prefered. The reason is that it allows:

1. A construction of the geometry of the total space of the bundle of higher order accelerations.
2. A clear definition, for the first time, of the notion of higher order Lagrange space.
3. The solving of the old problem of prolongation of Riemann spaces and by means of it the construction of some remarkable examples of Lagrange spaces of order $k > 1$.
4. The elaboration of the geometrical ground for variational calculus involving Lagrangians which depend on higher order accelerations.
5. Notable applications in Analytical Mechanics of Higher Order and Theoretical Physics.

For the readers interested in a general view on this book, we present here a summary description of it.

Just from the beginning the bundle of k–jets $(J_0^k M, \pi, M)$ is replaced by the k–osculator bundle $(\mathrm{Osc}^k M, \pi, M)$, equivalent to the preceeding one, for its more pregnant geometrical aspect. $(\mathrm{Osc}^1 M, \pi, M)$ is identified with the tangent bundle.

In this book the emphasis falls on the elucidation of the geometry of the differentiable manifold $\mathrm{Osc}^k M$. Thus one shows that the vertical distribution V_1 decomposes in k subdistributions from the sequence of inclusions $V_1 \supset V_2 \supset \cdots \supset V_k$, that there exist k independent Liouville vector fields $\overset{1}{\Gamma}, \overset{2}{\Gamma}, ..., \overset{k}{\Gamma}$ and a natural k–tangent structure $J : \mathcal{X}(\mathrm{Osc}^k M) \longrightarrow \mathcal{X}(\mathrm{Osc}^k M)$. Then, one defines the notion of k–spray S by the equation $JS = \overset{k}{\Gamma}$. This allows the construction of a nonlinear connection N derived from S only. More precisely, S uniquely determines the dual coefficients of N. But N implies a direct decomposition

$$T_u \mathrm{Osc}^k M = N_0(u) \oplus N_1(u) \oplus \cdots \oplus N_{k-1}(u) \oplus V_k \tag{1}$$

to which all geometrics objects on $\mathrm{Osc}^k M$ are reported.

Thus N leads to the study of the simplest linear connection D on $\mathrm{Osc}^k M$, which preserves by parallelism the distribution defining N and makes J covariant constant, i.e., $DJ = 0$. It comes out that D preserves by parallelism the distributions from (1), too. The local coefficients characterizing D are in the smallest possible number.

For D one studies the parallelism of vector fields, the torsion, the curvature, the structure equations, the geometrical models on the tangent bundle to $\mathrm{Osc}^k M$ etc. In such a way the geometry of differentiable manifold $\mathrm{Osc}^k M$ appears as a well–stated field.

As a first application the problem of the prolongation of the Riemann space $R^n = (M, \gamma_{ij}(x))$ to the k–osculator bundle is solved. This problem was formulated in the Italian school of L. Bianchi and E. Bompiani, but a solution for the general case was not found up today because of the enormous difficulties in calculating the direct local coefficients of the nonlinear connection N.

Our solution, published in [197],[198],[204] follows a different way based on dual coefficients of N, constructed only with the metric tensor $\gamma_{ij}(x)$. The space $\mathrm{Prol}^k R^n = (\mathrm{Osc}^k M, G)$, where G is the Sasaki lift of $\gamma_{ij}(x)$ with respect to N, is just the solution of the above mentioned problem. With this opportunity there also was solved the problem of prolongations to $\mathrm{Osc}^k M$ of Finsler spaces $F^n = (M, F(x,y))$ and Lagrange spaces $L^n = (M, L(x,y))$. The spaces $\mathrm{Prol}^k R^n$, $\mathrm{Prol}^k F^n$, $\mathrm{Prol}^k L^n$ are excellent examples of higher order Lagrange spaces.

The geometry of the total space $\mathrm{Osc}^k M$ of the k–osculator bundle offers geometrical support for the variational problem involving functionals of the form

$$I(c) = \int_0^1 L\left(x, \frac{dx}{dt}, \frac{1}{2}\frac{d^2x}{dt^2}, ..., \frac{1}{k!}\frac{d^kx}{dt^k}\right) dt. \tag{2}$$

One derives the Euler–Lagrange equation as well as the main invariants $\overset{1}{I}(L), ..., \overset{k}{I}(L)$, which show a geometrical meaning of Zermelo's conditions on the integral in (2). In the same context, one defines the energies of order $1, 2, ..., k$ of the Lagrangian L and a remarkable conservation law is found. Introducing the notion of symmetry for L, by requiring the preservation of the variational principle with respect to the functional (2), one proves two theorems of Noether type. The applications to higher–order Analytical Mechanics are evident.

A Lagrange space of order $k \geq 1$ is a pair $L^{(k)n} = (M, L)$, where M is a real differentiable manifold of dimension n and $L : \mathrm{Osc}^k M \longrightarrow R$ is a regular Lagrangian of order k whose fundamental tensor $g_{ij} = = \frac{1}{2} \frac{\partial^2 L}{\partial y^{(k)i} \partial y^{(k)j}}$ has a constant signature. Here, $(x^i, y^{(1)i}, ..., y^{(k)i})$ are canonical local coordinates on $\mathrm{Osc}^k M$.

The geometry of $L^{(k)n}$ is thought of as the geometry of the manifold $\mathrm{Osc}^k M$ in the presence of the fundamental function $L(x^i, y^{(1)}, ..., y^{(k)})$. It is desirable the determination of a canonical k–spray S depending on the Lagrangian L only. For $k = 1$, S is derived from the Euler–Lagrange equations. This is no longer possible for $k > 1$. Replacing L by $\phi(t) L(x^i, y^{(1)}, ..., y^{(k)})$ in the Euler–Lagrange operator $\overset{0}{E_i}(L)$, with ϕ an arbitrary function, one obtains

$$\overset{0}{E_i}(\phi L) = \phi \overset{0}{E_i}(L) + \frac{d\phi}{dt} \overset{1}{E_i}(L) + \cdots + \frac{d^k \phi}{dt^k} \overset{k}{E_i}(L).$$

As it is easy to see, $\overset{0}{E_i}(L), ..., \overset{k}{E_i}(L)$ are just the covectors discovered by Craig and Synge. A surprise was for us the fact that the equation $\overset{k-1}{E}_i(L) = 0$ determines a k–spray S for the Lagrange space $L^{(k)n}$.

Based on this canonical spray, the geometry of $L^{(k)n}$ can be edified. It is to be noted that S determines a nonlinear connection which depends only on L. This produces the decomposition (1). Then the metrical N–connection D has as coefficients $k+1$ generalized Christoffel symbols. The geometrical model on $T(\mathrm{Osc}^k M)$ of the Lagrange space of order k, $L^{(k)n}$ is no longer an almost Kählerian space as for $k = 1$. This is an almost contact Riemannian space. It follows that the applications of theory in Mechanics and Physics have to be rephrased in view of this model.

Generally, this monograph is based on the original construction of the geometry of the higher order Lagrange spaces given by Miron and Atanasiu [201],[202],[204]. The variational theory of the higher order Lagrangians and its applications to Analytical Mechanics and Physics belongs to different authors (see References).

In the final part of the book we have briefly presented a new theory of subspaces $L^{(k)m}$ in $L^{(k)n}$ $(m < n)$ and a gauge theory for Lagrange space $L^{(k)n}$ as an extension of the theory of G.S. Asanov for $k = 1$, [225].

The book is divided into three parts: Lagrange spaces of order 1, Lagrange spaces of order 2 and Lagrange spaces of order $k > 2$.

The readers can go quickly into the heart of the subject carefully studying only the first two parts. For this reason, the book is accesible for readers ranging from students to researchers in Mathematics, Mechanics, Physics, Biology, Informatics, etc.

Acknowledgements. First, I wish to express my deep gratitude to Professor P.L. Antonelli (Alberta University, Canada) who supported me in developing a part of this theory at his university. He is competently promoting the Lagrange geometries. The book benefited from many personal discussions with R.G. Beil, S.S. Chern, M. Crampin, R.S. Ingarden, D. Krupka, S. Kobayashi, M. Matsumoto, R.M. Santilli, Gr. Tsagas. I feel indebted to all of them.

My sincere thanks go to M. Anastasiei (University "Al.I.Cuza" Iaşi) for the immense support granted me in the elaboration of this book. I am also indebted to Gh. Munteanu and I. Bucătaru.

Special thanks are due to Mrs. Elena Mocanu–Cosma for having typeset this book.

Finally, I should like to thank the publishers for their co–operation and courtesy.

Chapter 1

Lagrange Spaces of Order 1

The Lagrange spaces of order 1 are the smooth manifolds M endowed with a regular Lagrangian L of order 1. These spaces have appeared twenty years ago and were studied because of their applications to Mechanics, Physics, Control theory etc. They lead to geometrical models more general than those provided by Riemannian or Finslerian structures. For the extensive presentation of the geometry of these spaces we refer to the books [15], [16], [17], [195], [196].

Here, the foundations of the differential geometry of Lagrange spaces of order 1 (Lagrange spaces, for brevity) are briefly expounded as an argument for a natural extension: the theory of higher order Lagrangians.

We study here the main problems related to the geometry of the higher order Lagrangians, using, in this section, the order 1 only. Namely, we shall study:

a) The geometry of total space of the 1–osculator bundle. It includes such subjects as: spray, nonlinear connections, N–linear connections, curvatures, torsions, parallelism and structure equations.

b) The prolongations of the Riemannian structures defined on the base manifold M, to the space $TM = \mathrm{Osc}^1 M$.

c) The notion of Lagrange spaces and canonical spray, as well as a canonical N–linear connection – derived from the variational problem involving the integral of action. The almost Kählerian model of a Lagrange space is constructed.

As applications, the geometry of Finsler spaces, particularly the theory of Randers spaces are sketched.

The line of presentation is chosen so that the extension of this geometry to that of the higher order Lagrange spaces, treated in the next chapters, to appear very naturally.

Throughout the text we assume that manifolds, mappings etc. are of C^∞–class.

1.1 The Manifold $\mathrm{Osc}^1 M$. Sprays

The general theory of Lagrange spaces of order 1 is based on the geometry of total space $\mathrm{Osc}^1 M$ of the 1-osculator bundle and $\mathrm{Osc}^1 M$ will be identified to the total space of tangent bundle TM.

So, let M be a real n–dimensional manifold. Two curves in M, $\rho, \sigma : I \to M$, which have a common point $x_0 \in M$, $x_0 = \rho(0) = \sigma(0)$, $(0 \in I)$, have at x_0 a "*contact of order 1*", if for any function $f \in \mathcal{F}(U)$, $x_0 \in U$, (U being an open set in M) we have

$$\frac{d}{dt}(f \circ \rho)(t)|_{t=0} = \frac{d}{dt}(f \circ \sigma)(t)|_{t=0}. \tag{1.1}$$

The relation "*contact of order 1*" is an equivalence on the set of smooth curves in M, which pass through the point x_0. Let $[\rho]_{x_0}$ be a class of equivalence. It will be called an "*1–osculator space*" in the point $x_0 \in M$. We denote by $\mathrm{Osc}^1_{x_0}$ the set of 1–osculator spaces in x_0 of M and put

$$\mathrm{Osc}^1 M = \bigcup_{x_0 \in M} \mathrm{Osc}^1_{x_0}. \tag{1.2}$$

One considers the mapping $\pi : \mathrm{Osc}^1 M \to M$, defined by $\pi([\rho]_{x_0}) = x_0$. Clearly, π is a surjection.

The set $\mathrm{Osc}^1 M$ is endowed with a natural differentiable structure, induced by that of the manifold M so that π is a differentiable mapping which will be described below.

If (U, φ) is a local chart of M, $x_0 \in U$ and the curve $\rho : I \to M$ is represented in (U, φ) by $x^i = x^i(t)$, $t \in I$, $x_0 = (x_0^1 = x^i(0))$, taking the function f from (1.1) succesively equal to the coordinate functions x^i (throughout in the following the indices $i, j, k, h, ...$ run over the set $\{1, 2, ..., n\}$), then a representative of the class $[\rho]_{x_0}$ is given by

$$x^{*i}(t) = x^i(0) + t\frac{dx^i}{dt}(0), \quad t \in (-\varepsilon, \varepsilon) \subset I. \tag{1.3}$$

The polynomial from (1.3) is determined by its coefficients

$$x_0^i = x^i(0), \quad y_0^i = \frac{dx^i}{dt}(0). \tag{1.4}$$

So, the pair $(\pi^{-1}(U, \phi))$ $\phi([\rho]_{x_0}) = (x_0^i, y_0^i) \in R^{2n}$, $\forall [\rho]_{x_0} \in \pi^{-1}(U)$ is a local chart on $\mathrm{Osc}^1 M$. Therefore a differentiable atlass $\mathcal{A}_M$ on the manifold M determines a differentiable atlass $\mathcal{A}_{\mathrm{Osc}^1 M}$ on $\mathrm{Osc}^1 M$. The latter one gives a differentiable structure on $\mathrm{Osc}^1 M$ and the triple $(\mathrm{Osc}^1 M, \pi, M)$ is a vector bundle.

By (1.4), a transformation of local coordinates $(x^i, y^i) \to (\tilde{x}^i, \tilde{y}^i)$ on $\mathrm{Osc}^1 M$ is given by

$$\begin{aligned} &\tilde{x}^i = \tilde{x}^i(x^1, ..., x^n), \ \det \left\| \frac{\partial \tilde{x}^i}{\partial x^j} \right\| \neq 0 \\ &\tilde{y}^i = \frac{\partial \tilde{x}^i}{\partial x^j} y^j. \end{aligned} \tag{1.5}$$

The manifold $\mathrm{Osc}^1 M$ is of dimension $2n$ and is orientable. One can see that if the manifold M is paracompact, then $\mathrm{Osc}^1 M$ is paracompact, too.

Based on the equalities (1.4) we can identify the point $[\rho]_{x_0} \in$ $\in \mathrm{Osc}^1 M$, with the tangent vector $y_0 \in T_{x_0} M$. Thus, the bundle $(\mathrm{Osc}^1 M,$ $\pi, M)$ can be identified to the tangent bundle (TM, π, M). In this case some notations are in order: $E = \mathrm{Osc}^1 M = TM$, $\tilde{E} = \mathrm{Osc}^1 M \setminus \{0\}$, 0 being the null–section of the projection π.

The coordinate transformations (1.5) determine the transformations of the natural basis of the tangent space $T_u E$ at the point $u \in \pi^{-1}(U \cap \tilde{U})$ as follows

$$\frac{\partial}{\partial x^i} = \frac{\partial \tilde{x}^j}{\partial x^i}\frac{\partial}{\partial \tilde{x}^j} + \frac{\partial \tilde{y}^j}{\partial x^i}\frac{\partial}{\partial \tilde{y}^j}, \quad \frac{\partial}{\partial y^i} = \frac{\partial \tilde{y}^j}{\partial y^i}\frac{\partial}{\partial \tilde{y}^j}. \tag{1.6}$$

Of course, from (1.5) we get $\dfrac{\partial \tilde{y}^j}{\partial y^i} = \dfrac{\partial \tilde{x}^j}{\partial x^i}$.

One remarks the existence of some natural geometrical object field on E. Namely, the tangent space V_u to the fibre $E_x = \pi^{-1}(x)$ at the point $u \in E_x$ is spanned by the vectors $\left(\dfrac{\partial}{\partial y^1}, \dots, \dfrac{\partial}{\partial y^n}\right)_u$. Therefore, the mapping $V : u \in E \to V_u \subset T_uE$ provides a regular distribution. It is called the *vertical distribution* on E. Its local dimension is n and $\left\{\dfrac{\partial}{\partial y^1}, \dots, \dfrac{\partial}{\partial y^n}\right\}$ is a local adapted basis. So, V is an integrable distribution on E.

Taking into account (1.5) and (1.6) we can see that

$$\overset{1}{\Gamma} = y^i \frac{\partial}{\partial y^i} \tag{1.7}$$

is a vector field on E, which does not vanish on $\tilde{E}$. Clearly $\overset{1}{\Gamma}$ is a vertical vector field. It is called the *Liouville vector field.*

Let us consider the $\mathcal{F}(E)$–linear mapping $J : \mathcal{X}(E) \to \mathcal{X}(E)$,

$$J\left(\frac{\partial}{\partial x^i}\right) = \frac{\partial}{\partial y^i}, \quad J\left(\frac{\partial}{\partial y^i}\right) = 0, \quad (i = 1, \dots, n). \tag{1.8}$$

Theorem 1.1.1 *The following properties hold:*

1° *J is globally defined on E.*

2° $J^2 = 0,\ \operatorname{Im} J = \operatorname{Ker} J = V,\ \operatorname{rank} \|J\| = n.$

3° *J is an integrable structure.*

4° $J\overset{1}{\Gamma} = 0.$

Proof. 1° From (1.5) and (1.6), (1.8) it follows that J is defined in every domain of the local chart of E and under a coordinate transformation we have $\tilde{y}^i \dfrac{\partial}{\partial \tilde{y}^i} = y^i \dfrac{\partial}{\partial y^i}$. 2° is immediate. 3° Since (1.8) holds, the Nijenhuis tensor N_J vanishes. 4° $J\overset{1}{\Gamma} = y^i J\left(\dfrac{\partial}{\partial y^i}\right) = 0.$ **q.e.d.**

We say that J is the *natural tangent structure* on E. It allows us to define the notion of spray.

Definition 1.1.1 A vector field $S \in \mathcal{X}(E)$ with the property

$$JS = \overset{1}{\Gamma} \tag{1.9}$$

is called a spray on E.

Note that here S is in no way homogeneous. But we prefer this nonstandard denomination, since it is convenient for higher order geometries. The notion of local spray is immediate.

The following theorem holds:

Theorem 1.1.2

1° *A spray S can be uniquely written in the form:*

$$S = y^i \frac{\partial}{\partial x^i} - 2G^i(x, y)\frac{\partial}{\partial y^i}. \tag{1.10}$$

2° *With respect to* (1.5) *the coefficients G^i are changed as follows*

$$2\tilde{G}^i = 2\frac{\partial \tilde{x}^i}{\partial x^j}G^j - \frac{\partial \tilde{y}^i}{\partial x^j}y^j. \tag{1.11}$$

3° *If the functions $G^i(x, y)$ are given on every domain of local chart of the manifold E, so that* (1.11) *holds, then the vector field $S = y^i \frac{\partial}{\partial x^i} - 2G^i(x, y)\frac{\partial}{\partial y^i}$ is a spray.*

Proof. 1° If a vector field $S = a^i(x, y)\frac{\partial}{\partial x^i} + b^i(x, y)\frac{\partial}{\partial y^i}$ is a spray, then $JS = \overset{1}{\Gamma}$ implies $a^i = y^i$. We denote $b^i(x, y) = -2G^i(x, y)$. 2° The transformations (1.5), (1.6) lead to (1.11). 3° The functions G^i being given on every $\pi^{-1}(U)$ and satisfying (1.11) it follows that S from (1.10) is a vector field, globally defined on E and $JS = \overset{1}{\Gamma}$ holds. **q.e.d.**

Let X be a vector field on the manifold M. A curve $\rho : I \to M$ satisfying the condition $\frac{d\rho}{dt} = X \circ \rho$ is called an integral curve of X. On

U, ρ can be represented by $x^i = x^i(t)$, $t \in I$ and $X = X^i(x^1, ..., x^n)\dfrac{\partial}{\partial x^i}$. Then ρ is an integral curve of X if, and only if, it verifies the differential equation

$$\frac{dx^i}{dt} = X^i(x^1(t), ..., x^n(t)), \quad (i = 1, ..., n).$$

Of course, the curve $\rho : I \to M$ can be prolonged to E, taking the map, $\left(\rho, \dfrac{d\rho}{dt}\right) : t \in I \to \left(x^i(t), \dfrac{dx^i}{dt}(t)\right) \in E$.

The curve ρ in M is called a *solution* or a *path* of the spray S if its extension $\left(\rho, \dfrac{d\rho}{dt}\right)$ is an integral curve of S. In this case $y^i = \dfrac{dx^i}{dt}$ and

$$\frac{d^2x^i}{dt^2} + 2G^i\left(x, \frac{dx}{dt}\right) = 0. \tag{1.12}$$

The equation (1.12) of the paths has geometrical meaning. That is (1.12) is preserved with respect to the coordinate transformations on the base manifold M. It is easy to see that the inverse statement is true. So we have:

Theorem 1.1.3 *If the equation* (1.12) *on the base manifold M has geometrical meaning, then the functions $G^i(x, y)$ are the coefficients of a local spray S on E.*

Now we can prove:

Theorem 1.1.4 *If the base manifold M is paracompact, then on $E = \mathrm{Osc}^1 M$ there exist the local sprays.*

Indeed, there is a Riemannian metric g on M. The equations of geodesics of g, in a fixed parametrization, is of the form (1.12), with $2G^i = \gamma^i_{jk}\dfrac{dx^j}{dt}\dfrac{dx^k}{dt}$, $\gamma^i_{jk}(x)$ being the Christoffel symbols of g and, in our case, this equation has a geometrical meaning. **q.e.d.**

Finally, for a spray S with the coefficients $G^i(x, y)$ we denote

$$N^i{}_j(x, y) = \frac{\partial G^i}{\partial y^j}. \tag{1.13}$$

We can prove, without difficulties:

Theorem 1.1.5 *If G^i are the coefficients of a spray S, then, with respect to* (1.5), *the set of functions $N^i{}_j(x,y)$ from* (1.13) *has the rule of transformation:*

$$\tilde{N}^i{}_m \frac{\partial \tilde{x}^m}{\partial x^j} = N^m{}_j \frac{\partial \tilde{x}^i}{\partial x^m} - \frac{\partial \tilde{y}^i}{\partial x^j}. \tag{1.14}$$

In the next section we shall prove that $N^i{}_j(x,y)$ are the coefficients of a nonlinear connection.

1.2 Nonlinear Connections

Geometrical object fields such as tensors or connections on $E = \mathrm{Osc}^1 M = TM$ have, with respect to (1.5), very complicated rules of transformation for their components. In order to avoid these complications, we introduce the notion of nonlinear connection and operate with the components of various geometrical objects in an adapted frame to it and to the vertical distribution V on E.

Definition 1.2.1 A nonlinear connection on E is a distribution $N : u \in E \to N_u \subset T_u E$, supplementary to the vertical distribution V. That is:

$$T_u E = N_u \oplus V_u, \quad \forall u \in E. \tag{2.1}$$

We say that N is a *horizontal* distribution.

Let N be a nonlinear connection on E. From (2.1) it follows that the local dimension of N is n. One proves, [195], that if M is a paracompact manifold, there exist on E the nonlinear connections.

Consider the projectors h and v determined by the direct decomposition (2.1) and we denote

$$X^H = hX, \; X^V = vX, \quad \forall X \in \mathcal{X}(E). \tag{2.2}$$

The horizontal lift is a $\mathcal{F}(M)$–linear mapping $\ell_h : \mathcal{X}(M) \to \mathcal{X}(E)$, for which

$$v \circ \ell_h = 0, \quad d\pi \circ \ell_h = \mathrm{id}, \tag{2.3}$$

where $d\pi$ is the differential of the projection $\pi : TM \to M$. Consequently, for any vector field $X \in \mathcal{X}(M)$, $\ell_h X$ is a vector field in the horizontal distribution N. Locally, $\ell_h X$ always exists.

So, we can prove:

Proposition 1.2.1

a. *There exists a unique local basis adapted to the distribution* N, *which is projected by* $d\pi$ *onto the natural basis of* $\mathcal{X}(M)$. *It is given by*

$$\frac{\delta}{\delta x^i} = \ell_h\left(\frac{\partial}{\partial x^i}\right), \quad (i = 1, .., n). \tag{2.4}$$

b. *The linearly independent vector fields* $\dfrac{\delta}{\delta x^1}, \dots, \dfrac{\delta}{\delta x^n}$ *can be uniquely written under the form*

$$\frac{\delta}{\delta x^i} = \frac{\partial}{\partial x^i} - N^j{}_i(x,y)\frac{\partial}{\partial y^j}. \tag{2.5}$$

c. *With respect to* (1.5) *we get*

$$\frac{\delta}{\delta x^i} = \frac{\partial \tilde{x}^j}{\partial x^i}\frac{\delta}{\delta \tilde{x}^j}. \tag{2.6}$$

The equality (2.6) is deduced from $\dfrac{\partial}{\partial x^i} = \dfrac{\partial \tilde{x}^j}{\partial x^i}\dfrac{\partial}{\partial \tilde{x}^j}$ (from $\mathcal{X}(M)$) by applying (2.4).

Note the simplicity of the transformation (2.6) in comparison with (1.6).

The functions $N^i{}_j(x,y)$ are called *coefficients* of the nonlinear connection N.

Theorem 1.2.1

1° *With respect to* (1.5) *the coefficients* $N^i{}_j(x,y)$ *of a nonlinear connection* N *are changed by the law* (1.14).

2° *If on every domain of chart,* $\pi^{-1}(U)$, *there is given a set of functions* $N^i{}_j(x,y)$ *so that, with respect to* (1.5), *we have* (1.8), *then there exists a unique nonlinear connection* N *on* E *which has the coefficients* $N^i{}_j(x,y)$.

The reader can see the proof in the book [195].

The theorems 1.1.5 and 1.2.1 have as a consequence:

Theorem 1.2.2 *A spray* S, *with the coefficients* G^i, *determines a nonlinear connection* N *on* E *with the local coefficients* $N^i{}_j = \dfrac{\partial G^i}{\partial y^j}\cdot$

The local vector fields $\left\{\dfrac{\delta}{\delta x^i}, \dfrac{\partial}{\partial y^i}\right\}$, $i = 1, ..., n$, gives a local basis of the $\mathcal{F}(E)$–module of vector fields $\mathcal{X}(E)$, adapted to the direct decomposition (2.1). Its dual basis $(dx^i, \delta y^i)$, $i = 1, ..., n$ has 1–forms δy^i given by

$$\delta y^i = dy^i + N^i{}_j(x,y)dx^j. \tag{2.7}$$

By (2.2), a vector field $X \in \mathcal{X}(E)$ can be uniquely written in the form

$$X = X^H + X^V, \tag{2.8}$$

where X^H belongs to the horizontal distribution N and X^V to the vertical distribution V. In the adapted basis we can write

$$X^H = X^i(x,y)\frac{\delta}{\delta x^i},\quad X^V = \dot{X}^i(x,y)\frac{\partial}{\partial y^i}\cdot \tag{2.8)'}$$

Proposition 1.2.2 *With respect to* (1.5) *the components* $X^i(x,y)$ *and* $\dot{X}^i(x,y)$ *of* X^H *and* X^V, *respectively, detain the rules of transformation*

$$\tilde{X}^i = \frac{\partial \tilde{x}^i}{\partial x^j}X^j,\quad \dot{\tilde{X}}^i = \frac{\partial \tilde{x}^i}{\partial x^j}\dot{X}^j. \tag{2.8)''}$$

A 1–form field $\omega \in \mathcal{X}^*(E)$ can be written in the form

$$\omega = \omega^H + \omega^V,\ \text{ with} \tag{2.9}$$

$$\omega^H = \omega_j(x,y)dx^j,\quad \omega^V = \dot{\omega}_j(x,y)\delta y^j. \tag{2.9)'}$$

We clearly get

$$\omega(X^H) = \omega^H(X),\quad \omega(X^V) = \omega^V(X), \forall X \in \mathcal{X}(E). \tag{2.9)''}$$

Proposition 1.2.3 *With respect to* (1.5) *the components* $\omega_j, \dot{\omega}_j$ *of* ω^H, ω^V *are, respectively, changed by the following rules*

$$\omega_j = \frac{\partial \tilde{x}^i}{\partial x^j}\tilde{\omega}_i, \quad \dot{\omega}_j = \frac{\partial \tilde{x}^i}{\partial x^j}\tilde{\dot{\omega}}_j. \tag{2.9)'''$$

A curve $c : I \to E$ has the tangent vector $\frac{dc}{dt} = \dot{c}$ given by

$$\dot{c} = \dot{c}^H + \dot{c}^V = \frac{dx^i}{dt}\frac{\delta}{\delta x^i} + \frac{\delta y^i}{dt}\frac{\partial}{\partial y^i}. \tag{2.10}$$

It is a *horizontal curve* if $\frac{\delta y^i}{dt} = 0$, $\forall t \in I$, $(i = 1, ..., n)$.

A horizontal curve c with the property $y^i = \frac{dx^i}{dt}$ is said to be an *autoparallel* curve of the nonlinear connection N.

Proposition 1.2.4 *An autoparallel curve of the nonlinear connection* N *is characterized by the system of differential equations*

$$\frac{dy^i}{dt} + N^i{}_j(x, y)\frac{dx^j}{dt} = 0, \quad \frac{dx^i}{dt} = y^i. \tag{2.11}$$

Definition 1.2.2 A tensor field T of type (r, s) on E is called distinguished tensor field (briefly, a d–tensor) if it has the property

$$T(\overset{1}{\omega}, ..., \overset{r}{\omega}, \underset{1}{X}, ..., \underset{s}{X}) = T(\overset{1}{\omega}{}^H, ..., \overset{r}{\omega}{}^V, \underset{1}{X}{}^H, ..., \underset{s}{X}{}^V),$$
$$\forall \underset{1}{X}, ..., \underset{s}{X} \in \mathcal{X}(E), \quad \forall \overset{1}{\omega}, ..., \overset{r}{\omega} \in \mathcal{X}^*(E).$$

For any vector field $X \in \mathcal{X}(E)$, the components X^H and X^V are d–vector fields. Also, the components ω^H and ω^V of a 1–form $\omega \in \mathcal{X}^*(E)$ are d–covector fields.

We consider the components of the d–tensor T in adapted basis $\left(\frac{\delta}{\delta x^i}, \frac{\partial}{\partial y^i}\right)$, $(i = \overline{1, n})$:

$$T^{i_1 \dots i_r}_{j_1 \dots j_s}(x, y) = T\left(dx^{i_1}, ..., \delta y^{i_r}, \frac{\delta}{\delta x^{j_1}}, ..., \frac{\partial}{\partial y^{j_s}}\right)$$

or

$$T = T^{i_1 \dots i_r}_{j_1 \dots j_s}(x,y)\frac{\delta}{\delta x^{j_1}} \otimes \dots \otimes \frac{\partial}{\partial y^{i_r}} \otimes dx^{j_1} \otimes \dots \otimes \delta y^{j_s}. \tag{2.12}$$

With respect to (1.5) we get

$$\tilde{T}^{i_1 \dots i_r}_{j_1 \dots j_s} = \frac{\partial \tilde{x}^{i_1}}{\partial x^{h_1}} \cdots \frac{\partial \tilde{x}^{i_r}}{\partial x^{h_r}} \frac{\partial x^{k_1}}{\partial \tilde{x}^{j_1}} \cdots \frac{\partial x^{k_s}}{\partial \tilde{x}^{j_s}} T^{h_1 \dots h_r}_{k_1 \dots k_s}. \tag{2.13}$$

Indeed, (2.12),(1.6),(2.5) and

$$d\tilde{x}^i = \frac{\partial \tilde{x}^i}{\partial x^j} dx^j, \quad \delta \tilde{y}^i = \frac{\partial \tilde{x}^i}{\partial x^j} \delta y^j, \tag{2.14}$$

have as consequences (2.13).

But (2.13) is just the classical law of transformation of the local coefficients of a tensor field on the base manifold M. Of course, (2.13) characterizes a d–tensor field of type (r, s) (up to the choice of the basis from (2.12)).

Using the local expression (2.12) of a d–tensor field it follows that $\{1, \frac{\delta}{\delta x^i}, \frac{\partial}{\partial y^i}\}$ generate the tensor algebra of d–tensor fields, over the ring of functions defined on E.

Notice that, if $N^i{}_j$ are the coefficients of a nonlinear connection N then

$$R^i{}_{jk} = \frac{\delta N^i{}_j}{\delta x^k} - \frac{\delta N^i{}_k}{\delta x^j}, \quad t^i{}_{jk} = \frac{\partial N^i{}_j}{\partial y^k} - \frac{\partial N^i{}_k}{\partial y^j} \tag{2.15}$$

are d–tensor fields of type $(1, 2)$.

Proposition 1.2.5 *The Lie brackets of the vectors* $\frac{\delta}{\delta x^i}, \frac{\partial}{\partial y^i}$, $(i = \overline{1,n})$, *are given by*

$$\left[\frac{\delta}{\delta x^j}, \frac{\delta}{\delta x^k}\right] = R^i{}_{jk}\frac{\partial}{\partial y^i}, \quad \left[\frac{\delta}{\delta x^j}, \frac{\partial}{\partial y^k}\right] = \frac{\partial N^i{}_j}{\partial y^k}\frac{\partial}{\partial y^i}. \tag{2.16}$$

It is not difficult to prove these formulae. They have an interesting consequence:

Theorem 1.2.3 *The horizontal distribution N is integrable if, and only if, the d–tensor field $R^i{}_{jk}$ vanishes.*

1.3 N–Linear Connections

Let N be a given nonlinear connection on TM, $\{\frac{\delta}{\delta x^i}, \frac{\partial}{\partial y^i}\}$, $(i = 1, ..., n)$ be the adapted basis to N, V, $\{dx^i, \delta y^i\}$ its dual basis, and J the natural tangent structure. We have:

Definition 1.3.1 A linear connection D (i.e. a Koszul connection or covariant derivative) on TM is called an N–linear connection if:

1° D preserves by parallelism the horizontal distribution N.

2° J is absolute parallel with respect to D, that is $D_X J = 0$, $\forall X \in \mathcal{X}(E)$.

Therefore, an N–linear connection D on TM has the properties:

(3.1) $$(D_X Y^H)^V = 0, \quad (D_X Y^V)^H = 0,$$

(3.1)′ $$D_X(JY^H) = J(D_X Y^H), \quad D_X(JY^V) = J(D_X Y^V),$$

(3.1)″ $$D_X h = 0, \quad D_X v = 0.$$

Consequently, setting

(3.2) $$D^H_X Y = D_{X^H} Y, \quad D^V_X Y = D_{X^V} Y,$$

we have

(3.3) $$D_X = D^H_X + D^V_X.$$

The operators D^H and D^V are special derivations in the algebra of d–tensor fields on TM. Of course they are not covariant derivations, but have similar properties. So, D^H and D^V are called the $h-$ and $v-$ covariant derivations, respectively.

For us it is important to know the local form of these operators (see [195]).

Proposition 1.3.1 *In the adapted basis, an N–linear connection D can be uniquely written in the form*

$$(3.4)\qquad D_{\frac{\delta}{\delta x^k}}\frac{\delta}{\delta x^j}=L^i{}_{jk}(x,y)\frac{\delta}{\delta x^i};\quad D_{\frac{\delta}{\delta x^k}}\frac{\partial}{\partial y^j}=L^i{}_{jk}(x,y)\frac{\partial}{\partial y^i},$$

$$(3.4)'\qquad D_{\frac{\partial}{\partial y^k}}\frac{\delta}{\delta x^j}=C^i{}_{jk}(x,y)\frac{\delta}{\delta x^i};\quad D_{\frac{\partial}{\partial y^k}}\frac{\partial}{\partial y^j}=C^i{}_{jk}(x,y)\frac{\partial}{\partial y^i}.$$

Indeed, we can uniquely write

$$D_{\frac{\delta}{\delta x^k}}\frac{\delta}{\delta x^j}=L^i{}_{jk}(x,y)\frac{\delta}{\delta x^i}+L'^i{}_{jk}(x,y)\frac{\partial}{\partial y^i};$$
$$D_{\frac{\delta}{\delta x^k}}\frac{\partial}{\partial y^j}=\underset{1}{L}{}^i{}_{jk}(x,y)\frac{\delta}{\delta x^i}+\underset{1}{L'}{}^i{}_{jk}(x,y)\frac{\partial}{\partial y^i}.$$

Applying (3.1) and (3.1)′ it follows $L^i{}_{jk}(x,y)=\underset{1}{L'}{}^i{}_{jk}(x,y)$, $L'^i{}_{jk}(x,y)=$ $=\underset{1}{L}{}^i{}_{jk}(x,y)=0$. So, (3.4) holds. Similarly we get (3.4)′. **q.e.d.**

The system of functions $L^i{}_{jk}(x,y)$ gives the *coefficients* of the h–covariant derivation D^H and the system of functions $C^i{}_{jk}(x,y)$ are the coefficients of v–covariant derivation D^V.

Proposition 1.3.2 *With respect to* (1.5), *the coefficients* $(L^i{}_{jk}(x,y)$, $C^i{}_{jk}(x,y))$ *are transformed as follows:*

$$(3.5)\qquad \begin{aligned}\tilde{L}^i{}_{rs}\frac{\partial\tilde{x}^r}{\partial x^j}\frac{\partial\tilde{x}^s}{\partial x^k}&=L^r{}_{jk}\frac{\partial\tilde{x}^i}{\partial x^r}-\frac{\partial^2\tilde{x}^i}{\partial x^j\partial x^k},\\ \tilde{C}^i{}_{rs}\frac{\partial\tilde{x}^r}{\partial x^j}\frac{\partial\tilde{x}^s}{\partial x^k}&=C^r{}_{jk}\frac{\partial\tilde{x}^i}{\partial x^r}.\end{aligned}$$

These rules of transformation are easily obtained from the formulae (3.4) and (3.4)′.

It is important to remark that $C^i{}_{jk}(x,y)$ are the coordinates of a d–tensor field of type $(1,2)$.

Proposition 1.3.3 *If on every domain of chart* $\pi^{-1}(\mathcal{U})$ *in TM a set of functions* $(L^i{}_{jk}(x,y),C^i{}_{jk}(x,y))$ *is given, so that* (3.5) *holds, then there exists a unique N–linear connection D on TM which satisfies* (3.4) *and* (3.4)′.

We denote $D\Gamma(N) = (L^i{}_{jk}, C^i{}_{jk})$.

If $N^i{}_j(x,y)$ are the coefficients of the nonlinear connection N, it is easily to see that $B\Gamma(N) = \left(\dfrac{\partial N^i{}_j}{\partial y^k}, 0\right)$ are the coefficients of an N–linear connection. It is the famous *Berwald connection* on TM.

In the case when $G^i(x,y)$ are the coefficients of a spray it follows that $B\Gamma(N) = \left(\dfrac{\partial^2 G^i}{\partial y^i \partial y^k}, 0\right)$ are the coefficients of an N–linear connection, N having the coefficients $N^i{}_j = \dfrac{\partial G^i}{\partial y^j}$.

From the previous remarks we can deduce:

Theorem 1.3.1 *If the base manifold M is paracompact, then on TM there exist N-linear connections.*

Let T be the d–tensor field given by (2.12) and $X = X^i(x,y)\dfrac{\delta}{\delta x^i}$ a horizontal vector field. Taking into account (3.4), the h–covariant derivation of T is expressed by

$$D^H_X T = X^m T^{i_1\dots i_r}_{j_1\dots j_s|m} \frac{\delta}{\delta x^{i_1}} \otimes \dots \otimes \frac{\partial}{\partial y^{i_r}} \otimes dx^{j_1} \otimes \dots \otimes \delta y^{j_s}, \tag{3.6}$$

where

$$\begin{aligned} T^{i_1\dots i_r}_{j_1\dots j_s|m} = \frac{\delta T^{i_1\dots i_r}_{j_1\dots j_s}}{\delta x^m} &+ T^{hi_2\dots i_r}_{j_1\dots j_s} L^{i_1}_{hm} + \dots + T^{i_1\dots i_{r-1}h}_{j_1\dots j_s} L^{i_r}_{hm} - \\ &- T^{i_1\dots i_r}_{hj_2\dots j_s} L^h_{j_1 m} - \dots - T^{i_1\dots i_r}_{j_1\dots j_{s-1}h} L^h_{j_s m}. \end{aligned} \tag{3.7}$$

The operator $|$ from the formula (3.7) is called the *h–covariant derivation*. It has similar properties with the ordinary operators of the covariant derivation on the base manifold M. But, in (3.7), the action of the operators $\dfrac{\delta}{\delta x^i}$ on the function $f \in \mathcal{F}(TM)$ has the form

$$f_{|i} = \frac{\delta f}{\delta x^i} = \frac{\partial f}{\partial x^i} - N^j{}_i(x,y)\frac{\partial}{\partial y^j}. \tag{3.8}$$

Proposition 1.3.4 *The h–covariant derivation $|$ has the properties:*

(1) *$T^{i_1\dots i_r}_{j_1\dots j_s|m}$ is a d–tensor field of type $(r, s+1)$.*

(2) $\left(\underset{(1)}{T}{}^{\cdots}_{\cdots} + \underset{(2)}{T}{}^{\cdots}_{\cdots}\right)_{|m} = \underset{(1)}{T}{}^{\cdots}_{\cdots|m} + \underset{(2)}{T}{}^{\cdots}_{\cdots|m}.$

(3) $(fT^{\cdots}_{\cdots})_{|m} = f_{|m}T^{\cdots}_{\cdots} + fT^{\cdots}_{\cdots|m}.$

(4) $\left(\underset{(1)}{T}{}^{\cdots}_{\cdots} \otimes \underset{(2)}{T}{}^{\cdots}_{\cdots}\right)_{|m} = \underset{(1)}{T}{}^{\cdots}_{\cdots|m} \otimes \underset{(2)}{T}{}^{\cdots}_{\cdots} + \underset{(1)}{T}{}^{\cdots}_{\cdots} \otimes \underset{(2)}{T}{}^{\cdots}_{\cdots|m}$

(this is the Leibniz formula).

(5) *The operator* $_|$ *commutates with the operation of contraction.*

Now, we consider $X = \dot{X}^i(x,y)\dfrac{\partial}{\partial y^i}$, a vertical vector field.

According to (3.4)′, the v–covariant derivation of the d–tensor T is:

$$D^V_X T = \dot{X}^m T^{i_1\dots i_r}_{j_1\dots j_s}\Big|_m \frac{\delta}{\delta x^{i_1}} \otimes \dots \otimes \frac{\partial}{\partial y^{i_r}} \otimes dx^{j_1} \otimes \dots \otimes \delta y^{j_s}, \tag{3.9}$$

where

$$T^{i_1\dots i_r}_{j_1\dots j_s}\Big|_m = \frac{\partial T^{i_1\dots i_r}_{j_1\dots j_s}}{\partial y^m} + T^{hi_2\dots i_r}_{j_1\dots j_s}C^{i_1}_{hm} + \dots + T^{i_1\dots i_{r-1}h}_{j_1\dots j_s}C^{i_r}_{hm} - \tag{3.10}$$
$$- T^{i_1\dots i_r}_{hj_2\dots j_s}C^h_{j_1 m} - \dots - T^{i_1\dots i_r}_{j_1\dots j_{s-1}h}C^h_{j_s m}.$$

The properties of the v–covariant operator $\Big|$ from (3.10) are similar to (1)–(5) from the previous proposition, with few modifications. For instance, instead of (3) we have $(fT^{\cdots}_{\cdots})\Big|_m = \dfrac{\partial f}{\partial y^m}T^{\cdots}_{\cdots} + fT^{\cdots}_{\cdots}\Big|_m$.

A first application of this theory is to the case of the Liouville vector field $\overset{1}{\Gamma} = y^i\dfrac{\partial}{\partial y^i}$. We obtain

$$D^H_X \overset{1}{\Gamma} = X^m y^i{}_{|m}\frac{\partial}{\partial y^i},\quad D^V_X \overset{1}{\Gamma} = \dot{X}^m y^i\Big|_m \frac{\partial}{\partial y^i},$$

from which we get two importan d–tensors:

$$D^i{}_m = y^i{}_{|m} = -N^i{}_m + y^h L^i{}_{hm};\quad d^i{}_m = y^i\Big|_m = \delta^i{}_m + y^h C^i{}_{hm}, \tag{3.11}$$

called the $h-$ and $v-$ *deflection* tensors of the N–linear connection D.

If

$$D^i{}_m = 0,\quad d^i{}_m = \delta^i{}_m, \tag{3.12}$$

D is called a *Cartan N–linear connection.*

1.4 *d*–Tensors of Torsion and Curvature

Let D be an N–linear connection on $E = TM$. Its torsion $\mathbb{T}$ is given by

$$(4.1)\qquad \mathbb{T}(X,Y) = D_X Y - D_Y X - [X,Y], \quad \forall X,Y \in \mathcal{X}(TM).$$

The vector field $\mathbb{T}(X,Y)$ can be uniquely written under the form

$$(4.2)\qquad \mathbb{T}(X,Y) = h\mathbb{T}(X,Y) + v\mathbb{T}(X,Y).$$

Then, we can deduce, [195]:

Theorem 1.4.1 *The torsion $\mathbb{T}$ of an N–linear connection D is completely determined by the following d–tensor fields:*

$$(4.2)'\qquad \begin{cases} h\mathbb{T}(X^H,Y^H) = D^H_X Y^H - D^H_Y X^H - [X^H,Y^H]^H, \\ v\mathbb{T}(X^H,Y^H) = -[X^H,Y^H]^V, \end{cases}$$

$$(4.2)''\qquad \begin{cases} h\mathbb{T}(X^H,Y^V) = -D^V_Y X^H - [X^H,Y^V]^H, \\ v\mathbb{T}(X^H,Y^V) = D^H_X Y^V - [X^H,Y^V]^V, \end{cases}$$

and

$$(4.2)'''\qquad v\mathbb{T}(X^V,Y^V) = D^V_X Y^V - D^V_Y X^V - [X^V,Y^V]^V, \quad \forall X,Y \in \mathcal{X}(TM).$$

Corollary 1.4.1 *The following properties hold good:*

a. $v\mathbb{T}(X^H,Y^H) = 0$, *if, and only if, the horizontal distribution N is integrable.*

b. $v\mathbb{T}(X^H,Y^V) = 0$, *if, and only if,* $D^H_X Y^V = v[X^H,Y^V]$,
$h\mathbb{T}(X^H,Y^V) = 0$, *if, and only if,* $D^V_X Y^H = -h[X^H,Y^V]$.

We shall say that $[\mathbb{T}(X^H,Y^H)]^H$ is $h(hh)$*–torsion* of D and $[\mathbb{T}(X^V,Y^V)]^V$ is $v(vv)$*–torsion* of D.

The local forms of these d–tensors of torsion are obtained from (4.2)′–(4.2)‴ setting, in adapted basis:

$$\begin{aligned}
&hI\!\!T\left(\frac{\delta}{\delta x^k},\frac{\delta}{\delta x^j}\right)=T^i{}_{jk}\frac{\delta}{\delta x^i},\quad vI\!\!T\left(\frac{\delta}{\delta x^k},\frac{\delta}{\delta x^j}\right)=\tilde{T}^i{}_{jk}\frac{\partial}{\partial y^i},\\
(4.3)\qquad &hI\!\!T\left(\frac{\partial}{\partial y^k},\frac{\delta}{\delta x^j}\right)=\tilde{P}^i{}_{jk}\frac{\delta}{\delta x^i},\quad vI\!\!T\left(\frac{\partial}{\partial y^k},\frac{\delta}{\delta x^j}\right)=P^i{}_{jk}\frac{\partial}{\partial y^i},\\
&vI\!\!T\left(\frac{\partial}{\partial y^k},\frac{\partial}{\partial y^j}\right)=S^i{}_{jk}\frac{\partial}{\partial y^i}.
\end{aligned}$$

By Theorem 1.4.1 and Proposition 1.2.5, one easily gets:

Theorem 1.4.2 *The d-tensors of torsion* (4.3) *of an N–linear connection D, with the coefficients* $D\Gamma(N)=(L^i{}_{jk},C^i{}_{jk})$ *have the expressions:*

$$(4.4)\qquad \begin{aligned}
&T^i{}_{jk}=L^i{}_{jk}-L^i{}_{kj},\quad \tilde{T}^i{}_{jk}=R^i{}_{jk},\quad \tilde{P}^i{}_{jk}=C^i{}_{jk},\\
&P^i{}_{jk}=\frac{\partial N^i{}_j}{\partial y^k}-L^i{}_{kj},\quad S^i{}_{jk}=C^i{}_{jk}-C^i{}_{kj}.
\end{aligned}$$

Corollary 1.4.2 *The N–linear connection with the coefficients* $D\Gamma(N)=(L^i{}_{jk},C^i{}_{jk})$ *is without torsion if, and only if:*

a. *The horizontal distribution N is integrable;*

b. $D\Gamma(N)=B\Gamma(N)$, *where* $B\Gamma(N)$ *is the Berwald connection of the nonlinear connection N.*

Corollary 1.4.3 *The N–linear connection with the coefficients* $D\Gamma(N)=(L^i{}_{jk},C^i{}_{jk})$ *is without* $h(hh)-$ *and* $v(vv)-$ *torsions if the coefficients* $L^i{}_{jk}$ *and* $C^i{}_{jk}$ *are symmetric:*

$$(4.4)'\qquad L^i{}_{jk}=L^i{}_{kj},\quad C^i{}_{jk}=C^i{}_{kj}.$$

In the following paragraph we shall deal with the curvature of an N–linear connection D. This is given by

$$(4.5)\qquad I\!\!R(X,Y)Z=D_XD_YZ-D_YD_XZ-D_{[X,Y]}Z.$$

We can prove [195]:

Theorem 1.4.3 *The curvature tensor field $I\!R$ of the N–linear connection D has the properties:*

a. $I\!R(X,Y)(JZ) = J[I\!R(X,Y)Z]$,

b. $v\{I\!R(X,Y)Z^H\} = 0, \quad h\{I\!R(X,Y)Z^V\} = 0$,

c. $I\!R(X,Y)Z = h\{I\!R(X,Y)Z^H\} + v\{I\!R(X,Y)Z^V\}$.

Therefore, we deduce

Theorem 1.4.4 *The curvature tensor field $I\!R$ of the N–linear connection D is completely determined by the following three d–tensor fields:*

$$
\begin{aligned}
&I\!R(X^H,Y^H)Z^H &&= D^H_X D^H_Y Z^H - D^H_Y D^H_X Z^H - D^H_{[X^H,Y^H]} Z^H - \\
& &&- D^V_{[X^H,Y^H]} Z^H, \\
(4.6)\quad &I\!R(X^V,Y^H)Z^H &&= D^V_X D^H_Y Z^H - D^H_Y D^V_X Z^H - D^H_{[X^V,Y^H]} Z^H - \\
& &&- D^V_{[X^V,Y^H]} Z^H, \\
&I\!R(X^V,Y^V)Z^H &&= D^V_X D^V_Y Z^H - D^V_Y D^V_X Z^H - D^V_{[X^V,Y^V]} Z^H.
\end{aligned}
$$

In the adapted basis, the local coefficients of these d–tensors are as follows:

$$
(4.7)\qquad
\begin{aligned}
I\!R\left(\frac{\delta}{\delta x^k},\frac{\delta}{\delta x^j}\right)\frac{\delta}{\delta x^h} &= R_h{}^i{}_{jk}\,\frac{\delta}{\delta x^i},\\
I\!R\left(\frac{\partial}{\partial y^k},\frac{\delta}{\delta x^j}\right)\frac{\delta}{\delta x^h} &= P_h{}^i{}_{jk}\,\frac{\delta}{\delta x^i},\\
I\!R\left(\frac{\partial}{\partial y^k},\frac{\partial}{\partial y^j}\right)\frac{\delta}{\delta x^h} &= S_h{}^i{}_{jk}\,\frac{\delta}{\delta x^i},
\end{aligned}
$$

By a direct computation, using (2.16) and (3.4),(3.4)′ we get:

Theorem 1.4.5 *The local coefficients $R_h{}^i{}_{jk}, P_h{}^i{}_{jk}$ and $S_h{}^i{}_{jk}$ of the curvature tensor $\mathbb{R}$ are expressed in the form*

$$
\begin{aligned}
R_h{}^i{}_{jk} &= \frac{\delta L^i{}_{hj}}{\delta x^k} - \frac{\delta L^i{}_{hk}}{\delta x^j} + L^s{}_{hj}L^i{}_{sk} - L^s{}_{hk}L^i{}_{sj} + C^i{}_{hs}R^s{}_{jk},\\
P_h{}^i{}_{jk} &= \frac{\partial L^i{}_{hj}}{\partial y^k} - C^i{}_{hk|j} + C^i{}_{hs}P^s{}_{jk},\\
S_h{}^i{}_{jk} &= \frac{\partial C^i{}_{hj}}{\partial y^k} - \frac{\partial C^i{}_{hk}}{\partial y^j} + C^s{}_{hj}C^i{}_{sk} - C^s{}_{hk}C^i{}_{sj}.
\end{aligned}
\tag{4.8}
$$

As usual, here the small bar denotes the h–covariant derivation.

The torsions and curvatures of an N–linear connection D are not independent. They verify a number of the Bianchi identities (see [195]), obtained from the general Bianchi identities

$$
\begin{aligned}
&\Sigma[D_X\mathbb{T}(Y,Z) - \mathbb{R}(X,Y)Z + \mathbb{T}(\mathbb{T}(X,Y)Z)] = 0,\\
&\Sigma[(D_X\mathbb{R})(U,Y,Z) + \mathbb{R}(\mathbb{T}(X,Y),Z)U] = 0,
\end{aligned}
\tag{4.9}
$$

where Σ means the cyclic sum over X, Y, Z.

If we rewrite (4.6) in another manner, then we deduce:

Theorem 1.4.6 *For an N-linear connection D with the coefficients $D\Gamma(N) = (L^i{}_{jk}, C^i{}_{jk})$ the following Ricci identities hold:*

$$
\begin{aligned}
X^i{}_{|k|h} - X^i{}_{|h|k} &= X^m R_m{}^i{}_{kh} - X^i{}_{|m}T^m{}_{kh} - X^i|_m R^m{}_{kh},\\
X^i{}_{|k}\big|_h - X^i\big|_{h|k} &= X^m P_m{}^i{}_{kh} - X^i{}_{|m}C^m{}_{kh} - X^i|_m P^m{}_{kh},\\
X^i|_k|_h - X^i|_h|_k &= X^m S_m{}^i{}_{kh} - X^i|_m S^m{}_{kh},
\end{aligned}
\tag{4.10}
$$

where X^i is an arbitrary d–vector field.

Of course, these Ricci identities can be extended to any d–tensor field of type (r, s).

Applying the Ricci identities (4.10) to the Liouville vector field, $\overset{1}{\Gamma}$ from (1.7) and taking into account the expressions (3.11) of the $h-$ and $v-$ deflection tensors $D^i{}_j$ and $d^i{}_j$, we obtain the following important identities:

Theorem 1.4.7 *The deflection tensors $D^i{}_j$ and $d^i{}_j$ of an N–linear connection D satisfy the identities:*

$$
(4.11)\qquad
\begin{aligned}
&D^i{}_{j|k} - D^i{}_{k|j} = y^m R_m{}^i{}_{jk} - D^i{}_m T^m{}_{jk} - d^i{}_m R^m{}_{jk},\\
&D^i{}_{j}\big|_k - d^i{}_{k|j} = y^m P_m{}^i{}_{jk} - D^i{}_m C^m{}_{jk} - d^i{}_m P^m{}_{jk},\\
&d^i{}_{j}\big|_k - d^i{}_{k}\big|_j = y^m S_m{}^i{}_{jk} - d^i{}_m S^m{}_{jk}.
\end{aligned}
$$

When D is a Cartan N–linear connection the previous theorem implies:

Proposition 1.4.1 *A Cartan N–linear connection D has the following properties:*

$$
(4.12)\qquad
\begin{aligned}
R^i{}_{jk} &= y^m R_m{}^i{}_{jk},\\
P^i{}_{jk} &= y^m P_m{}^i{}_{jk},\\
S^i{}_{jk} &= y^m S_m{}^i{}_{jk}.
\end{aligned}
$$

We will see that in a Finsler space, for the so–called Cartan connection, (4.12) holds.

Another application is obtained writing the Ricci identities for an arbitrary function $L : E \to R$. So we have

$$
(4.13)\qquad
\begin{aligned}
&L_{|i|j} - L_{|j|i} = -L_{|m} T^m{}_{ij} - L\big|_m R^m{}_{ij},\\
&L_{|i}\big|_j - L\big|_{j|i} = -L_{|m} C^m{}_{ij} - L\big|_m P^m{}_{ij},\\
&L\big|_i\big|_j - L\big|_j\big|_i = -L\big|_m S^m{}_{ij}.
\end{aligned}
$$

Considering the partial differential equations

$$
(4.14)\qquad L_{|j} = \frac{\delta L}{\delta x^j} = \frac{\partial L}{\partial x^j} - N^i{}_j \frac{\partial L}{\partial y^i} = 0,
$$

then the necessary conditions for the existence of the solutions $L(x, y)$ of (4.14) are given by:

$$
(4.15)\quad L\big|_m R^m{}_{ij} = 0,\ L\big|_{j|i} + L\big|_m P^m{}_{ij} = 0,\ L\big|_i\big|_j - L\big|_j\big|_i + L\big|_m S^m{}_{ij} = 0.
$$

Remark. The equations (4.14) are frequently used in the Finsler spaces or Lagrange spaces. For instance, in a Finsler space the fundamental function satisfies (4.14).

1.5 Parallelism. Structure Equations

Let D be an N–linear connection on TM and $D\Gamma(N) = (L^i{}_{jk}, C^i{}_{jk})$ its coefficients in the adapted basis $\left(\frac{\delta}{\delta x^i}, \frac{\partial}{\partial y^i}\right)$.

If c is a parametrized curve in TM, $c : t \in I \to c(t) \in TM$ then on a domain of chart $\pi^{-1}(U)$ on TM, the tangent vector $\dot{c} = \frac{dc}{dt}$ can be represented in the form (2.10). The curve c is horizontal if $\frac{\delta y^i}{dt} = 0$ and it is an autoparallel of the nonlinear connection N if $\frac{\delta y^i}{dt} = 0$, $y^i = \frac{dx^i}{dt}$.

We denote

$$\frac{DX}{dt} = D_{\dot{c}}X, \quad DX = \frac{DX}{dt} \cdot dt, \quad \forall X \in \mathcal{X}(TM). \tag{5.1}$$

$\frac{DX}{dt}$ is the covariant differential along the curve c.

Setting $X = X^H + X^V$, $X^H = X^i \frac{\delta}{\delta x^i}$, $X^V = \dot{X}^i \frac{\partial}{\partial y^i}$, we have

$$\begin{aligned}\frac{DX}{dt} &= \frac{DX^H}{dt} + \frac{DX^V}{dt} = \left\{X^i{}_{|k} \frac{dx^k}{dt} + X^i\big|_k \frac{\delta y^k}{dt}\right\} \frac{\delta}{\delta x^i} + \\ &+ \left\{\dot{X}^i{}_{|k} \frac{dx^k}{dt} + \dot{X}^i\big|_k \frac{\delta y^k}{dt}\right\} \frac{\partial}{\partial y^i}.\end{aligned} \tag{5.2}$$

Let us consider

$$\omega^i{}_j = L^i{}_{jk} dx^k + C^i{}_{jk} \delta y^k. \tag{5.3}$$

$\omega^i{}_j$ are called 1–forms connection of D.

Then the equation (5.2) can be written under the form:

$$\frac{DX^i}{dt} = \left\{\frac{dX^i}{dt} + X^m \frac{\omega^i{}_m}{dt}\right\} \frac{\delta}{\delta x^i} + \left\{\frac{d\dot{X}^i}{dt} + \dot{X}^m \frac{\omega^i{}_m}{dt}\right\} \frac{\partial}{\partial y^i}. \tag{5.4}$$

The vector field X on TM is said to be *parallel* along the curve c, with respect to N–connection D, if $\frac{DX}{dt} = 0$. This is equivalent to $\frac{DX^H}{dt} = 0$, $\frac{DX^V}{dt} = 0$. Looking to (5.4), one finds the following result:

Proposition 1.5.1 *The vector field* $X = X^i \dfrac{\delta}{\delta x^i} + \dot{X}^i \dfrac{\partial}{\partial y^i}$ *is parallel along the parametrized curve c, if, and only if, its coefficients* $X^i(x,y)$, $\dot{X}^i(x,y)$ *are solutions of the linear system of differential equations*

$$\frac{dZ^i}{dt} + Z^m(x(t),y(t))\,\frac{\omega^i{}_m(x(t),y(t))}{dt} = 0.$$

A theorem for the existence and uniqueness for the parallel vector fields along to a given curve on TM can be formulated. A *horizontal path* of an N–linear connection D on TM is a horizontal parametrized curve c with the property $D_{\dot{c}}\dot{c} = 0$.

Using (5.4), with $X^i = \dfrac{dx^i}{dt}$, and taking into account the previous proposition, we get the next theorem:

Theorem 1.5.1 *The horizontal paths of the N–linear connection D are characterized by the system of differential equations*

$$\text{(5.5)} \qquad \frac{d^2x^i}{dt^2} + L^i{}_{jk}(x,y)\,\frac{dx^j}{dt}\,\frac{dx^k}{dt} = 0, \quad \frac{dy^i}{dt} + N^i{}_j(x,y)\,\frac{dx^j}{dt} = 0.$$

Now we can consider a curve $c^v_{x_0}$ in the fiber $\pi^{-1}(x_0)$ of the vertical bundle VTM. It can be represented by

$$x^i = x_0{}^i, \; y^i = y^i(t), \quad t \in I.$$

$c^v_{x_0}$ is called a vertical curve in the point $x_0 \in M$.

A vertical curve $c^v_{x_0}$ is called a *vertical path* with respect to D if $D_{\dot{c}^v_{x_0}}\dot{c}^v_{x_0} = 0$.

Again, the equation (5.4) leads to:

Theorem 1.5.2 *The vertical paths in the point* $x_0 \in M$*, with respect to the N–linear connection D, is characterized by the system of differential equations*

$$\text{(5.6)} \qquad x^i = x_0{}^i, \quad \frac{d^2y^i}{dt^2} + C^i{}_{jk}(x_0,y)\,\frac{dy^j}{dt}\,\frac{dy^k}{dt} = 0.$$

Obviously, the local existence and uniqueness of horizontal paths are assured if the initial conditions for (5.5) are given. The same considerations can be made for the vertical paths.

Considering the 1–form connection (5.3) we can formulate the following result:

Theorem 1.5.3 *The structure equations of an N–linear connection D on TM are given by*

$$
(5.7)\qquad
\begin{aligned}
d(dx^i) - dx^m \wedge \omega^i{}_m &= -\overset{(0)}{\Omega}{}^i\\
d(\delta y^i) - \delta y^m \wedge \omega^i{}_m &= -\overset{(1)}{\Omega}{}^i\\
d\omega^i{}_j - \omega^m{}_j \wedge \omega^i{}_m &= -\Omega^i{}_j,
\end{aligned}
$$

where $\overset{(0)}{\Omega}{}^i$, $\overset{(1)}{\Omega}{}^i$ *are the 2–forms of torsion,*

$$
(5.8)\qquad
\begin{aligned}
\overset{(0)}{\Omega}{}^i &= \frac{1}{2} T^i{}_{jk} dx^j \wedge dx^k + C^i{}_{jk} dx^j \wedge \delta y^k,\\
\overset{(1)}{\Omega}{}^i &= \frac{1}{2} R^i{}_{jk} dx^j \wedge dx^k + P^i{}_{jk} dx^j \wedge \delta y^k + \frac{1}{2} S^i{}_{jk} \delta y^j \wedge \delta y^k,
\end{aligned}
$$

and the 2–forms of curvature $\Omega^i{}_j$ *are expressed by*

$$
(5.9)\qquad
\begin{aligned}
\Omega^i{}_j &= \frac{1}{2} R_j{}^i{}_{kh} dx^k \wedge dx^h + P_j{}^i{}_{kh} dx^k \wedge \delta y^h +\\
&+ \frac{1}{2} S_j{}^i{}_{kh} \delta y^k \wedge \delta y^h.
\end{aligned}
$$

One can get the proof by taking the exterior differential of the 1–forms $\delta y^i = dy^i + N^i{}_j dx^j$, as well as the exterior differential of the 1–forms connection $\Omega^i{}_j$.

Now, the Bianchi identities of an N–linear connection D can be obtained from (5.7) by calculating the exterior differential of (5.7), moduling the same system (5.7) and using the exterior differential of the 2–forms $\overset{(0)}{\Omega}{}^i$, $\overset{(1)}{\Omega}{}^i$ and $\Omega^i{}_j$.

Of course, these identities can be also obtained from (4.9).

1.6 Lagrange Spaces

In the last twenty five years, many geometrical models in Mechanics, Physics, Theory of control, Biology were based on the notion of Lagrangian or Hamiltonian. Thus, the concepts of Lagrange space or Hamilton space were introduced. The differential geometry of Lagrange spaces and Hamilton spaces is now considerably developed and used in various fields to study the natural processes where the dependence on position, velocity or momentum is involved, [185], [195].

We start with the following definition:

Definition 1.6.1 A differentiable Lagrangian, on a real n–dimensional manifold M, is a mapping

$$L : (x, y) \in TM \to L(x, y) \in R, \quad \forall u = (x, y) \in TM, \tag{6.1}$$

of class C^∞ on $\widetilde{TM}$ and continuous on null section 0 of the projection $\pi : TM \to M$.

Clearly, the conditions imposed on differentiable Lagrangian have a geometrical character.

Observing that, with respect to (1.5) we have $\dfrac{\partial}{\partial y^i} = \dfrac{\partial \tilde{x}^j}{\partial x^i} \dfrac{\partial}{\partial \tilde{y}^j}$, it follows that

$$g_{ij}(x, y) = \frac{1}{2} \frac{\partial^2 L(x, y)}{\partial y^i \partial y^j} \tag{6.2}$$

is a d–tensor field on $\widetilde{TM}$, of type $(0, 2)$, symmetric.

A differential Lagrangian L on M is said to be *regular* if

$$rank \, \|g_{ij}(x, y)\| = n \quad \text{on } \widetilde{TM}. \tag{6.3}$$

Of course, in this case, we will use its contravariant d–tensor $g^{ij}(x, y)$ from

$$g_{ih}(x, y) g^{hj}(x, y) = \delta_i^j. \tag{6.2$'$}$$

Now we can set the following definition.

Definition 1.6.2 A Lagrange space is a pair $L^n = (M, L)$ formed by a smooth real n–dimensional manifold M and a regular differentiable Lagrangian L on M, for which the d–tensor field g_{ij}, from (6.2), has a constant signature on $\widetilde{TM}$.

Sometimes L is called the *fundamental function* and $g_{ij}(x, y)$ is called the *fundamental tensor field* (or *metric tensor field*).

We can prove, without difficulties:

Theorem 1.6.1 *If the manifold M is paracompact, there exist the regular Lagrangians L on M, so that the pair $L^n = (M, L)$ is a Lagrange space.*

Of course, we can repeat the previous definitions for an open set of M and obtain the notion of Lagrange space over that open set.

An example is given by electrodynamics.

Let us consider the Lagrangian

$$L(x, y) = mc\gamma_{ij}(x)y^i y^j + \frac{2e}{m} A_i(x)y^i + U(x), \tag{6.4}$$

where $\gamma_{ij}(x)$ is a pseudo–Riemannian metric tensor on M (we assume the existence of this metric on M), $A_i(x)$ is a covector field on M and $U(x)$ a real function, and $m \neq 0$, c, e are the well–known constants from Physics. The coefficients $\gamma_{ij}(x)$ are called gravitational potentials and those of $A_i(x)$ are called electromagnetic potentials.

The Lagrangian (6.4) is regular, since $g_{ij}(x) = mc\gamma_{ij}(x)$ and $g_{ij}(x)$ has a constant signature on M. So the pair $L^n = (M, L)$ is a Lagrange space. We say that L^n is the *Lagrange space of electrodynamics.*

1.7 Variational Problem

Let $L : TM \to R$ be a differentiable Lagrangian on the manifold M, which is not obligatorly regular.

A curve $c : t \in [0, 1] \to (x^i(t)) \in U \subset M$ (with a fixed parametrization) having the image in a domain of a chart U on M, has the extension to TM given by $c^* : t \in [0, 1] \to (x^i(t), \dfrac{dx^i}{dt}(t)) \in \pi^{-1}(U)$. Since nowhere the tangent vector field $\frac{dx^i}{dt}(t)$, $t \in [0,1]$ vanishes, the image of c^* belongs to $\widetilde{T}M$.

The integral of action of the Lagrangian L on the curve c is given by the functional

$$I(c) = \int_0^1 L(x, \frac{dx}{dt})dt. \tag{7.1}$$

Consider the curves $c_\varepsilon : t \in [0,1] \to (x^i(t) + \varepsilon V^i(t)) \in M$, which have the same endpoints $x^i(0), x^i(1)$ as the curve c, $V^i(t) = V^i(x(t))$ being a regular vector field on the curve c, with the property $V^i(0) = V^i(1) = 0$ and ε a real number, sufficiently small in absolute value, so that $Im\ c_\varepsilon \subset U$.

The extension of the curve c_ε to TM is

$$c^*_\varepsilon : t \in [0,1] \to \left(x^i(t) + \varepsilon V^i(t), \frac{dx^i}{dt} + \varepsilon \frac{dV^i}{dt}\right) \in \pi^{-1}(U).$$

The integral of action of the Lagrangian L on the curve c_ε is

$$I(c_\varepsilon) = \int_0^1 L\left(x + \varepsilon V, \frac{dx}{dt} + \varepsilon \frac{dV}{dt}\right) dt. \tag{7.1$'$}$$

A necessary condition for $I(c)$ to be an extremal value of $I(c_\varepsilon)$ is

$$\left.\frac{dI(c_\varepsilon)}{d\varepsilon}\right|_{\varepsilon=0} = 0. \tag{7.2}$$

In our conditions of differentiability, the operator $\frac{d}{d\varepsilon}$ is permuting with the operator of integration. We obtain from (7.1)$'$

$$\frac{dI(c_\varepsilon)}{d\varepsilon} = \int_0^1 \frac{d}{d\varepsilon} L\left(x + \varepsilon V, \frac{dx}{dt} + \varepsilon \frac{dV}{dt}\right) dt. \tag{7.3}$$

A straightforward calculation leads to

$$\left.\frac{d}{d\varepsilon} L\left(x + \varepsilon V, \frac{dx}{dt} + \varepsilon \frac{dV}{dt}\right) dt\right|_{\varepsilon=0} = \frac{\partial L}{\partial x^i} V^i + \frac{\partial L}{\partial y^i}\frac{dV^i}{dt} =$$

$$= \left\{\frac{\partial L}{\partial x^i} - \frac{d}{dt}\frac{\partial L}{\partial y^i}\right\} V^i + \frac{d}{dt}\left\{\frac{\partial L}{\partial y^i} V^i\right\}, \quad y^i = \frac{dx^i}{dt}.$$

Substituting the last expression in (7.3) and taking into account that $\int_0^1 \frac{d}{dt}\left\{\frac{\partial L}{\partial y^i}V^i\right\}dt = 0$, the equation (7.2) and the fact that $V^i(x(t))$ is arbitrary, we obtain the following property:

Theorem 1.7.1 *In order that the functional $I(c)$ be an extremal value of $I(c_\varepsilon)$ it is necessary that c be the solution of the Euler–Lagrange equations*

$$E_i(L) \overset{\text{def}}{=} \frac{\partial L}{\partial x^i} - \frac{d}{dt}\frac{\partial L}{\partial y^i} = 0, \quad y^i = \frac{dx^i}{dt}. \tag{7.4}$$

The curves c which are the solutions of the equations (7.4) are called the *extremal curves* of the Lagrangian L.

We can see that the next theorem holds.

Theorem 1.7.2 *The following properties hold:*

1° *$E_i(L)$ is a d–covector field.*

2° $E_i(L+L') = E_i(L) + E_i(L')$, $E_i(aL) = aE_i(L)$, $a \in R$.

3° $E_i\left(\frac{dF}{dt}\right) = 0$, $\forall F \in \mathcal{F}(TM)$, *with* $\frac{\partial F}{\partial y^i} = 0$.

The notion of *energy* can be introduced as in the Theoretical Mechanics (Santilli [263], [264]), by

$$E_L = y^i\frac{\partial L}{\partial y^i} - L. \tag{7.5}$$

Theorem 1.7.3 *On a smooth curve c in M the following formula holds*

$$\frac{dE_L}{dt} = -\frac{dx^i}{dt}E_i(L), \quad \left(y^i = \frac{dx^i}{dt}\right). \tag{7.6}$$

Proof. By (7.5), $\frac{dE_L}{dt} = \frac{dy^i}{dt}\frac{\partial L}{\partial y^i} + y^i\frac{d}{dt}\left(\frac{\partial L}{\partial y^i}\right) - y^i\frac{\partial L}{\partial x^i} - \frac{dy^i}{dt}\frac{\partial L}{\partial y^i} =$
$= -y^i\left(\frac{\partial L}{\partial x^i} - \frac{d}{dt}\frac{\partial L}{\partial y^i}\right), \quad y^i = \frac{dx^i}{dt}.$ **q.e.d.**

Theorem 1.7.4 *For any differentiable Lagrangian $L(x,y)$ the energy E_L is conserved along to every solution curve c of the Euler–Lagrange equations $E_i(L) = 0$, $\left(y^i = \frac{dx^i}{dt}\right)$.*

1.8 A Noether Theorem

In the previous section, Theorem 7.2 shows that the integral of action $I(c) = \int_0^1 L\left(x, \frac{dx}{dt}\right) dt$ and the integral of action

(8.1) $$I'(c) = \int_0^1 \left\{L\left(x, \frac{dx}{dt}\right) + \frac{dF(x)}{dt}\right\} dt,$$

for any function $F(x)$ give rise to the same Euler-Lagrange equations $E_i(L) = 0$ only depending on the Lagrangian L. Based on this reason we can introduce:

Definition 1.8.1 A symmetry of the differentiable Lagrangian $L(x, y)$ is a C^∞ diffeomorphism $\varphi : M \times R \to M \times R$ which transforms the integral of action $I'(c)$ of L, from (8.1), into the integral of action $I(c) = \int_0^1 L\left(x, \frac{dx}{dt}\right) dt.$

For us, it is convenient to study the infinitesimal symmetries of $L(x, y)$. Therefore, we start with an infinitesimal transformation on $M \times R$ of the form

(8.2) $$x'^i = x^i + \varepsilon V^i(x, t), \quad t' = t + \varepsilon\tau(x, t),$$

where ε is a real number, sufficiently small as absolute value so that the points (x^i, t) and (x'^i, t') belong to the same local chart $U \times (a, b) \subset \subset M \times R$.

In the following considerations, the terms of order greater than 1 in ε will be neglected and $V^i(x, t)$ will be a vector field on $U \times (a, b)$.

The inverse of the diffeomorphism (8.2) is

$$x^i = x'^i - \varepsilon V^i(x, t), \quad t = t' - \varepsilon\tau(x, t).$$

Of course, a curve $c : t \in [0, 1] \to (x^i(t), t) \in U \times (a, b)$ has the endpoints $c(0)$ and $c(1)$. Therefore, looking at (8.1), the infinitesimal transformation (8.2) is a symmetry for the Larangian $L(x, y)$ if and only if, for any C^∞–function $F(x)$, the following equation holds

(8.3) $$L\left(x', \frac{dx'}{dt'}\right) dt' = \left\{L\left(x, \frac{dx}{dt}\right) + \frac{dF}{dt}(x)\right\} dt.$$

From (8.2) we deduce

$$\frac{dt'}{dt} = 1 + \varepsilon \frac{d\tau}{dt}, \quad \frac{dx'^i}{dt'} = \frac{dx^i}{dt} + \varepsilon \left(\frac{dV^i}{dt} - \frac{dx^i}{dt}\frac{d\tau}{dt} \right). \tag{8.4}$$

The equality (8.3), by virtue of (8.4), and neglecting the terms in $\varepsilon^2, \varepsilon^3, \ldots$, leads to

$$L\frac{d\tau}{dt} + \frac{\partial L}{\partial x^i} V^i + \frac{\partial L}{\partial y^i}\left(\frac{dV^i}{dt} - \frac{dx^i}{dt}\frac{d\tau}{dt} \right) = \frac{d\phi}{dt}, \; y^i = \frac{dx^i}{dt}, \tag{8.5}$$

where we set $\phi(x) = \varepsilon F(x)$.

Conversely, if (8.5) is verified, when L, V^i, τ are given, it is easily to see that (8.3) holds, up to the terms of higher order in ε.

The equations (8.5) can be written under the form

$$V^i E_i(L) - E_L \frac{d\tau}{dt} + \frac{d}{dt}\left(\frac{\partial L}{\partial y^i} V^i \right) = \frac{d\phi}{dt}. \tag{8.5)'}$$

So, we have the following Noether theorem:

Theorem 1.8.1 *For any infinitesimal symmetry (8.2) of the Lagrangian $L(x, y)$ and for any function $\phi(x)$ the following function*

$$\mathcal{F}(L, \phi) \overset{\text{def}}{=} V^i \frac{\partial L}{\partial y^i} - \tau E_L - \phi(x)$$

is conserved on the solution curves of the Euler–Lagrange equations $E_i(L) = 0$, $y^i = \dfrac{dx^i}{dt}$.

Proof. The equations $E_i(L) = 0$, $y^i = \dfrac{dx^i}{dt}$, Theorem 1.7.4 and (8.5)′ imply the conclusion of the Theorem.

Of course, all previous results are valid in a Lagrange space $L^n = (M, L)$.

1.9 Canonical Nonlinear Connection of a Lagrange Space

Let $L^n = (M, L)$ be a Lagrange space. Taking into account that $E_i(L)$ is a covector field and that the fundamental tensor $g_{ij}(x, y)$ is nondegenerate we can deduce:

Theorem 1.9.1 *If $L^n = (M, L)$ is a Lagrange space, then the system of differential equations*

$$g^{ij} E_j(L) = 0, \quad y^i = \frac{dx^i}{dt} \tag{9.1}$$

determines a spray, whose coefficients depend only on the Lagrangian $L(x, y)$.

Indeed, we have

$$E_j(L) = \frac{\partial L}{\partial x^j} - \frac{d}{dt}\frac{\partial L}{\partial y^j} = \frac{\partial L}{\partial x^j} - \left\{ \frac{\partial^2 L}{\partial y^j \partial x^h} y^h + 2g_{jh}\frac{dy^h}{dt} \right\}, \ \frac{dx^i}{dt} = y^i.$$

Hence (9.1) is equivalent to the system of differential equations

$$\frac{d^2 x^i}{dt^2} + 2G^i\left(x, \frac{dx}{dt}\right) = 0, \tag{9.2}$$

where

$$2G^i(x, y) = \frac{1}{2} g^{ij} \left\{ \frac{\partial^2 L}{\partial y^j \partial x^h} y^h - \frac{\partial L}{\partial x^j} \right\}. \tag{9.3}$$

The system of differential equations (9.2) has a geometrical meaning, since the system (9.1) has this property. So, (9.2) determines a spray with the coefficients (9.3). **q.e.d.**

Corollary 1.9.1 *The extremal curves of a Lagrange space $L^n = (M, L)$ are given by the system of differential equations (9.2), whose coefficients are given by (9.3).*

Applying Theorem 1.1.5, we obtain an important result:

Theorem 1.9.2 *For any Lagrange space $L^n = (M, L)$, on $\widetilde{TM}$, there exist nonlinear connections N which depend only on the fundamental function L. One of them has the coefficients*

$$N^i{}_j = \frac{\partial G^i}{\partial y^j} = \frac{1}{4}\frac{\partial}{\partial y^j}\left\{g^{ik}\left(\frac{\partial^2 L}{\partial y^k \partial x^h}y^h - \frac{\partial L}{\partial x^k}\right)\right\}. \tag{9.4}$$

Looking at (9.4), or using the property $E_i\left(L(x,y) + \frac{dF(x)}{dt}\right) = E_i(L(x,y))$, we get:

Proposition 1.9.1 *The nonlinear connection N, with the coefficients (9.4), is invariant to the transformations of Lagrangians $L(x,y)$ of the form*

$$L'(x,y) = L(x,y) + \frac{\partial \varphi(x)}{\partial x^i}y^i. \tag{9.5}$$

The nonlinear connection N mentioned in Theorem 1.9.2 is called *canonical* for the Lagrange space $L^n = (M, L)$.

Example 1.9.1 The Lagrange space of Electrodynamics. Let us consider the Lagrange space of electrodynamics, $L^n = (M, L)$, where L is the Lagrangian (6.4), with $U(x) = 0$, $\forall x \in M$. We denote $\gamma^i{}_{jk}(x)$ the Christoffel symbols of the metric tensor $\gamma_{ij}(x)$. The coefficients of the spray (9.2) are given by

$$G^i(x,y) = \frac{1}{2}\gamma^i{}_{jk}(x)y^j y^k - g^{ij}F_{jk}(x)y^k, \tag{9.6}$$

where $F_{jk}(x)$ is the electromagnetic tensor:

$$F_{jk}(x) = \frac{e}{2m}\left(\frac{\partial A_k}{\partial x^j} - \frac{\partial A_j}{\partial x^k}\right). \tag{9.7}$$

Therefore, the extremal curves of this Lagrange space are given by the *Lorentz equation:*

$$\frac{d^2x^i}{dt^2} + \gamma^i{}_{jk}(x)\frac{dx^j}{dt}\frac{dx^k}{dt} = g^{ij}(x)F_{jk}(x)\frac{dx^k}{dt}. \tag{9.8}$$

The canonical nonlinear connection of L^n has the coefficients

$$N^i{}_j(x,y) = \gamma^i{}_{jk}(x)y^k - g^{ik}(x)F_{kj}(x). \tag{9.9}$$

The fact that $N^i{}_j$ are linear in the variables y^i is important in applications.

Indeed, the tensor $R^i{}_{jk} = \dfrac{\delta N^i{}_j}{\delta x^k} - \dfrac{\delta N^i{}_k}{\delta x^j}$ is expressed by the curvature tensor of the metric tensor γ_{ij} and the Berwald connection $B\Gamma(N) = (B^i{}_{jk}, 0)$ has the coefficients

$$B^i{}_{jk} = \gamma^i{}_{jk}(x). \tag{9.10}$$

Finally, we remark:

Theorem 1.9.3 *In the Lagrange space of Electrodynamics $L^n = (M, L)$, having the fundamental function* (6.4), $(U(x) = 0)$, *the solution curves of Euler–Lagrange equations and the autoparallel curves of the canonical nonlinear connection are the solutions of the Lorentz equation* (9.8).

1.10 Canonical Metrical Connection

Let N be the canonical nonlinear connection of a Lagrange space $L^n = (M, L)$. It will be the basic ingredient in the notion of the canonical metrical connection of the space L^n which we introduce now.

Theorem 1.10.1 *The following properties hold:*

1° *There exists a unique N–linear connection D on $\widetilde{TM}$ verifying the axioms:*

$$g_{ij|k} = 0, \quad g_{ij}|_k = 0, \quad \text{and} \tag{10.1}$$

$$T^i{}_{jk} = 0, \quad S^i{}_{jk} = 0. \tag{10.2}$$

2° *This connection has as coefficients the generalized Christoffel symbols:*

$$(10.3)\qquad \begin{aligned} L^i{}_{jk} &= \frac{1}{2}g^{im}\left(\frac{\delta g_{mk}}{\delta x^j} + \frac{\delta g_{jm}}{\delta x^k} - \frac{\delta g_{jk}}{\delta x^m}\right), \\ C^i{}_{jk} &= \frac{1}{2}g^{im}\left(\frac{\partial g_{mk}}{\partial y^j} + \frac{\partial g_{jm}}{\partial y^k} - \frac{\partial g_{jk}}{\partial y^m}\right). \end{aligned}$$

3° *The previous connection depends only on the fundamental function $L(x, y)$ of the Lagrange space.*

Indeed, the N–linear connection D with the coefficients (10.3) satisfies (10.1), (10.2). The uniqueness of this connection can be obtained by contradiction. It is clearly that the coefficients $L^i{}_{jk}$ and $C^i{}_{jk}$ from (10.3) depend only on the fundamental function L, since the canonical nonliner connection N has the coefficients (9.4). **q.e.d.**

The connection D from the previous theorem will be called *canonical metrical connection* of the space L^n, and denoted with $C\Gamma(N)$.

Of course, by means of the canonical metrical connection we can study the geometrical problems of the Lagrange space L^n, for instance h–paths, v–paths, structure equations, parallelism, etc.

Let us consider the electromagnetic tensor of the space L^n endowed with the canonical metrical connection $C\Gamma(N)$. Taking into account the covariant deflection tensor $D_{ij} = g_{ik}D^k{}_j$, $d_{ij} = g_{ik}d^k{}_j$, we can introduce the so–called $h-$ and $v-$ electromagnetic tensors

$$(10.4)\qquad F_{ij} = \frac{1}{2}(D_{ij} - D_{ji}),\quad f_{ij} = \frac{1}{2}(d_{ij} - d_{ji}).$$

For the Lagrange space L^n we have $f_{ij} = 0$.

Applying the Ricci identities to the Liouville vector field, and using Theorem 1.4.7 and the definition of the tensors F_{ij} and f_{ij} we get:

Theorem 1.10.2

1° *The $h-$ and $v-$ electromagnetic tensors of the Lagrange space L^n, endowed with the connection $C\Gamma(N)$, are given by*

$$(10.5)\qquad F_{ij} = \frac{1}{2}\left(g_{io|j} - g_{jo|i}\right),\quad f_{ij} = 0,\quad (g_{io} = g_{ij}y^j).$$

2° *The electromagnetic tensors F_{ij} satisfy the Maxwell equations*

$$(10.6)\quad \begin{aligned} &F_{ij|k} + F_{jk|i} + F_{ki|j} = -(C_{oim}R^m{}_{jk} + C_{ojm}R^m{}_{ki} + C_{okm}R^m{}_{ij}) \\ &F_{ij}\big|_k + F_{jk}\big|_i + F_{ki}\big|_j = 0 \quad (C_{oij} = y^m C_{mij}). \end{aligned}$$

Corollary 1.10.1 *If the fundamental tensor $g_{ij}(x,y)$ of the Lagrange space L^n is 0–homogeneous with respect to y^i, then the h-electromagnetic tensor F_{ij} satisfies the following Maxwell equation*

$$(10.6)' \qquad F_{ij|k} + F_{jk|i} + F_{ki|j} = 0, \quad F_{ij}\big|_k + F_{jk}\big|_i + F_{ki}\big|_j = 0.$$

Of course, we can introduce the $h-$ and $v-$ charge–current densities and study their properties [211].

If L^n is the Lagrange space of electrodynamics, studied in the Example 1.9.1, then F_{jk} from (10.4) is the same with the classical electromagnetic tensor and the equations (10.6)′ are reduced to the classical Maxwell equations, since $F_{ij}\big|_k = 0$ and, on the other hand, $F_{ij|k}$ is the covariant derivation with respect to Levi–Civita connection of $\gamma_{ij}(x)$.

One can also prove the following

Theorem 1.10.3 *For $n > 2$, the Einstein equations of a Lagrange space L^n endowed with the canonical metrical connection $C\Gamma(N)$ are as follows*

$$(10.7)\quad \begin{aligned} &R_{ij} - \frac{1}{2}Rg_{ij} = \kappa \overset{H}{T}_{ij}, \quad 'P_{ij} = \kappa \overset{1}{T}_{ij}, \\ &S_{ij} - \frac{1}{2}Sg_{ij} = \kappa \overset{V}{T}_{ij}, \quad ''P_{ij} = -\kappa \overset{2}{T}_{ij}, \end{aligned}$$

where

$$(10.8)\quad \begin{aligned} &R_{ij} = R_i{}^m{}_{jm}, \quad S_{ij} = S_i{}^m{}_{jm}, \quad 'P_{ij} = P_i{}^m{}_{jm}, \quad ''P_{ij} = P_i{}^m{}_{mj}, \\ &R = g^{ij}R_{ij}, \qquad S = g^{ij}S_{ij}; \end{aligned}$$

κ is a constant and $\overset{H}{T}_{ij}$, $\overset{V}{T}_{ij}$, $\overset{1}{T}_{ij}$, $\overset{2}{T}_{ij}$ are the components of the energy–momentum tensor.

Of course, the d–tensors of energy–momentum satisfy some laws of conservation [195].

In the Lagrange space of electrodynamics, the equation (10.7) is reduced to the classical Einstein equations.

1.11 Finsler Spaces

An important class of Lagrange spaces is provided by the so–called Finsler spaces [179], [188], [261].

Definition 1.11.1 A Finsler space is a pair $F^n = (M, F)$, formed by a real n–dimensional manifold M and a scalar positive function F on TM, differentiable on $\widetilde{TM}$ and continuous on the null section, which has the properties:

1° $F(x, y)$ is positively homogeneous of degree 1, with respect to y^i on $\widetilde{TM}$.

2° The pair (M, F^2) is a Lagrange space.

Thus, to a Finsler space $F^n = (M, F)$ a Lagrange space $L^n = (M, F^2)$ corresponds. So we can apply the previous theory of Lagrange spaces to the Finsler spaces.

The function F is called the *fundamental* or *metric function* and the d–tensor field

(11.1)
$$g_{ij}(x,y) = \frac{1}{2}\frac{\partial^2 F^2}{\partial y^i \partial y^j}$$

is called the *fundamental* or *metric tensor* of the Finsler space F^n. The tensor field $g_{ij}(x, y)$ is 0–homogeneous with respect to y^i and it is nondegenerate. The Cartan tensor field

(11.2)
$$C_{ijk} = \frac{1}{4}\frac{\partial^3 F^2}{\partial y^i \partial y^j \partial y^k}$$

is totally symmetric and $C_{oij} = y^m C_{mij}$ has the property

(11.2)′
$$C_{oij} = 0.$$

If we denote by $\gamma^i{}_{jk}(x, y)$ the Christoffel symbols of the fundamental tensor field g_{ij} of the Finsler space F^n, after a straightforward calculus in the formula (9.3) we find the coefficients of the canonical spray (9.2):

(11.3)
$$G^i = \frac{1}{2}\gamma^i{}_{oo} = \frac{1}{2}\gamma^i{}_{jk}y^j y^k.$$

Therefore, from Theorem 1.9.2, the canonical nonlinear connection of the Finsler space F^n has the coefficients given by E. Cartan

$$N^i{}_j = \frac{1}{2}\frac{\partial \gamma^i{}_{oo}}{\partial y^j}. \tag{11.4}$$

The canonical metrical connection $C\Gamma(N)$ of F^n is also obtained from Theorem 1.10.1. It has the coefficients (10.3), denoted by $C\Gamma(N) = (F^i{}_{jk}, C^i{}_{jk})$, where $C^i{}_{jk} = g^{im}C_{mjk}$. $C\Gamma(N)$ is the famous *Cartan connection* of the Finsler space F^n.

It is not difficult to prove:

Theorem 1.11.1 *The Cartan connection $C\Gamma(N)$ of a Finsler space $F^n = (M, F)$ has the properties:*

1° $F_{|k} = 0,\ F^2{}_{|k} = 0,\ F^2\big|_k = \dfrac{\partial F^2}{\partial y^k}$.

2° *F is constant on the autoparallel curve of the canonical nonlinear connection.*

3° *The connection $C\Gamma(N)$ is of the Cartan type, i.e.* $D^i{}_j = 0$, $d^i{}_j = \delta^i{}_j$.

4° $y^m R_m{}^i{}_{jk} = R^i{}_{jk},\ y^m P_m{}^i{}_{jk} = P^i{}_{jk},\ y^m S_m{}^i{}_{jk} = 0$.

5° *The electromagnetic tensors F_{ij} and f_{ij} of $C\Gamma(N)$ vanish.*

6° *$C\Gamma(N)$ is a metrical connection:* $g_{ij|k} = 0,\ g_{ij}\big|_k = 0$.

7° *The $h-$ and $v-$ torsions $T^i{}_{jk}$, $S^i{}_{jk}$ of $C\Gamma(N)$ vanish.*

8° *The tensors $P_{ijk} = g_{im}P^m{}_{jk}$ are totally symmetric.*

9° $R_{ijhk} + R_{jihk} = P_{ijhk} + P_{jihk} = S_{ijhk} + S_{jihk} = 0$,
where $R_{ijhk} = g_{jm}R_i{}^m{}_{hk}$, *etc.*

10° $R_{ijk} + R_{jki} + R_{kij} = 0$, where $R_{ijk} = g_{im}R^m{}_{jk}$.

Examples.

11.1 Let $\gamma_{ij}(x)$ be a Riemannian structure on the base manifold M. The function $F(x,y) = \{\gamma_{ij}(x)y^iy^j\}^{1/2}$ is a fundamental function of a Finsler space. In fact, this is a Riemannian space, since its fundamental tensor is $g_{ij}(x,y) = \gamma_{ij}(x)$. The Cartan tensor (11.2) vanishes and this condition characterizes the Riemannian spaces in the class of Finsler spaces. So, for the classes of Riemannian spaces $\{R^n\}$, Finsler spaces $\{F^n\}$ and Lagrange spaces $\{L^n\}$ we have the inclusions

$$\{R^n\} \subset \{F^n\} \subset \{L^n\}. \tag{11.5}$$

11.2 The function $F(x,y)=\{\gamma_{ij}(x)y^iy^j\}^{1/2}+A_i(x)y^i$, with $A_i(x)y^i>0$, on $\widetilde{TM}$, is a fundamental function of a Finsler space. This is called a *Randers* space. It was introduced as Finsler space by R.Ingarden [15].

11.3 The function $F(x,y)=\dfrac{\gamma_{ij}(x)y^iy^j}{A_i(x)y^i}$, $A_i(x)\in\mathcal{X}^*(M)$ with $A_i(x)y^i>$ 0,
on $\widetilde{TM}$, gives us a fundamental function of a Finsler space, called *Kropina* space [15].

11.4 The function $F(x,y)|1 = e^{2\alpha_i x^i}\{(y^1)^m+\cdots+(y^n)^m\}^{1/m}$, m integer, $m\geq 3$, where $\alpha_i = const. \neq 0$ and F is expressed in a preferential local chart on $\widetilde{TM}$. F is a fundamental function of a Finsler space, called *Antonelli's Ecological metric*, [15], [16], [17].

1.12 Generalized Lagrange Spaces

A first natural generalization of the notion of Lagrange space is provided by a notion which we call a generalized Lagrange space. This notion was given by the author of the present book in the paper [195], [196] in the following form:

Definition 1.12.1 A generalized Lagrange space is a pair $GL^n =$ $= (M, g_{ij}(x,y))$, where $g_{ij}(x,y)$ is a d-tensor field on $\widetilde{TM}$, of type $(0,2)$, symmetric, of rank n and having a constant signature on $\widetilde{TM}$.

We continue to call $g_{ij}(x,y)$ the *fundamental tensor* of GL^n.

One easily can see that any Lagrange space $L^n = (M, L)$ is a generalized Lagrange space, with the fundamental tensor field:

$$g_{ij}(x,y) = \frac{1}{2}\frac{\partial^2 L}{\partial y^i \partial y^j}. \tag{12.1}$$

But not any GL^n–space is an L^n–space. Indeed, if $g_{ij}(x,y)$ is given, it may happen that the system of partial differential equations (12.1) does not admit any solution $L(x,y)$.

Proposition 1.12.1

1° *A necessary condition in order that the system* (12.1) *admit a solution* $L(x,y)$ *is that d–tensor field* $\dfrac{\partial g_{ij}}{\partial y^k}$ *be completely symmetric.*

2° *If the condition* 1° *is verified and* $g_{ij}(x,y)$ *is* 0*–homogeneous with respect to* y^i*, then the function*

$$L(x,y) = g_{ij}(x,y)y^i y^j + A_i(x)y^i + U(x) \tag{12.2}$$

is a solution of (12.1). *Here* $A_i(x)$ *is an arbitrary covector field and* U *is an arbitrary function on* M.

In the case when (12.1) does not admit solutions we say that the generalized Lagrange space $GL^n = (M, g_{ij}(x,y))$ is *not reducible* to a Lagrange space.

So, the inclusions (11.5) can be extended as follows

$$\{R^n\} \subset \{F^n\} \subset \{L^n\} \subset \{GL^n\}. \tag{12.3}$$

Remark 1.12.1 The Lagrange spaces with the fundamental function $L(x,y)$ from (12.2) give us an important class of Lagrange spaces which have several properties, similar to those of Finsler spaces.

Examples.

12.1 The pair $GL^n = (M, g_{ij})$, with

$$g_{ij}(x,y) = e^{2\sigma(x,y)}\gamma_{ij}(x), \tag{12.4}$$

when the covector field $\frac{\partial \sigma}{\partial y^i}$ does not vanish and $\gamma_{ij}(x)$ is a Lorentz metric, is a generalized Lagrange space. It is not reducible to a Lagrange space. R. Miron and R. Tavakol [213], showed that M endowed with the metric tensor (12.4) supports the Ehlers–Pirani–Shield axioms of General Relativity.

12.2 The pair $GL^n = (M, g_{ij})$, with

$$g_{ij}(x,y) = \gamma_{ij}(x) + \left(1 - \frac{1}{n^2(x,y)}\right) y_i y_j, \quad y_i = \gamma_{ij}(x) y^j, \tag{12.5}$$

where $\gamma_{ij}(x)$ is a Riemannian (or pseudo–Riemannian) metric tensor and $n(x,y) > 1$ is a refractive index, gives us a generalized Lagrange space.

The metric tensor $g_{ij}(x,y)$ from (12.5) restricted to the section $S_V : x^i = x^i,\ y^i = V^i(x)$ of the mapping $\pi : TM \to M$, where $V^i(x)$ is a vector field on M (assuming that V exists), gives us the known Synge's metric tensor of the relativistic geometrical optics. It was intensively studied in the papers [206], [207], [208].

For a generalized Lagrange space $GL^n = (M, g_{ij}(x,y))$, the first important notions associated to it are given by:

(a) the *absolute energy*

$$\mathcal{E}(x,y) = g_{ij}(x,y) y^i y^j \tag{12.6}$$

(b) the *energy* of $\mathcal{E}(x,y)$:

$$L_{\mathcal{E}} = y^i \frac{\partial \mathcal{E}}{\partial y^i} - \mathcal{E}. \tag{12.6$'$}$$

If $\mathcal{E}(x,y)$ is a regular Lagrangian, then we can determine a nonlinear connection, depending only on $g_{ij}(x,y)$. In this case we can develop the geometry of the generalized Lagrange space GL^n step by step using the same methods as in the case of the geometry of Lagrange spaces. In the contrary case, we give apriori a nonlinear connection N and study the pair (GL^n, N).

For instance, in Example 12.1 and 12.2 we can use the nonlinear connection N with the coefficients $N^i{}_j = \gamma^i{}_{jk}(x) y^k$.

1.13 Almost Kählerian Model of the Space L^n

We shall see that the Lagrange (or Finsler) space $L^n = (M, L)$ endowed with the canonical metrical connection $C\Gamma(N)$ can be thought of as an almost Kähler space on the smooth manifold $\widetilde{TM}$. We say that such a space is an almost Kählerian model of the Lagrange space L^n. Moreover, a Lagrangian theory of gravitational and electromagnetic fields can be geometrically studied much better on such a model, since the almost symplectic structure of the space is a symplectic one and the nonlinear connection is essential included into the mentioned model.

Let $L^n = (M, L(x,y))$ be a Lagrange space, having $g_{ij}(x,y)$ as fundamental tensor field, and $C\Gamma(N)$ as canonical nonlinear connection. Adapted basis to the horizontal distribution N and to the vertical distribution V is denoted as usual by $\left(\frac{\delta}{\delta x^i}, \frac{\partial}{\partial y^i}\right)$ and its dual by $(dx^i, \delta y^i)$, where

$$(13.1)\qquad \frac{\delta}{\delta x^i} = \frac{\partial}{\partial x^i} - N^j{}_i\frac{\partial}{\partial y^j}, \quad \delta y^i = dy^i + N^i{}_j dx^j \quad (i = 1, ..., n).$$

Let us define the $\mathcal{F}(E)$–linear mapping $\mathbb{F}$:

$$(13.2)\qquad \mathbb{F}\left(\frac{\delta}{\delta x^i}\right) = -\frac{\partial}{\partial y^i}, \quad \mathbb{F}\left(\frac{\partial}{\partial y^i}\right) = \frac{\delta}{\delta x^i}, \quad (i = 1, ..., n).$$

It is obvious that $\mathbb{F}$ is well–defined on $\widetilde{TM}$ and that it has the property:

$$(13.3)\qquad \mathbb{F}^2 = -I,$$

I being Kronecker's tensor field. Therefore, we have

Theorem 1.13.1

1° *The mapping $\mathbb{F}$ from (13.2) is an almost complex structure on $\widetilde{TM}$.*
2° *The structure $\mathbb{F}$ is integrable, if, and only if, the d–tensor field $R^i{}_{jk}$ vanishes.*

The metric tensor $g_{ij}(x,y)$ of the Lagrange space L^n induces a pseudo–Riemannian structure on $\widetilde{TM}$, given by

$$G = g_{ij}(x,y)dx^i \otimes dx^j + g_{ij}(x,y)\delta y^i \otimes \delta y^j. \tag{13.4}$$

With respect to G, the horizontal distribution and vertical distribution are orthogonal.

In this context we have, too:

Theorem 1.13.2

1° *The pair $(G, \mathbb{F})$ is an almost Hermitian structure on $\widetilde{TM}$.*

2° *The 2–form associated to the structure $(G, \mathbb{F})$ is given by*

$$\theta = g_{ij}(x,y)\delta y^i \wedge dx^j. \tag{13.5}$$

But θ is the Poincaré–Cartan 2–form associated to the Lagrangian L. Therefore we can consider the Poincaré–Cartan 1–form:

$$\omega = \frac{1}{2}\frac{\partial L}{\partial y^i}dx^i. \tag{13.5$'$}$$

So, we can prove:

Theorem 1.13.3 *The following equations hold:*

$$\theta = d\omega, \quad d\theta = 0, \tag{13.6}$$

d being the exterior differential operator.

Corollary 1.13.1 *The Poincaré–Cartan 2–form θ gives rise to a symplectic structure on $\widetilde{TM}$.*

Thus, we have

Theorem 1.13.4 *For any Lagrange space L^n, the corresponding Hermitian space $H^{2n} = (\widetilde{TM}, G, \mathbb{F})$ is an almost Kählerian space.*

The space $K^{2n} = (\widetilde{TM}, G, \mathbb{F})$ from the previous theorem will be called *the almost Kählerian model* of the Lagrange space L^n.

We know that K^{2n} becomes a Kähler space if the almost complex structure $\mathbb{F}$ is integrable. Hence we have:

Theorem 1.13.5 *The almost Kählerian model K^{2n} of the Lagrange space L^n is a Kählerian space if, and only if, the canonical nonlinear connection N is integrable.*

If we define the notion of N–linear connection of Lagrange type as an N–linear connection on $\widetilde{TM}$, denoted by D, which is metrical with respect to the metric structure G, then, by means of the property $D_X J = 0$, it follows $D_X \mathbb{F} = 0$. So, we have:

Theorem 1.13.6 *Any N–linear connection of Lagrange type is an almost Kählerian connection (i.e. $D_X G = 0$, $D_X \mathbb{F} = 0$).*

Theorem 1.13.7 *There exists a unique N–linear connection on $\widetilde{TM}$ of Lagrange type having h– and v– torsions zero. The coefficients $(L^i{}_{jk}, C^i{}_{jk})$ of D, in the adapted basis $\left(\frac{\delta}{\delta x^i}, \frac{\partial}{\partial y^i}\right)$ are just the coefficients $C\Gamma(N)$ of the canonical metrical connection.*

It is now clear that we can use the almost Kählerian model K^{2n} in order to study:

1° The Einstein equations of the Lagrange space L^n, defined as the Einstein equations of the space K^{2n} endowed with the canonical metrical connection of Lagrange type.

2° The electromagnetic field of the space L^n, using the same model K^{2n} endowed with the above mentioned connection.

Remark. We can construct an almost Hermitian model for a generalized Lagrange space $GL^n = (M, g_{ij}(x, y))$. Generally, this is not reducible to an almost Kählerian space as in the case of the Lagrange space L^n.

1.14 Problems

1. Determine the canonical nonlinear connection and the canonical metrical connection of the Finsler space with the fundamental function $F(x, y) = \{\gamma_{ij}(x) y^i y^i\}^{1/2}$, where $\gamma_{ij}(x)$ is a Riemannian structure on the base manifold.

2. Prove that in a Finsler space $F^n = (M, F(x,y))$ the extremal curves of the Lagrangian $L(x,y) = F^2(x,y)$, the autoparallel curves of the canonical non-linear connection and the solution curves of the variational problem on the functional $I(c) = \int_0^1 F\left(x, \frac{dx}{dt}\right) dt$ (in the canonical parametrization $F\left(x, \frac{dx}{dt}\right) = 1$) are solutions of the same system of differential equations:
$$\frac{d^2x^i}{dt^2} + \gamma^i{}_{jk}(x,y)\frac{dx^j}{dt}\frac{dx^k}{dt} = 0,$$
$\gamma^i{}_{jk}(x,y)$ being the Christoffel symbols of the fundamental tensor of space F^n.

3. Let $L^n = (M, L)$ be the so-called *almost Finsler–Lagrange* space i.e. its fundamental function is
$$L(x,y) = F^2(x,y) + A_i(x)y^i + U(x),$$
where $F(x,y)$ is the fundamental function of a Finsler space. Prove that:

   (a) L^n is not a Finsler space.
   (b) The fundamental tensor field $g_{ij}(x,y)$ of the space L^n coincides with the fundamental tensor field of the Finsler space $F^n = (M, F(x,y))$.

   Determine:

   (a)′ The canonical nonlinear connection of L^n.
   (b)′ The extremal curves of L^n.
   (c)′ The canonical metrical connection $C\Gamma(N)$.
   (d)′ Its almost Kählerian model.

4. Considering a generalized Lagrange space $GL^n = (M, g_{ij})$ endowed with a nonlinear connection N, determine its almost Hermitian model and establish when this space is

   (a) Hermitian,
   (b) Almost Kählerian,

(c) Kählerian,
(d) Apply the results to the case when $g_{ij} = e^{2\sigma(x,y)}\gamma_{ij}(x)$ and N has the coefficients $N^i{}_j = \gamma^i{}_{jk}(x)y^k$.

5. A generalized Randers space [15], [16] is a Lagrange space L^n with the fundamental function: $L(x,y) = (F(x,y) + A_i(x)y^i)^2$, where $F(x,y)$ is a Finsler metric function and $A_i(x)$ a d–covector field. Prove:

   (a) $F'(x,y) = F(x,y) + A_i(x)y^i$ is a Finsler metric function and $L^n = (M,L)$ is its associated Lagrange space.
   (b) Determine the fundamental tensor field of L^n.
   (c) Find the canonical nonlinear connection and the canonical metrical connection of L^n.

Chapter 2

The Geometry of 2–Osculator Bundle

The geometry of Lagrange spaces, studied in the previous chapter can be extended step by step to the higher order Lagrange spaces. In this case the base manifold is the so–called k–osculator bundle. It is a natural extension of the notion of 1–osculator bundle. So it is necessary to study the total space of the k–osculator bundle $(\mathrm{Osc}^k M, \pi, M)$.

In order to maximize the clarity of the next chapters we consider first the particular case $k = 2$. Therefore, we begin with the study of the geometry of the total space $\mathrm{Osc}^2 M$ of the 2–osculator bundle.

We show that on the manifold $\mathrm{Osc}^2 M$ there exist two independent Liouville vector fields, two special vertical distributions, a 2–tangent structure and 2–sprays. The fundamental concept of nonlinear connection and its relations with 2–sprays is pointed out.

This theory is based on Miron–Atanasiu's papers [197]–[204].

2.1 The Fibre Bundle $\mathrm{Osc}^2 M$

We shall introduce the 2–osculator bundle as a natural extension of the notion of 1–osculator bundle described in the first chapter.

Of course, we assume again that the geometrical object fields and mappings, used here are of class C^∞.

Let M be a real n–dimensional manifold. Two curves in M, ρ, σ :

$I \to M$, which have a common point $x_0 \in M$, $x_0 = \rho(0) = \sigma(0)$, $(0 \in I)$, have in x_0 "*a contact of order* 2" if for any function $f \in \mathcal{F}(U)$, $x_0 \in U$ and U is an open set in M, we have

$$(1.1)\qquad \begin{aligned} \frac{d}{dt}(f \circ \rho)(t)|_{t=0} &= \frac{d}{dt}(f \circ \sigma)(t)|_{t=0}, \\ \frac{d^2}{dt^2}(f \circ \rho)(t)|_{t=0} &= \frac{d^2}{dt^2}(f \circ \sigma)(t)|_{t=0}. \end{aligned}$$

The relation "*to have a contact of order* 2" is an equivalence on the set of curves in M, which pass through the point x_0. We denote by $[\rho]_{x_0}$ a class of equivalence. It will be called a "*2–osculator space*" in the point x_0 of M.

Let $\mathrm{Osc}^2_{x_0}$ be the set of 2–osculator spaces in the point x_0 of M and let us consider the set

$$(1.2)\qquad \mathrm{Osc}^2 M = \bigcup_{x_0 \in M} \mathrm{Osc}^2_{x_0},$$

together with the mapping

$$(1.3)\qquad \pi : \mathrm{Osc}^2 M \to M, \quad \pi([\rho]_{x_0}) = x_0, \quad \forall [\rho]_{x_0} \in \mathrm{Osc}^2 M.$$

The set $\mathrm{Osc}^2 M$ has a natural differentiable structure induced by that of M, so that π becomes a differentiable mapping. This can be described as follows.

Let (U, φ) be a local chart of M, $x_0 \in U$ and a curve $\rho : I \to M$ represented in (U, φ) by

$$(1.4)\qquad x^i = x^i(t),\ t \in I,\ \ x^i_0 = x^i(0),\ (0 \in I),$$

x^i_0 being the coordinates of the point x_0.

Taking the function f from (1.1) succesively equal to the coordinate functions, a representative of class $[\rho]_{x_0}$ is given by

$$(1.5)\qquad x^{*i}(t) = x^i(0) + t\frac{dx^i}{dt}(0) + \frac{1}{2}t^2\frac{d^2x^i}{dt^2}(0),\ t \in (-\varepsilon, \varepsilon) \subset I.$$

The polynomial function from the right hand side of (1.5) is determined by its coefficients:

$$(1.6)\qquad x^i_0 = x^i(0),\ y^{(1)i}_0 = \frac{dx^i}{dt}(0),\ y^{(2)i}_0 = \frac{1}{2}\frac{d^2x^i}{dt^2}(0),$$

and it results that the pair $(\pi^{-1}(U), \phi)$ is a local chart on $\mathrm{Osc}^2 M$, where $\phi([\rho]_{x_0}) = (x_0^i, y_0^{(1)i}, y_0^{(2)i}) \in R^{3n}$.

Indeed, $\pi^{-1}(U)$ is an open set and it is homeomorphic to the open set $U \times R^{2n}$, by means of the homeomorphism ϕ.

Therefore, a differentiable atlass $\mathcal{A}_M$ of the differentiable structure of the manifold M determines a differentiable atlass $\mathcal{A}_{\mathrm{Osc}^2 M}$ on $\mathrm{Osc}^2 M$. So that the triple $(\mathrm{Osc}^2 M, \pi, M)$ is a differentiable bundle. The mapping $\pi : \mathrm{Osc}^2 M \to M$ is a differentiable submersion.

By (1.6), a transformation of local coordinates $(x^i, y^{(1)i}, y^{(2)i}) \to$ $\to (\tilde{x}^i, \tilde{y}^{(1)i}, \tilde{y}^{(2)i})$ on the manifold $\mathrm{Osc}^2 M$ is given by

$$\text{(1.7)} \quad \begin{aligned} &\tilde{x}^i = \tilde{x}^i(x^1, .., x^n), \ \det \left\| \frac{\partial \tilde{x}^i}{\partial x^j} \right\| \neq 0, \\ &\tilde{y}^{(1)i} = \frac{\partial \tilde{x}^i}{\partial x^j} y^{(1)j}, \ 2\tilde{y}^{(2)i} = \frac{\partial \tilde{y}^{(1)i}}{\partial x^j} y^{(1)j} + 2 \frac{\partial \tilde{y}^{(1)i}}{\partial y^{(1)j}} y^{(2)j}. \end{aligned}$$

It is important to remark that, from (1.7), it follows

$$\text{(1.7)}' \qquad \frac{\partial \tilde{x}^i}{\partial x^j} = \frac{\partial \tilde{y}^{(1)i}}{\partial y^{(1)j}} = \frac{\partial \tilde{y}^{(2)i}}{\partial y^{(2)j}}; \ \frac{\partial \tilde{y}^{(1)j}}{\partial x^j} = \frac{\partial \tilde{y}^{(2)i}}{\partial y^{(1)j}}.$$

A point $u \in \mathrm{Osc}^2 M$ with the coordinates $(x^i, y^{(1)i}, y^{(2)i})$ will be denoted also by $u = (x, y^{(1)}, y^{(2)})$.

Therefore, we can consider the projections

$$\text{(1.8)} \quad \begin{aligned} &\pi = \pi_0^2 : (x, y^{(1)}, y^{(2)}) \in \mathrm{Osc}^2 M \longrightarrow (x) \in M, \\ &\pi_1^2 : (x, y^{(1)}, y^{(2)}) \in \mathrm{Osc}^2 M \longrightarrow (x, y^{(1)}) \in \mathrm{Osc}^1 M. \end{aligned}$$

Clearly, π_1^2 is also a submersion.

A mapping $s : M \longrightarrow \mathrm{Osc}^2 M$ is said to be a *section* of the projection $\pi : \mathrm{Osc}^2 M \longrightarrow M$ if $\pi \circ s = 1_M$ and called a *local section* if $\pi \circ s\big|_U = 1_U$, for an open set $U \subset M$. Of course, if $c : I \longrightarrow M$ is a curve on M, we can define the notion of section of the projection π along the curve c. It will be a mapping $s : M \longrightarrow \mathrm{Osc}^2 M$ with the property $\pi \circ s\big|_c = 1_c$.

Let us consider a curve $c : I \longrightarrow M$, represented in a local chart (U, φ) by $x^i(t)$, $t \in I$. Thus, the mapping $\tilde{c} : I \longrightarrow \mathrm{Osc}^2 M$, given on $\pi^{-1}(U)$ by

$$\text{(1.9)} \qquad x^i = x^i(t), \ y^{(1)i} = \frac{dx^i}{dt}(t), \ y^{(2)i} = \frac{1}{2} \frac{d^2 x^i}{dt^2}(t), \ t \in I,$$

is a section of the projection π along the curve c. $\tilde{c}$ is a curve in $\mathrm{Osc}^2 M$, called the *extension* to $\mathrm{Osc}^2 M$ of the curve c.

More general, let us consider a vector field V on $U \subset M$ and its restriction $V_{|c}$ to the curve $c : I \longrightarrow U$. V can be given by its local coordinates $V^i(x(t)) = V^i(t),\ t \in I$.

Thus, the mapping $S_V : c \longrightarrow \mathrm{Osc}^2 M$ defined by

$$(1.10) \qquad S_V : x^i = x^i(t),\ y^{(1)i} = V^i(t),\ y^{(2)i} = \frac{1}{2}\frac{dV^i}{dt}(t),\ t \in I$$

is a section of the projection π along the curve c.

Indeed, a transformation of local coordinates on M gives:

$$\tilde{V}^i = \frac{\partial \tilde{x}^i}{\partial x^j} V^j, \quad \frac{d\tilde{V}^i}{dt} = \frac{\partial \tilde{y}^{(1)i}}{\partial x^j} V^j + \frac{\partial \tilde{y}^{(1)i}}{\partial y^{(1)j}} \frac{dV^j}{dt}.$$

So, the equalities (1.10) are preserved by a transformation of local coordinates (1.7) on $\mathrm{Osc}^2 M$.

In particular, $S_{\dot{c}} : c \longrightarrow \mathrm{Osc}^2 M$ is the extension $\tilde{c}$ of the curve c.

We can see that $\mathrm{Osc}^2 : \mathrm{Man} \longrightarrow \mathrm{Man}$ is a covariant functor from the category of differentiable manifolds, Man, to itself. Namely:

$$\mathrm{Osc}^2 : M \in \mathrm{Ob\,Man} \longrightarrow \mathrm{Osc}^2 M \in \mathrm{Ob\,Man},$$
$$\mathrm{Osc}^2 : \{f : M \longrightarrow N\} \longrightarrow \{\mathrm{Osc}^2 f : \mathrm{Osc}^2 M \longrightarrow \mathrm{Osc}^2 N\}.$$

If $f : x \in M \longrightarrow f(x) \in N$, in local coordinates is represented by

$$(1.11) \qquad x^{i'} = x^{i'}(x^1, ..., x^n),\ (i', j', ... = 1, ..., m = \dim N),$$

then $\mathrm{Osc}^2 f$ is given by

$$(1.12) \qquad \begin{aligned} & x^{i'} = x^{i'}(x^1, ..., x^n) \\ & y^{(1)i'} = \frac{\partial x^{i'}}{\partial x^j} y^{(1)j},\ 2y^{(2)i'} = \frac{\partial y^{(1)i'}}{\partial x^j} y^{(1)j} + 2\frac{\partial y^{(1)i'}}{\partial y^{(1)j}} y^{(2)j}. \end{aligned}$$

It follows that Osc^2 satisfies the properties of a covariant functor. We use these considerations in the theory of submanifolds of the manifold $\mathrm{Osc}^2 M$.

In order to assure the global existence of the various geometrical objects on $\mathrm{Osc}^2 M$ the following result is useful.

Theorem 2.1.1 *If the differentiable manifold M is paracompact, then* $\mathrm{Osc}^2 M$ *is a paracompact differentiable manifold, too.*

Proof. Let $\{(U_\alpha, \varphi_\alpha)_{\alpha\in I}\}$ be the complete atlass of M. Thus, $\{(\pi^{-1}(U_\alpha)_{\alpha\in I}\}$ give us the domains of local charts on $\mathrm{Osc}^2 M$. But, every open set $\pi^{-1}(U_\alpha)$ is homeomorphic to the open set $U_\alpha \times R^{2n}$ from $M \times R^{2n}$. Thus, we can determine a base $\mathcal{B} = \{U_\alpha \times W_{\alpha'}\}$, $\alpha \in I$, $\alpha' \in I'$ of the topology of $\mathrm{Osc}^2 M$, where $\{W_{\alpha'}\}_{\alpha'\in I'}$ is a base of topology of R^{2n}. For any two distinct points $(x, y^{(1)}, y^{(2)})$ and $(x', y'^{(1)}, y'^{(2)})$ of $\mathrm{Osc}^2 M$, if we have $\pi(x, y^{(1)}, y^{(2)}) \neq \pi(x', y'^{(1)}, y'^{(2)})$, there exist two open sets $U_\alpha \ni x$, $U_\beta \ni x'$, with the property $U_\alpha \cap U_\beta = \emptyset$. It follows that $\pi^{-1}(U_\alpha) \cap \pi^{-1}(U_\beta) = \emptyset$. In the case $x = x'$, there exist disjoint open sets $W_{\alpha'}, W_{\beta'}$ such that $(y^{(1)}, y^{(2)}) \in W_{\alpha'}$, $(y'^{(1)}, y'^{(2)}) \in W_{\beta'}$ and an open set $U_\alpha \ni x = x'$. It follows again that $\pi^{-1}(U_\alpha) \cap \pi^{-1}(U_\beta) = \emptyset$. Consequently, the topological space $\mathrm{Osc}^2 M$ is a separate Hausdorff one.

Let $\mathcal{A}$ be any open covering of the topological space $\mathrm{Osc}^2 M$. Since $\mathcal{B}$ is a base of the topology of $\mathrm{Osc}^2 M$ there exists an open covering $\mathcal{B}' \in$ $\in \{(U_\alpha \times W_{\alpha'}\}_{\alpha\in J, \alpha'\in J'}$; $J \subset I$, $J' \subset I'$ of $\mathrm{Osc}^2 M$, which refines $\mathcal{A}$ so that $\{U_\alpha\}_{\alpha\in J}$ be an open local finite covering of M, and $\{W_{\alpha'}\}_{\alpha'\in J'}$ be an open local finite covering of R^{2n}. Consequently, $\mathcal{B}'$ is an open local finite covering which refines $\mathcal{A}$. Thus, $\mathrm{Osc}^2 M$ is paracompact.

q.e.d.

2.2 Vertical Distributions. Liouville Vector Fields

As usually, we set

$$(2.1)\qquad \begin{aligned} &E = \mathrm{Osc}^2 M, \\ &\tilde{E} = \mathrm{Osc}^2 M \setminus \{0\} = \{(x, y^{(1)}, y^{(2)}) \in \mathrm{Osc}^2 M \mid rank\|y^{(1)i}\| = 1\}. \end{aligned}$$

Note that the condition $rank\|y^{(1)i}\| = 1$ has a geometrical character. So, $\tilde{E}$ is an open set of E. It will be endowed with the differentiable structure induced by that of E.

The differential of the projection $\pi : E \longrightarrow M$ is a mapping $d\pi : TE \rightarrow TM$ of tangent bundles. Let $V(E)$ be the kernel of $d\pi$. Thus,

$V(E)$ is a subbundle of the tangent bundle TE, called the vertical bundle. The fibres of $V(E)$ provide the vertical distribution V_1 on E:

$$V_1 : u \in E \longrightarrow V_1(u) \in T_uE. \tag{2.2}$$

Therefore, a local basis of the vertical distribution is $\left(\dfrac{\partial}{\partial y^{(1)i}}, \dfrac{\partial}{\partial y^{(2)i}}\right)$, $(i = 1, ..., n)$. Consequently, the vertical distribution V_1 is an integrable distribution of the local dimension $2n$.

Similarly, the projection $\pi_1^2 : E = \mathrm{Osc}^2 M \longrightarrow \mathrm{Osc}^1 M$ gives rise to the bundle mapping $d\pi_1^2 : TE \longrightarrow T(\mathrm{Osc}^1 M)$. Its kernel determines a vector subbundle whose fibres provide a new vertical distribution:

$$V_2 : u \in E \longrightarrow V_2(u) \subset T_uE. \tag{2.2$'$}$$

A local basis of V_2 is $\left\{\dfrac{\partial}{\partial y^{(2)1}}, ..., \dfrac{\partial}{\partial y^{(2)n}}\right\}$. So, V_2 is an integrable distribution, has the local dimension n and is a subdistribution of V_1.

Therefore, in every point $u \in E$, we have the vector spaces $V_2(u)$, $V_1(u)$, T_uE of dimensions n, $2n$, $3n$, respectively, and satisfying the inclusions:

$$V_2(u) \subset V_1(u) \subset T_uE, \quad \forall u \in E. \tag{2.2$''$}$$

By calculating the Jacobian of the transformation of coordinates (1.7), one finds the transformation for the local natural basis $\left(\dfrac{\partial}{\partial x^i}, \dfrac{\partial}{\partial y^{(1)i}}, \dfrac{\partial}{\partial y^{(2)i}}\right)$ of the module of vector fields $\mathcal{X}(E)$, in the form

$$\begin{aligned}
\frac{\partial}{\partial x^j} &= \frac{\partial \tilde{x}^i}{\partial x^j}\frac{\partial}{\partial \tilde{x}^i} + \frac{\partial \tilde{y}^{(1)i}}{\partial x^j}\frac{\partial}{\partial \tilde{y}^{(1)i}} + \frac{\partial \tilde{y}^{(2)i}}{\partial x^j}\frac{\partial}{\partial \tilde{y}^{(2)i}},\\
\frac{\partial}{\partial y^{(1)j}} &= \frac{\partial \tilde{y}^{(1)i}}{\partial y^{(1)j}}\frac{\partial}{\partial \tilde{y}^{(1)i}} + \frac{\partial \tilde{y}^{(2)i}}{\partial y^{(1)j}}\frac{\partial}{\partial \tilde{y}^{(2)i}},\\
\frac{\partial}{\partial y^{(2)j}} &= \frac{\partial \tilde{y}^{(2)i}}{\partial y^{(2)j}}\frac{\partial}{\partial \tilde{y}^{(2)i}}.
\end{aligned} \tag{2.3}$$

Taking into account $(1.7)'$, we deduce that $\left\{\dfrac{\partial}{\partial y^{(2)i}}\right\}$, $(i = 1, ..., n)$ span locally the distribution V_2 and $\left\{\dfrac{\partial}{\partial y^{(1)i}}, \dfrac{\partial}{\partial y^{(2)i}}\right\}$, $(i = 1, ..., n)$ span locally the distribution V_1, and $(2.2)''$ holds.

Based on (2.3) we can prove without difficulties:

Theorem 2.2.1
1° *The following operators in the algebra of functions on E, $\mathcal{F}(E)$:*

$$\overset{1}{\Gamma} = y^{(1)i} \frac{\partial}{\partial y^{(2)i}}, \quad \overset{2}{\Gamma} = y^{(1)i} \frac{\partial}{\partial y^{(1)i}} + 2y^{(2)i} \frac{\partial}{\partial y^{(2)i}} \tag{2.4}$$

are two vector fields, globally defined on E and linearly independent of $\check{E}$.

2° $\overset{1}{\Gamma}$ *belongs to the distribution V_2 and* $\overset{2}{\Gamma}$ *belongs to the distribution V_1.*

Indeed, with respect to (1.7), (2.3) we get that $\overset{1}{\Gamma}$ and $\overset{2}{\Gamma}$ have a geometrical meaning. Evidently, $\overset{1}{\Gamma}$ and $\overset{2}{\Gamma}$ are linearly independent of $\check{E}$ and $\overset{1}{\Gamma} \subset V_2$, $\overset{2}{\Gamma} \subset V_1$.

The vector fields $\overset{1}{\Gamma}, \overset{2}{\Gamma}$ are called the *Liouville vector fields.* Their presence is essential in our construction of the theory of Lagrangians of second order.

In applications we shall use also the operator

$$\Gamma = y^{(1)i} \frac{\partial}{\partial x^i} + 2y^{(2)i} \frac{\partial}{\partial y^{(1)i}}. \tag{2.5}$$

By a direct calculation one can check the following lemma:

Lemma 2.2.1
a. *Under a coordinate transformation* (1.7) *on E, Γ changes as follows*

$$\Gamma = \widetilde{\Gamma} + \left(y^{(1)j} \frac{\partial \tilde{y}^{(2)m}}{\partial x^j} + 2y^{(2)j} \frac{\partial \tilde{y}^{(2)m}}{\partial y^{(1)j}} \right) \frac{\partial}{\partial \tilde{y}^{(2)m}}. \tag{2.5$'$}$$

b. *For any function $f \in \mathcal{F}(E)$ having the property* $\dfrac{\partial f}{\partial y^{(2)i}} = 0$, *with respect to* (1.7), *we have*

$$\Gamma f = \widetilde{\Gamma}\, \tilde{f}. \tag{2.5)''}$$

Then, Γ *is not* a vector field.

2.3 2–Tangent Structure. 2–Sprays

An extremely important structure on E is the so–called 2–*tangent structure*, introduced by Eliopoulous [85], [86]. It is defined as a $\mathcal{F}(E)$–linear mapping $J : \mathcal{X}(E) \longrightarrow \mathcal{X}(E)$ given on the natural basis of $\mathcal{X}(E)$ by:

$$J\left(\frac{\partial}{\partial x^i}\right) = \frac{\partial}{\partial y^{(1)i}}, \; J\left(\frac{\partial}{\partial y^{(1)i}}\right) = \frac{\partial}{\partial y^{(2)i}}, \; J\left(\frac{\partial}{\partial y^{(2)i}}\right) = 0, \quad (i = 1, ..., n). \tag{3.1}$$

Theorem 2.3.1 *The following properties hold:*
1° *The 2–tangent structure J is globally defined on E;*
2° *J is a tensor field of type* (1,1) *on E;*
3° *J is an integrable structure;*
4° $\operatorname{Im} J = V_1$, $\operatorname{Ker} J = V_2$, $J(V_1) = V_2$;
5° $\operatorname{rank} \|J\| = 2n$;
6° $J \overset{2}{\Gamma} = \overset{1}{\Gamma}$, $\quad J \overset{1}{\Gamma} = 0$;
7° $J^3 = 0$.

Proof. 1° Assuming (3.1) holds in the coordinates $(\tilde{x}^i, \tilde{y}^{(1)i}, \tilde{y}^{(2)i})$ and applying J to both members of (2.3) we get $J\left(\dfrac{\partial}{\partial x^j}\right) = \dfrac{\partial \tilde{x}^i}{\partial x^j}\dfrac{\partial}{\partial \tilde{y}^{(1)j}} + \dfrac{\partial \tilde{y}^{(1)i}}{\partial x^j}\dfrac{\partial}{\partial \tilde{y}^{(2)i}} = \dfrac{\partial}{\partial y^{(1)j}}$; $\quad J\left(\dfrac{\partial}{\partial y^{(1)j}}\right) = \dfrac{\partial \tilde{y}^{(1)i}}{\partial y^{(1)j}}\dfrac{\partial}{\partial \tilde{y}^{(2)i}} = \dfrac{\partial}{\partial y^{(2)j}}$ and $J\left(\dfrac{\partial}{\partial y^{(2)j}}\right) = 0$. Consequently, J is preserved by the changes of local coordinates on E and is defined in every point on E.

2° Follows from the definition of J.
3° The Nijenhuis tensor N_J of J vanishes, by virtue of (3.1).
4°–7° Follow from (3.1), too. **q.e.d.**

Now, we can introduce the following:

Definition 2.3.1 *A* 2–spray on E is a vector field $S \in \mathcal{X}(E)$ with the property

$$JS = \overset{2}{\Gamma} . \tag{3.2}$$

We repeat the remark made in §1, Ch.1. Here, a 2–spray S is not homogeneous in any way [195]. So, this is a nonstandard denomination. But it is convenient in our study of the higher order Lagrangians.

Obviously, there not always exists a vector field S on E with the property (3.2). Therefore, the notion of local spray must be introduced. In this case, $S \in \mathcal{X}(U)$, U is an open set in E, with the property (3.2) that gives a 2–spray on U. We get:

Theorem 2.3.2

1° *A 2–spray S can be uniquely written in local coordinates in the form:*

$$S = \Gamma - 3G^i(x, y^{(1)}, y^{(2)})\frac{\partial}{\partial y^{(2)}} \quad \text{or} \tag{3.3}$$

$$S = y^{(1)i}\frac{\partial}{\partial x^i} + 2y^{(2)i}\frac{\partial}{\partial y^{(1)i}} - 3G^i(x, y^{(1)}, y^{(2)})\frac{\partial}{\partial y^{(2)i}}. \tag{3.3'}$$

2° *With respect to* (1.7) *the coefficients* G^i *change as follows:*

$$3\widetilde{G}^i = 3\frac{\partial \tilde{x}^i}{\partial x^j}G^j - \left(y^{(1)j}\frac{\partial \tilde{y}^{(2)i}}{\partial x^j} + 2y^{(2)j}\frac{\partial \tilde{y}^{(2)i}}{\partial y^{(1)j}}\right). \tag{3.4}$$

3° *If the functions* $G^i(x, y^{(1)}, y^{(2)})$ *are given on every domain of local chart of* E, *so that* (3.4) *holds, then the vector field* $S = \Gamma - 3G^i(x, y^{(1)i}, y^{(2)i})\frac{\partial}{\partial y^{(2)i}}$ *is a 2–spray.*

Proof. 1° If $S = a^i \dfrac{\partial}{\partial x^i} + b^i \dfrac{\partial}{\partial y^{(1)i}} + c^i \dfrac{\partial}{\partial y^{(2)i}}$ is a 2–spray, the condition (3.2) leads to $a^i = y^{(1)i}$, $b^i = 2y^{(2)i}$ and we put $c^i = -3G^i$.
2° The transformations (1.7) and (2.3) give us $S = \widetilde{S} = \widetilde{\Gamma} - 3\widetilde{G}^i \dfrac{\partial}{\partial \tilde{y}^{(2)i}} =$
$= \Gamma - 3G^i \dfrac{\partial}{\partial y^{(2)i}}$. Now, using (2.5)′, we obtain (3.4).
3° Being given the functions G^i being given on every $\pi^{-1}(U)$ and satisfying (3.4), it follows that S from (3.3) is a vector field defined on E and $JS = \overset{2}{\Gamma}$. **q.e.d.**

Let $c : I \to M$ be a curve, represented in a coordinate neighbourhood U by $x^i = x^i(t)$, $t \in I$, and $\tilde{c} : I \to E$ be its extension to $\mathrm{Osc}^2 M$, represented on $\pi^{-1}(U)$ by

$$(*) \qquad x^i = x^i(t),\ y^{(1)i} = \frac{dx^i}{dt}(t),\ y^{(2)i} = \frac{1}{2}\frac{d^2x^i}{dt^2}(t),\quad t \in I,$$

The curve c is called a *path* of a 2–spray S (from (3.3)′) if $\tilde{c}$ is an integral curve of S.

Clearly, the notion of path of the 2–spray S has a geometrical character. We obtain:

Theorem 2.3.3

1° *The paths of the 2–spray S in (3.3) are given by the differential equations*

$$(3.5) \qquad \frac{d^3x^i}{dt^3} + 3!G^i\left(x, \frac{dx}{dt}, \frac{1}{2}\frac{d^2x}{dt^2}\right) = 0.$$

2° *If on the base manifold M, the equations (3.5) are given and they have a geometrical character (i.e. the first member is a vector field), then the function $G^i(x, y^{(1)}, y^{(2)})$ are the coefficients of a local 2–spray.*

Proof. 1° $\tilde{c}$ from $(*)$ is an integral curve of S if, and only if,

$$(3.5)' \qquad \frac{dx^i}{dt} = y^{(1)i},\ \frac{dy^{(1)i}}{dt} = 2y^{(2)i},\ \frac{dy^{(2)i}}{dt} = -3G^i(x, y^{(1)}, y^{(2)}).$$

Consequently (3.5) holds.

2° If (3.5) is preserved by the transformations of local coordinates (1.7) and $\tilde{t} = t$, then it follows that (G^i) is changed by the rule (3.4). Applying the Theorem 2.3.2, it follows that G^i are the coefficients of a 2–spray S of the form (3.3)′. **q.e.d.**

Using the previous theorem we prove:

Theorem 2.3.4 *If the base manifold M is paracompact, then on $E = \mathrm{Osc}^2 M$ there exist 2–sprays.*

Proof. Theorem 2.1.1 shows that E is a paracompact manifold. Let g be a Riemannian metric on M with local coefficients $\gamma_{ij}(x)$ and $\gamma^i_{jk}(x)$ as its Christoffel symbols. It is easy to prove that

$$z^{(2)i} = y^{(2)i} + \frac{1}{2}\gamma^i{}_{jk}(x) y^{(1)j} y^{(1)k} \tag{3.6}$$

is a distinguished vector field. Thus, with respect to (1.7), we have $\tilde{z}^{(2)i} = \dfrac{\partial \tilde{x}^i}{\partial x^j} z^{(2)j}$. It follows that the function

$$L(x, y^{(1)}, y^{(2)}) = \gamma_{ij}(x) z^{(2)i} z^{(2)j} \tag{3.7}$$

does not depend on the transformations of coordinates (1.7). Therfore, we can consider the functions

$$\begin{aligned} 3G^i &= \frac{1}{2}\gamma^{ij}(x)\left\{ y^{(1)m}\frac{\partial}{\partial x^m}\left(\frac{\partial L}{\partial y^{(2)i}}\right) + \right. \\ &\left. +2y^{(2)m}\frac{\partial}{\partial y^{(1)m}}\left(\frac{\partial L}{\partial y^{(2)j}}\right) - \frac{\partial L}{\partial y^{(1)j}}\right\}. \end{aligned} \tag{3.8}$$

It is not difficult to prove that, with respect to (1.7), the functions $G^i(x, y^{(1)}, y^{(2)})$ from (3.8) obey the transformation law (3.4). From Theorem 2.3.2 we deduce that G^i are the coefficients of a 2–spray.
q.e.d.

Finally, if we consider a 2–spray S with the coefficients $G^i(x, y^{(1)}, y^{(2)})$ and we denote

$$\underset{(1)}{N}{}^i{}_j = \frac{\partial G^i}{\partial y^{(2)j}} \tag{3.9}$$

we can prove:

Theorem 2.3.5 *The functions $\underset{(1)}{N}{}^i{}_j$ from (3.9), determined by a 2–spray S, give us the components of a geometrical object on E. With respect to the transformations (1.7), $\underset{(1)}{N}{}^i{}_j$ is transformed by the rule*

$$\underset{(1)}{\tilde{N}}{}^i{}_m \frac{\partial \tilde{x}^m}{\partial x^j} = \underset{(1)}{N}{}^m{}_j \frac{\partial \tilde{x}^i}{\partial x^m} - \frac{\partial \tilde{y}^{(1)i}}{\partial x^j}. \tag{3.10}$$

Proof. We get (3.10) from (3.4) by applying to it the operator $\dfrac{\partial}{\partial y^{(2)j}} = \dfrac{\partial \tilde{x}^m}{\partial x^j}\dfrac{\partial}{\partial \tilde{y}^{(2)m}}$ and remarking that $\dfrac{\partial}{\partial y^{(2)j}}\left(\dfrac{\partial \tilde{y}^{(2)i}}{\partial x^m}\right) = \dfrac{\partial^2 \tilde{x}^i}{\partial x^j \partial x^m}$. But (3.10) gives us the rule of transformation of the components of a geometrical object field on E. **q.e.d.**

The system of functions $\underset{(1)}{N}{}^i{}_j$ is important to define the notion of nonlinear connection.

2.4 Nonlinear Connections

The tangent bundle of $E = \mathrm{Osc}^2 M$ is (TE, τ_E, E). We also take into account the vector bundles (TM, τ, M) and $(TE, d\pi, TM)$, where $d\pi : TE \to TM$ is he tangent mapping of the projection $\pi : E \to M$.

Let us denote by $(\pi^*(TM), \pi^*(\tau), E)$ the pull–back of the bundle $\tau : TM \to M$ through π.

Let the map $\overline{\pi}^* : TE \longrightarrow \pi^*(TM)$ be defined by $\overline{\pi}^*(X_u) = (u, d\pi_u(X_u))$, $\forall X_u \in T_u E$. As it is known, the following sequence of vector bundles over E is exact:

$$0 \longrightarrow VE \xrightarrow{\ i\ } TE \xrightarrow{\ \overline{\pi}^*\ } \pi^*(TM) \longrightarrow 0, \tag{4.1}$$

where i is the canonical injection.

Definition 2.4.1 A nonlinear connection on E is a splitting C on the left of the exact sequence (4.1).

It means that $C : TE \longrightarrow VE$ is a vector bundle morphism with the property $C \circ i = \mathrm{Id}$.

The kernel of the morphism C is a vector subbundle N of TE. So that it follows:

Theorem 2.4.1 *There exists a nonlinear connection on the 2–osculator bundle E if, and only if, there exists a vector subbundle NE of TE, so that the Whitney sum*

$$TE = NE \oplus VE \tag{4.2}$$

holds.

NE is called a *horizontal subbundle* of TE.

The fibres of NE on E will be denoted by N_u, $u \in E$.

It follows that the mapping $N : u \in E \longrightarrow N_u \subset T_uE$ is a differentiable distribution on E. From (4.2) we get that the linear spaces N_u and V_u are supplementary in T_uE. We have

$$T_uE = N_u \oplus V_u, \quad \forall u \in E. \tag{4.3}$$

Corollary 2.4.1 *A nonlinear connection on E is a distribution on E,* $N : u \in E \longrightarrow N_u \subset T_uE$ *so that* (4.3) *holds.*

Generally, we consider a nonlinear connection on E from the point of view of Corollary 2.4.1. We denote it by N and call it a horizontal distribution. According to (4.3) we deduce that the local dimension of N is n ($n = \dim M$).

We are looking for the conditions under which the nonlinear connections on E exist.

Theorem 2.4.2 *If the base manifold M is paracompact, then on* $E = \mathrm{Osc}^2 M$ *there exist nonlinear connections.*

Proof. The base manifold M being paracompact, in conformity with Theorem 2.1.1, E is a paracompact manifold. There exists at least a Riemannian metric G on E. Considering N as the orthogonal distribution to the vertical distribution V with respect to G, (4.3) is true. Thus, N is a nonlinear connection on E. **q.e.d.**

Let us consider a nonlinear connection$N$ on E and the corresponding direct sum (4.3) and denote by h and v the horizontal and vertical projectors, given by (4.3). We have

$$h + v = I, \ h^2 = h, \ v^2 = v, \ hv = vh = 0. \tag{4.4}$$

For simplicity, we denote

$$X^H = hX,\ X^V = vX,\quad \forall X \in \mathcal{X}(E). \tag{4.5}$$

Therefore, we have

$$X = X^H + X^V,\quad \forall X \in \mathcal{X}(E). \tag{4.6}$$

A *horizontal lift* is a $\mathcal{F}(M)$–linear map $\ell_h : \mathcal{X}(M) \longrightarrow \mathcal{X}(E)$ with the properties

$$v \circ \ell_h = 0,\quad d\pi \circ \ell_h = \text{Id}. \tag{4.7}$$

Consequently, locally, for any vector field $X \in \mathcal{X}(M)$ it follows that $\ell_h X$ is a uniquely determined vector field in the horizontal distribution N.

One can prove, without difficulties, the following proposition:

Proposition 2.4.1

a. *There exists a unique local basis, adapted to the horizontal distribution N which is projected by $d\pi$ onto the natural basis $\left\{\dfrac{\partial}{\partial x^i}\right\}$, $(i = 1, ..., n)$ of $\mathcal{X}(M)$. It is given by*

$$\frac{\delta}{\delta x^i} = \ell_h\left(\frac{\partial}{\partial x^i}\right),\ (i = 1, ..., n). \tag{4.8}$$

b. *The linearly independent vector fields $\dfrac{\delta}{\partial x^1}, ..., \dfrac{\delta}{\partial x^n}$ can be uniquely written in the form*

$$\frac{\delta}{\delta x^i} = \frac{\partial}{\partial x^i} - \underset{(1)}{N^j}{}_i \frac{\partial}{\partial y^{(1)j}} - \underset{(2)}{N^j}{}_i \frac{\partial}{\partial y^{(2)j}}. \tag{4.9}$$

c. *With respect to (1.7) the basis $\left(\dfrac{\delta}{\delta x^i}\right)$ is transformed as follows*

$$(4.10)\qquad \frac{\delta}{\delta x^i} = \frac{\partial \tilde{x}^j}{\partial x^i}\frac{\delta}{\delta \tilde{x}^j}.$$

The set of functions $\underset{(1)}{N}{}^j{}_i(x, y^{(1)}, y^{(2)}), \underset{(2)}{N}{}^j{}_i(x, y^{(1)}, y^{(2)})$, from (4.9), are called the *coefficients of the nonlinear connection* N and $\left\{\frac{\delta}{\delta x^1}, ..., \frac{\delta}{\delta x^n}\right\}$ is called the *adapted basis* to N.

Theorem 2.4.3 *The transformations of coordinates* (1.7) *on* E *produce the transformations of the coefficients* $\underset{(1)}{N}{}^i{}_j$ *and* $\underset{(2)}{N}{}^i{}_j$ *of the nonlinear connection* N *in the form:*

$$(4.11)\qquad \begin{aligned} \underset{(1)}{\tilde{N}}{}^i{}_m \frac{\partial \tilde{x}^m}{\partial x^j} &= \frac{\partial \tilde{x}^i}{\partial x^m}\underset{(1)}{N}{}^m{}_j - \frac{\partial \tilde{y}^{(1)i}}{\partial x^j},\\ \underset{(2)}{\tilde{N}}{}^i{}_m \frac{\partial \tilde{x}^m}{\partial x^i} &= \frac{\partial \tilde{x}^i}{\partial x^m}\underset{(2)}{N}{}^m{}_j + \frac{\partial \tilde{y}^{(1)i}}{\partial x^m}\underset{(1)}{N}{}^m{}_j - \frac{\partial \tilde{y}^{(2)i}}{\partial x^j} \end{aligned}$$

Proof. The equalities (4.11) result from (4.9) and (4.10) by means of (2.3). The converse property is true, too:

Theorem 2.4.4 *If on each local chart of* $E = \mathrm{Osc}^2 M$ *a set of functions* $\underset{(1)}{N}{}^i{}_j, \underset{(2)}{N}{}^i{}_j$ *is given so that, according to* (1.7), *the equalities* (4.11) *hold, then there exists on* E *a unique nonlinear connection* N *which has as coefficients just the given set of functions.*

Proof. Replacing $\underset{(1)}{N}{}^i{}_j, \underset{(2)}{N}{}^i{}_j$ in (4.9) and taking into account (4.11), we deduce the transformation (4.10) for $\frac{\delta}{\delta x^i}$. So, $\frac{\delta}{\delta x^i}$, $(i = 1, ..., n)$ giving us n linear independent local vector fields which span a distribution N of local dimension n. The distribution N is supplementary to the vertical distribution V. So, N is a nonlinear connection on E, having the coefficients $\underset{(1)}{N}{}^i{}_j, \underset{(2)}{N}{}^i{}_j$, the given set of functions. **q.e.d.**

We remark that $\underset{(1)}{N}{}^i{}_j$ from (3.9) possesses the rule of transformation, with respect to (1.7), the same as that of $\underset{(1)}{N}{}^i{}_j$ from (4.11).

2.5 J–Vertical Distributions

Let N be a nonlinear connection on $E = \mathrm{Osc}^2 M$. The 2–tangent structure J, defined by (3.1), applies to the horizontal distribution N in a vertical subdistribution N_1 from V_1 of local dimension n, supplementary to the subdistribution V_2. Setting $N_0 = N$, $J(N_0) = N_1$, we obtain from Theorem 2.3.1:

Theorem 2.5.1 *The following direct decomposition of linear spaces holds*

$$T_u E = N_0(u) \oplus N_1(u) \oplus V_2(u), \quad \forall u \in E, \tag{5.1}$$

N_1 is called the *J–vertical distribution.* $N_0(u), N_1(u)$ and $V_2(u)$ are the linear spaces of the distributions N_0, N_1, V_2, respectively, in the point $u \in E$.

The adapted basis $\left\{\dfrac{\delta}{\delta x^i}\right\}$, $(i = 1, ..., n)$ to the distribution N_0 determines the adapted basis $\left\{\dfrac{\delta}{\delta y^{(1)i}}\right\}$, $(i = 1, ..., n)$, in the J–vertical distribution N_1 as follows

$$\frac{\delta}{\delta y^{(1)i}} = J\left(\frac{\delta}{\delta x^i}\right), \quad (i = 1, ..., n). \tag{5.2}$$

In the same way, $\left\{\dfrac{\delta}{\delta y^{(2)i}}\right\}$ determines the adapted basis of the distribution V_2 :

$$\frac{\delta}{\delta y^{(2)i}} = J\left(\frac{\delta}{\delta y^{(1)i}}\right), \quad (i = 1, ..., n). \tag{5.3}$$

Proposition 2.5.1 *The adapted basis to the distributions N_0, N_1, V_2 are given, respectively, by*

$$\begin{aligned} &\frac{\delta}{\delta x^i} = \frac{\partial}{\partial x^i} - \underset{(1)}{N}{}^j{}_i \frac{\partial}{\partial y^{(1)j}} - \underset{(2)}{N}{}^j{}_i \frac{\partial}{\partial y^{(2)j}}, \\ &\frac{\delta}{\delta y^{(1)i}} = \frac{\partial}{\partial y^{(1)i}} - \underset{(1)}{N}{}^j{}_i \frac{\partial}{\partial y^{(2)j}}, \quad \frac{\delta}{\delta y^{(2)i}} = \frac{\partial}{\partial y^{(2)i}}, \end{aligned} \tag{5.4}$$

where $\underset{(1)}{N}{}^i{}_j$ and $\underset{(2)}{N}{}^i{}_j$ are the coefficients of the nonlinear connection N.

Indeed, (4.9) and (3.1),(5.2) and (5.3) imply (5.4). Consequently,

$$\left\{\frac{\delta}{\delta x^i}, \frac{\delta}{\delta y^{(1)i}}, \frac{\partial}{\partial y^{(2)i}}\right\} \tag{5.5}$$

is a *local basis adapted to the direct decomposition* (5.1).

Proposition 2.5.2 *The transformations* (1.7) *of coordinates on E imply the transformations of the adapted basis* (5.5), *as follows:*

$$\frac{\delta}{\delta x^i} = \frac{\partial \tilde{x}^j}{\partial x^i}\frac{\delta}{\delta \tilde{x}^j},\ \frac{\delta}{\delta y^{(1)i}} = \frac{\partial \tilde{x}^j}{\partial x^i}\frac{\delta}{\delta \tilde{y}^{(1)j}},\ \frac{\partial}{\partial y^{(2)i}} = \frac{\partial \tilde{x}^j}{\partial x^i}\frac{\partial}{\partial \tilde{y}^{(2)j}}. \tag{5.5)'$$

Indeed, (4.10) is transformed by J into (5.5)'.

The direct decomposition (5.1) is important in the geometrical study of the total space of the 2–osculator bundle $E = \mathrm{Osc}^2 M$. It allows to express the main geometrical object fields: tensors, connections etc. with respect to this decomposition. The new components of these objects, which are to be obtained, have some simple geometrical meanings.

If we consider the projectors h, v_1, v_2 determined by (5.1) and denote $v_1 X = X^{V_1}, v_2 X = X^{V_2}$, we can uniquely write:

$$X = X^H + X^{V_1} + X^{V_2},\quad \forall X \in \mathcal{X}(E). \tag{5.6}$$

Thus, in the adapted basis (5.4) we have:

$$X^H = X^{(0)i}\frac{\delta}{\delta x^i},\quad X^{V_1} = X^{(1)i}\frac{\delta}{\delta y^{(1)i}},\quad X^{V_2} = X^{(2)i}\frac{\delta}{\delta y^{(2)i}}. \tag{5.6)'$$

With respect to (1.7), each of these coordinates have the same rule of transformation:

$$\tilde{X}^{(\alpha)i} = \frac{\partial \tilde{x}^i}{\partial x^j}X^{(\alpha)j},\quad (\alpha = 0,1,2). \tag{5.6)''$$

Each of X^H, X^{V_1}, X^{V_2} or $X^{(0)i}, X^{(1)i}, X^{(2)i}$ is called a distinguished vector field. Briefly, we can say they are *d–vector fields.*

Of course, the projectors h, v, v_1, v_2 have the properties

$$\begin{gathered} h + v = I,\ h^2 = h,\ v^2 = v,\ hv = vh = 0,\ v = v_1 + v_2,\\ v_1 v_1 = v_1,\ v_2 v_2 = v_2,\ v_1 v_2 = v_2 v_1 = 0,\ hv_\alpha = v_\alpha h = 0,\\ (\alpha = 1,2). \end{gathered} \tag{5.7}$$

A first appliction is given by

Theorem 2.5.2 *The nonlinear connection N is integrable if, and only if, for any $X, Y \in \mathcal{X}(E)$ we have*

$$[X^H, Y^H]^{V_1} = [X^H, Y^H]^{V_2} = 0.$$

Indeed, the Lie bracket of any two horizontal vector fields X^H, Y^H belongs to the horizontal distribution N if, and only if, the last two equations hold.

We get also:

Theorem 2.5.3 *A J-vertical distribution N_1 is integrable if, and only if, for any vector fields $X, Y \in \mathcal{X}(E)$ we have*

$$[X^{V_1}, Y^{V_1}]^H = [X^{V_1}, Y^{V_1}]^{V_2} = 0.$$

2.6 The Dual Coefficients of a Nonlinear Connection

The dual basis (or adapted cobasis) of the adapted basis (5.5) will be denoted by

$$\left\{dx^i, \delta y^{(1)i}, \delta y^{(2)i}\right\}, \quad (i = 1, ..., n). \tag{6.1}$$

The scalar product of the covector fields (6.1) and vector fields (5.5) are expressed as follows:

$$\begin{array}{lll} \dfrac{\delta}{\delta x^i} \rfloor dx^j = \delta^j{}_i, & \dfrac{\delta}{\delta x^i} \rfloor \delta y^{(1)j} = 0, & \dfrac{\delta}{\delta x^i} \rfloor \delta y^{(2)j} = 0, \\ \dfrac{\delta}{\delta y^{(1)i}} \rfloor dx^j = 0, & \dfrac{\delta}{\delta y^{(1)i}} \rfloor \delta y^{(1)j} = \delta^i{}_j, & \dfrac{\delta}{\delta y^{(1)i}} \rfloor \delta y^{(2)j} = 0, \\ \dfrac{\partial}{\partial y^{(2)i}} \rfloor dx^j = 0, & \dfrac{\partial}{\partial y^{(2)i}} \rfloor \delta y^{(1)j} = 0, & \dfrac{\partial}{\partial y^{(2)i}} \rfloor \delta y^{(2)j} = \delta^i{}_j. \end{array} \tag{6.2}$$

By a straightforward calculus we obtain

Theorem 2.6.1 *The dual basis* (6.1) *of the adapted basis* (5.5) *is given by*

(6.3)
$$\delta x^i = dx^i,\ \delta y^{(1)i} = dy^{(1)i} + \underset{(1)}{M}{}^i{}_j dx^j,$$
$$\delta y^{(2)i} = dy^{(2)i} + \underset{(1)}{M}{}^i{}_j dy^{(1)j} + \underset{(2)}{M}{}^i{}_j dy^{(2)j},\quad \textit{where}$$

(6.4)
$$\underset{(1)}{M}{}^i{}_j = \underset{(1)}{N}{}^i{}_j,\ \underset{(2)}{M}{}^i{}_j = \underset{(2)}{N}{}^i{}_j + \underset{(1)}{N}{}^i{}_m \underset{(1)}{N}{}^m{}_j.$$

Conversely,

Theorem 2.6.2 *If the adapted cobasis* (6.1) *is given in the form* (6.3), *then the adapted basis* (5.5) *is expressed in the form* (5.4), *where*

(6.4)′
$$\underset{(1)}{N}{}^i{}_j = \underset{(1)}{M}{}^i{}_j,\ \underset{(2)}{N}{}^i{}_j = \underset{(2)}{M}{}^i{}_j - \underset{(1)}{M}{}^i{}_m \underset{(1)}{M}{}^m{}_j.$$

So, the coefficients $\left(\underset{(1)}{M}{}^i{}_j, \underset{(2)}{M}{}^i{}_j\right)$ of the adapted cobasis (6.1) are uniquely determined by the coefficients $\left(\underset{(1)}{N}{}^i{}_j, \underset{(2)}{N}{}^i{}_j\right)$ of the adapted basis, and conversely. Therefore, $\left(\underset{(1)}{M}{}^i{}_j, \underset{(2)}{M}{}^i{}_j\right)$ will be called the *dual coefficients* of the nonlinear connection N.

It is sometimes preferable and easier to determine the dual coefficients, instead of the direct coefficients of a nonlinear connection.

From (4.10) and (6.2), we deduce:

Proposition 2.6.1 *With respect to* (1.7) *the covector fields of the adapted cobasis* (6.1) *are transformed as follows:*

(6.5)
$$d\tilde{x}^i = \frac{\partial \tilde{x}^i}{\partial x^j} dx^j,\ \delta\tilde{y}^{(1)i} = \frac{\partial \tilde{x}^i}{\partial x^j}\delta y^{(1)j},\ \delta\tilde{y}^{(2)i} = \frac{\partial \tilde{x}^i}{\partial x^j}\delta y^{(2)j}.$$

By means of the previous proposition, we can prove by a straightforward calculus:

Theorem 2.6.3

1° *A tranformation of coordinates* (1.7) *on the differentiable manifold* $E = \mathrm{Osc}^2 M$ *implies the following transformation of the dual coefficients*

$$
(6.6)\quad
\begin{aligned}
&\frac{\partial \tilde{x}^i}{\partial x^m} \underset{(1)}{M}{}^m{}_j = \underset{(1)}{\widetilde{M}}{}^i{}_m \frac{\partial \tilde{x}^m}{\partial x^j} + \frac{\partial \tilde{y}^{(1)i}}{\partial x^j}\\
&\frac{\partial \tilde{x}^i}{\partial x^m} \underset{(2)}{M}{}^m{}_j = \underset{(2)}{\widetilde{M}}{}^i{}_m \frac{\partial \tilde{x}^m}{\partial x^j} + \underset{(1)}{\widetilde{M}}{}^i{}_m \frac{\partial \tilde{y}^{(1)m}}{\partial x^j} + \frac{\partial \tilde{y}^{(2)i}}{\partial x^j}.
\end{aligned}
$$

2° *If a set of functions* $\left(\underset{(1)}{M}{}^i{}_j, \underset{(2)}{M}{}^i{}_j\right)$ *is given on each domain of local chart on* E, *so that, according to* (1.7), *the equations* (6.6) *hold, then there exists on* E *a unique nonlinear connection* N, *which has as dual coefficients just the given set of functions.*

The second part of this theorem can be reduced to the Theorem 2.4.4, since (6.6) and (6.4)′ give us the rule (4.11) of transformation of the coefficients $\left(\underset{(1)}{N}{}^i{}_j, \underset{(2)}{N}{}^i{}_j\right)$.

A field of 1–form $\omega \in \mathcal{X}^*(E)$ can be written, uniquely, in the form

$$(6.7)\quad \omega = \omega^H + \omega^{V_1} + \omega^{V_2} \text{ where}$$

$$(6.7)'\quad \omega^H = \omega \circ h,\ \omega^{V_1} = \omega \circ v_1,\ \omega^{V_2} = \omega \circ v_2.$$

In the adapted cobasis (6.1), we get

$$(6.7)''\quad \omega^H = \omega_i^{(0)} dx^i,\ \omega^{V_1} = \omega_i^{(1)} \delta y^{(1)i},\ \omega^{V_2} = \omega_i^{(2)} \delta y^{(2)i}$$

and, with respect to (1.7):

$$(6.7)'''\quad \omega_i^{(\alpha)} = \frac{\partial \tilde{x}^j}{\partial x^i} \tilde{\omega}_j^{(\alpha)} \quad (\alpha = 0, 1, 2).$$

Each one of $\omega^H, \omega^{V_1}, \omega^{V_2}$ or of $\omega_i^{(0)}, \omega_i^{(1)}, \omega_i^{(2)}$ is called a *distinguished covector field* (or *d*–1– *form field*) or, briefly, a *d–covector field.*

In particular, if $f \in \mathcal{F}(E)$, then the 1–form df can be written:

$$df = (df)^H + (df)^{V_1} + (df)^{V_2}, \quad \text{where} \tag{6.8}$$

$$(df)^H = \frac{\delta f}{\delta x^i}\, dx^i, \; (df)^{V_1} = \frac{\delta f}{\delta y^{(1)i}}\, \delta y^{(1)i}, \; (df)^{V_2} = \frac{\delta f}{\delta y^{(2)i}}\, \delta y^{(2)i}. \tag{6.8)'}$$

Thus,

$$df = \frac{\delta f}{\delta x^i}\, dx^i + \frac{\delta f}{\delta y^{(1)i}} \delta y^{(1)i} + \frac{\delta f}{\delta y^{(2)i}} \delta y^{(2)i}. \tag{6.8)''}$$

Indeed, we have

$$df = \frac{\partial f}{\partial x^i}\, dx^i + \frac{\partial f}{\partial y^{(1)i}} dy^{(1)i} + \frac{\partial f}{\partial y^{(2)i}} dy^{(2)i}, \tag{*}$$

and taking into account (5.4), we deduce

$$\begin{aligned} \frac{\partial}{\partial x^i} &= \frac{\delta}{\delta x^i} + \underset{(1)}{M}{}^j{}_i \frac{\delta}{\delta y^{(1)j}} + \underset{(2)}{M}{}^j{}_i \frac{\delta}{\delta y^{(2)j}}, \\ \frac{\partial}{\partial y^{(1)i}} &= \frac{\delta}{\delta y^{(1)i}} + \underset{(1)}{M}{}^j{}_i \frac{\delta}{\delta y^{(2)j}}, \quad \frac{\partial}{\partial y^{(2)i}} = \frac{\delta}{\delta y^{(2)i}}. \end{aligned} \tag{6.9}$$

We get (6.8)–(6.8)″.

Proposition 2.6.2 *Each one of the equations*

$$\frac{\delta f}{\delta x^i} = 0, \quad \frac{\delta f}{\delta y^{(1)i}} = 0, \quad \frac{\delta f}{\delta y^{(2)i}} = 0 \tag{6.10}$$

has a geometrical meaning.

Indeed, the first member of these equations is a d–covector field.

q.e.d.

The natural cobasis $\{dx^i, dy^{(1)i}, dy^{(2)i}\}$ can be expressed by means of the adapted cobasis $\{\delta x^i, \delta y^{(1)i}, \delta y^{(2)i}\}$ as follows:

$$\begin{aligned} & dx^i = \delta x^i, \; dy^{(1)i} = \delta y^{(1)i} - \underset{(1)}{N}{}^i{}_j\, \delta x^j, \\ & dy^{(2)i} = \delta y^{(2)i} - \underset{(1)}{N}{}^i{}_j\, \delta y^{(1)j} - \underset{(2)}{N}{}^i{}_j\, \delta x^j. \end{aligned} \tag{6.11}$$

These formulae are obtained from (6.3) and (6.4)$'$.

We apply the previous considerations to establish the d–vector field of a tangent vector to a curve in E.

Let γ be a parametrized curve in E.

In a local chart $\pi^{-1}(U)$ in E we represent γ in the form:

$$x^i = x^i(t),\ y^{(1)i} = y^{(1)i}(t),\ y^{(2)i} = y^{(2)i}(t),\quad t \in I. \tag{6.12}$$

The tangent vector field $\dfrac{d\gamma}{dt}$ can be written, by means of its horizontal and vertical components as follows:

$$\frac{d\gamma}{dt} = \left(\frac{d\gamma}{dt}\right)^H + \left(\frac{d\gamma}{dt}\right)^{V_1} + \left(\frac{d\gamma}{dt}\right)^{V_2} \quad \text{where} \tag{6.13}$$

$$\begin{aligned} &\left(\frac{d\gamma}{dt}\right)^H = \frac{dx^i}{dt}\frac{\delta}{\delta x^i}, \\ &\left(\frac{d\gamma}{dt}\right)^{V_1} = \frac{\delta y^{(1)i}}{dt}\frac{\delta}{\delta y^{(1)i}},\quad \left(\frac{d\gamma}{dt}\right)^{V_2} = \frac{\delta y^{(2)i}}{dt}\frac{\delta}{\delta y^{(2)i}}, \end{aligned} \tag{6.13$'$}$$

or in the form

$$\frac{d\gamma}{dt} = \frac{dx^i}{dt}\frac{\delta}{\delta x^i} + \frac{\delta y^{(1)i}}{dt}\frac{\delta}{\delta y^{(1)i}} + \frac{\delta y^{(2)i}}{dt}\frac{\delta}{\delta y^{(2)i}}. \tag{6.13$''$}$$

Indeed, using (6.9) and (6.11), we deduce from (6.13)$''$

$$\frac{d\gamma}{dt} = \frac{dx^i}{dt}\frac{\partial}{\partial x^i} + \frac{dy^{(1)i}}{dt}\frac{\partial}{\partial y^{(1)i}} + \frac{dy^{(2)i}}{dt}\frac{\partial}{\partial y^{(2)i}}.$$

Proposition 2.6.3 *Each of these equations*

$$\frac{dx^i}{dt} = 0,\quad \frac{\delta y^{(1)i}}{dt} = 0,\quad \frac{\delta y^{(2)i}}{dt} = 0 \tag{6.14}$$

has a geometrical meaning.

Indeed, the first member of these equations is a d–vector fields.

Definition 2.6.1 A parametrized curve γ is calleu a horizontal curve, if the vertical components of its tangent vector field $\frac{d\gamma}{dt}$ vanish.

Theorem 2.6.4 *The parametrized curve $\gamma : I \longrightarrow E$, from (6.3), is horizontal if, and only if, the functions $\{x^i(t), y^{(1)i}(t), y^{(2)i}(t)\}$ are solutions of the system of differential equations*

$$\frac{\delta y^{(1)i}}{dt} = 0, \quad \frac{\delta y^{(2)i}}{dt} = 0. \tag{6.15}$$

Of course, if the functions $x^i(t)$, $t \in I$ are given, then, locally, there exist the horizontal curves, on E.

Let $c : I \longrightarrow M$ be a parametrized curve on the base manifold M, given by $x^i = x^i(t)$, $t \in I$. Let us also consider the extension $\tilde{c}$ to E of the curve c.

Definition 2.6.2 The parametrized curve $c : I \longrightarrow M$ on the base manifold M is called an autoparallel curve of the nonlinear connection N if its extension $\tilde{c}$ to E is a horizontal curve.

Theorem 2.6.5 *The autoparallel curves of the nonlinear connection N with the dual coefficients $\left(\underset{(1)}{M}{}^i{}_j, \underset{(2)}{M}{}^i{}_j\right)$ are characterized by the system of differential equations*

$$\begin{aligned} & y^{(1)i} = \frac{dx^i}{dt}, \quad y^{(2)i} = \frac{1}{2}\frac{d^2x^i}{dt^2}; \\ & \frac{\delta y^{(1)i}}{dt} = \frac{dy^{(1)i}}{dt} + \underset{(1)}{M}{}^i{}_j \frac{dx^j}{dt} = 0; \\ & \frac{\delta y^{(2)i}}{dt} = \frac{dy^{(2)i}}{dt} + \underset{(1)}{M}{}^i{}_j \frac{dy^{(1)j}}{dt} + \underset{(2)}{M}{}^i{}_j \frac{dx^j}{dt} = 0. \end{aligned} \tag{6.16}$$

A theorem of existence and uniqueness of the autoparallel curves of a nonlinear connection can now be easily formulated.

Finally, we can represent the Liouville vector fields $\overset{1}{\Gamma}$ and $\overset{2}{\Gamma}$ from (2.4) in the adapted basis (5.4). We get

$$\overset{1}{\Gamma} = z^{(1)i}\frac{\delta}{\delta y^{(2)i}}, \quad \overset{2}{\Gamma} = z^{(1)i}\frac{\delta}{\delta y^{(1)i}} + 2z^{(2)i}\frac{\delta}{\delta y^{(2)i}}, \quad \text{where} \tag{6.17}$$

$$z^{(1)i} = y^{(1)i}, \quad z^{(2)i} = y^{(2)i} + \frac{1}{2} \underset{(1)}{M}{}^i{}_j \, y^{(1)j}. \tag{6.18}$$

Therefore, $z^{(1)i}$ and $z^{(2)i}$ are d–vector fields. They are called the *Liouville d–vector fields.*

2.7 Determination of a Nonlinear Connection from a 2–Spray

One of the important problems concerning the notion of nonlinear connection consists in its determinations from a 2–spray. We give here a solution of this problem.

Let S be a 2–spray, with the coefficients $G^i(x, y^{(1)}, y^{(2)})$. According to $(3.3)'$, locally, S is expressed by

$$S = y^{(1)m} \frac{\partial}{\partial x^m} + 2y^{(2)m} \frac{\partial}{\partial y^{(1)m}} - 3G^m \frac{\partial}{\partial y^{(2)m}}. \tag{7.1}$$

We can formulate:

Theorem 2.7.1 *The set of functions*

$$\underset{(1)}{M}{}^i{}_j = \frac{\partial G^i}{\partial y^{(2)j}}, \quad \underset{(2)}{M}{}^i{}_j = \frac{1}{2}\left(S \underset{(1)}{M}{}^i{}_j + \underset{(1)}{M}{}^i{}_m \underset{(1)}{M}{}^m{}_j \right), \tag{7.2}$$

gives the dual coefficients of a nonlinear connection N, determined by the spray S only with the coefficients G^i.

Proof. According to Theorem 2.3.5, with respect to (1.7), $\underset{(1)}{M}{}^i{}_j$ are transformed by the rule:

$$\underset{(1)}{M}{}^m{}_j \frac{\partial \tilde{x}^i}{\partial x^m} = \underset{(1)}{\widetilde{M}}{}^i{}_m \frac{\partial \tilde{x}^m}{\partial x^j} + \frac{\partial \tilde{y}^{(1)i}}{\partial x^j}. \tag{a}$$

It is necessary to prove that the second equality of (6.6) holds.

Because S from (7.1) is invariant to the transformations (1.7), from (a) we obtain

$$\left(S\underset{(1)}{M}{}^m{}_j\right)\frac{\partial\tilde{x}^i}{\partial x^m}+\underset{(1)}{M}{}^m{}_j\left(S\,\frac{\partial\tilde{x}^i}{\partial x^m}\right)=$$
$$=\left(\widetilde{S}\underset{(1)}{\widetilde{M}}{}^i{}_m\right)\frac{\partial\tilde{x}^m}{\partial x^j}+\underset{(1)}{\widetilde{M}}{}^i{}_m\left(S\,\frac{\partial\tilde{x}^m}{\partial x^j}\right)+S\,\frac{\partial\tilde{y}^{(1)i}}{\partial x^j}$$

or

$$(b)\qquad \begin{aligned}&\left(S\underset{(1)}{M}{}^m{}_j\right)\frac{\partial\tilde{x}^i}{\partial x^m}+\underset{(1)}{M}{}^m{}_j\frac{\partial\tilde{y}^{(1)i}}{\partial x^m}=\\&=\left(\widetilde{S}\underset{(1)}{\widetilde{M}}{}^i{}_m\right)\frac{\partial\tilde{x}^m}{\partial x^j}+\underset{(1)}{\widetilde{M}}{}^i{}_m\frac{\partial\tilde{y}^{(1)m}}{\partial x^j}+2\,\frac{\partial\tilde{y}^{(2)i}}{\partial x^j}.\end{aligned}$$

Using (a), we find

$$(c)\qquad \begin{aligned}&\left(\underset{(1)}{M}{}^m{}_r\underset{(1)}{M}{}^r{}_j\right)\frac{\partial\tilde{x}^i}{\partial x^m}=\underset{(1)}{\widetilde{M}}{}^i{}_s\left(\underset{(1)}{\widetilde{M}}{}^s{}_r\frac{\partial\tilde{x}^r}{\partial x^j}+\frac{\partial\tilde{y}^{(1)s}}{\partial x^j}\right)+\\&+\frac{\partial x^m}{\partial\tilde{x}^s}\,\frac{\partial\tilde{y}^{(1)i}}{\partial x^m}\left(\underset{(1)}{\widetilde{M}}{}^s{}_r\,\frac{\partial\tilde{x}^r}{\partial x^j}+\frac{\partial\tilde{y}^{(1)s}}{\partial x^j}\right).\end{aligned}$$

Adding (b) and (c) and using the expression of $\underset{(2)}{M}{}^i{}_j$ from (7.2), we get

$$2\underset{(2)}{M}{}^m{}_j\,\frac{\partial\tilde{x}^i}{\partial x^m}+\underline{\underset{(1)}{M}{}^m{}_j\,\frac{\partial\tilde{y}^{(1)i}}{\partial x^m}}=2\underset{(2)}{\widetilde{M}}{}^i{}_m\,\frac{\partial\tilde{x}^m}{\partial x^j}+2\underset{(1)}{M}{}^i{}_m\,\frac{\partial\tilde{y}^{(1)m}}{\partial x^j}+2\,\frac{\partial\tilde{y}^{(2)i}}{\partial x^j}+$$
$$\underline{+\frac{\partial x^m}{\partial\tilde{x}^s}\,\frac{\partial\tilde{y}^{(1)i}}{px^m}\left(\underset{(1)}{\widetilde{M}}{}^s{}_r\,\frac{\partial\tilde{x}^r}{\partial x^j}+\frac{\partial\tilde{y}^{(1)s}}{\partial x^j}\right)}.$$

According to (a) the underlined terms vanish and so we have the second equality of (6.6). Consequently, $\left(\underset{(1)}{M}{}^i{}_j,\underset{(2)}{M}{}^i{}_j\right)$ from (7.2), with respect to (7.1), obey the transformation (6.6). Applying Theorem 2.6.3, second part, it follows that $\left(\underset{(1)}{M}{}^i{}_j,\underset{(2)}{M}{}^i{}_j\right)$ are the dual coefficients of a nonlinear connection N. They depend on the 2–spray S, only. **q.e.d.**

This result is important in the geometry of the Lagrange spaces of order 2 for the construction of a canonical nonlinear connection.

2.8 The Almost Product Structure $I\!P$. The Almost n–Contact Structure $I\!F$

In this last section of the present chapter we shall introduce two important geometrical structure on E.

Let N be a nonlinear connection on the total space E of the 2–osculator bundle. Taking into account the distributions

$$N_0 = N,\ N_1 = J(N_0),\ V_2 = J(N_1),$$

and the direct decomposition (5.1), any vector field X on E can be written as

$$X = X^H + X^{V_1} + X^{V_2}. \tag{8.1}$$

The $\mathcal{F}(E)$–linear mapping $I\!P : \mathcal{X}(E) \longrightarrow \mathcal{X}(E)$ defined by

$$\begin{aligned} &I\!P(X^H) = X^H,\ I\!P(X^{V_1}) = -X^{V_1},\ I\!P(X^{V_2}) = -X^{V_2},\\ &\forall X \in \mathcal{X}(E) \end{aligned} \tag{8.2}$$

determines a tensor field of type $(1,1)$ on E, with the property

$$I\!P \circ I\!P = I, \tag{8.3}$$

where I is the Kronecher tensor.

Consequently, $I\!P$ is an *almost product structure* on E.

Writing the vector field X in the form $X = X^H + X^V$, where $X^V = X^{V_1} + X^{V_2}$, it follows

$$I\!P(X^H) = X^H,\ I\!P(X^V) = -X^V,\ \forall X \in \mathcal{X}(E). \tag{8.2)'$$

Looking at the last equalities we notice that $I\!P$ has two eigenvalues $+1$ and -1. The linear eigenspace at a point $u \in E$, corresponding to the eigenvalue $+1$ is the horizontal space $N(u)$ and the linear space corresponding to the eigenvalue -1 is the vertical space $V(u)$.

We have also

Proposition 2.8.1

(a) *The almost product structure $I\!P$ can be expressed by means of the projectors h and v as follows*

$$I\!P = I - 2v; \quad I\!P = 2h - I. \tag{8.4}$$

(b) $\mathrm{rank}\|I\!P\| = 3n$.

The previous properties can be easily proved.
Now, we may state:

Theorem 2.8.1 *A nonlinear connection N on E is characterized by the existence of an almost product structure $I\!P$ on E, whose eigenspaces corresponding to the eigenvalue -1 coincide with the linear spaces of the vertical distribution V on E.*

Proof. Let N be the horizontal distribution of a given nonlinear connection. The vertical projector v is uniquely determined by $v = I - h$. Setting $I\!P = I - 2v$ it follows $I\!P \circ I\!P = I$ and $I\!P(X^V) = -X^V$, $\forall X \in \mathcal{X}(E)$. Thus, $I\!P$ is an almost product structure with the properties ennounced in the above theorem. Conversely, if $I\!P$ is a tensor field on E, of type $(1,1)$ and $I\!P \circ I\!P = I$, $I\!P(X^V) = -X^V, \forall X \in \mathcal{X}(E)$, then setting $v = \frac{1}{2}(I - I\!P)$, it follows $v^2 = v$ and $vX^V = X^V$. Consequently, $N = \mathrm{Ker}\, v$ is a horizontal distribution of local dimension n.

q.e.d.

Other important result is given by:

Theorem 2.8.2 *The almost product structure $I\!P$ determined by the nonlinear connection N is integrable if, and only if, the nonlinear connection N is integrable.*

Proof. The Nijenhuis tensor of structure $I\!P$ is

$$N_{I\!P}(X,Y) = [I\!PX, I\!PY] + I\!P^2[X,Y] - I\!P[I\!PX, Y] - I\!P[X, I\!PY],$$

for any vector fields X, Y on E.

Taking successively $X = X^H$, $Y = Y^H, ...,$ we get $N_{\mathbb{P}}(X^H, Y^H) = = 4[X^H, Y^H]$, $N_{\mathbb{P}}(X^H, Y^V) = N_{\mathbb{P}}(X^V, Y^V) = 0$. Thus, $N_{\mathbb{P}}(X, Y) = 0$ for any X, Y is equivalent to$[X^H, Y^H] = 0$. Theorem 2.5.2 is applicable and the enunciated property is valid. **q.e.d.**

The relation between the almost product structure $\mathbb{P}$ and the 2–tangent structure J is formulated in the next proposition:

Proposition 2.8.2 *The structures $\mathbb{P}$ and J have the property*

$$\mathbb{P} \circ J = -J. \tag{8.5}$$

As we have seen, on the total space of the 1–osculator bundle there exists an almost complex structure determined by a nonlinear connection (§13, Ch.1). The problem is how to extend this notion to the 2–osculator bundle.

Let us consider the $\mathcal{F}(E)$–linear mapping $\mathbb{F} : \mathcal{X}(E) \longrightarrow \mathcal{X}(E)$ defined on the adapted basis to the direct decomposition (5.1), by

$$\mathbb{F}\left(\frac{\delta}{\delta x^i}\right) = -\frac{\partial}{\partial y^{(2)i}},\ \mathbb{F}\left(\frac{\delta}{\delta y^{(1)i}}\right) = 0,\ \mathbb{F}\left(\frac{\delta}{\delta y^{(2)i}}\right) = \frac{\delta}{\delta x^i} \tag{8.6}$$
$$(i = 1, ..., n).$$

Taking into account the rules (5.5)$'$ of transformations of the adapted basis, it follows that $\mathbb{F}$ is globally defined on the manifold E. Now, we can prove, without difficulties, the following theorem:

Theorem 2.8.3 *The mapping $\mathbb{F}$ has the following properties:*

1° *$\mathbb{F}$ is globally defined on E.*

2° *$\mathbb{F}$ is a tensor field on E of type* $(1, 1)$.

3° *$Ker\,\mathbb{F} = N_1$,* $\operatorname{Im} \mathbb{F} = N_0 + V_2$.

4° *$rank\,\|\mathbb{F}\| = 2n$.*

5° $\mathbb{F}^3 + \mathbb{F} = 0$.

Thus, $\mathbb{F}$ is a $\mathbb{F}(3, 1)$ structure. This structure is also an *almost n–contact structure* on E [87], [107], [182].

Indeed, the dimension of the manifold $E = \operatorname{Osc}^2 M$ is $3n = 2n + n$. Let us consider a local basis $\{\underset{1a}{\xi}\}$, $(a = 1, ..., n)$ of the distribution

N_1 and $(\overset{1a}{\eta})$ its dual. Then the set $\left(\mathbb{F}, \underset{1a}{\xi}, \overset{1a}{\eta}\right)$, determine an *almost n–contact structure.* Namely, we have:

$$(8.7)\qquad \begin{aligned} &\mathbb{F}(\underset{1a}{\xi}) = 0, \quad \overset{1a}{\eta}(\underset{1b}{\xi}) = \delta^a_b, \\ &\mathbb{F}^2(X) = -X + \sum_{a=1}^{n} \overset{1a}{\eta}(X)\underset{1a}{\xi}, \quad \forall X \in \mathcal{X}(\widetilde{E}). \end{aligned}$$

From the last formulae we deduce the property

$$(8.7)' \qquad \overset{1a}{\eta} \circ \mathbb{F} = 0, \quad (a = 1, ..., n).$$

The structure $\left(\mathbb{F}, \underset{1a}{\xi}, \overset{1a}{\eta}\right)$ is said to be normal if

$$(8.8)\qquad N_{\mathbb{F}}(X,Y) + \sum_{a=1}^{n} d\overset{1a}{\eta}(X,Y)\underset{1a}{\xi} = 0,$$

where $N_{\mathbb{F}}$ is he Nijenhuis tensor of $\mathbb{F}$:

$$N_{\mathbb{F}}(X,Y) = [\mathbb{F}X, \mathbb{F}Y] + \mathbb{F}^2[X,Y] - \mathbb{F}[\mathbb{F}X, Y] - \mathbb{F}[X, \mathbb{F}Y].$$

Theorem 2.8.4 *The almost n–contact structure* $\left(\mathbb{F}, \underset{1a}{\xi}, \overset{1a}{\eta}\right)$ *is normal if, and only if, the following equation holds:*

$$(8.9)\qquad N_{\mathbb{F}}(X,Y) + \sum_{a=1}^{n} d(\delta y^{(1)a})(X,Y)\frac{\delta}{\delta y^{(1)a}} = 0, \quad \forall X, Y \in \mathcal{X}(\widetilde{E}).$$

Proof. Indeed, $\left(\underset{1a}{\xi}\right)$ being an arbitrary basis of N_1, we can take $\underset{1a}{\xi} = \frac{\delta}{\delta y^{(1)a}}$, $(a = 1, ..., n)$. Then its dual is $(\delta y^{(1)a})$. The condition (8.8) becomes the condition (8.9).

In a next section we shall express the equation (8.9) by means of d–tensor fields, using the expression of the exterior differentials in an adapted basis. We use the structure $\left(\mathbb{F}, \underset{1a}{\xi}, \overset{1a}{\eta}\right)$ in the case when we have a Riemannian structure G on E so that the set $\left(\mathbb{F}, \underset{1a}{\xi}, \overset{1a}{\eta}, G\right)$ will be an almost n–contact Riemannian structure on $\widetilde{E}$. The manifold E endowed with this structure gives us the geometrical model of a Lagrange space of order 2.

2.9 Problems

Let $\gamma_{ij}(x)$ be a Riemann structure on the base manifold M. Prove that:

1. $z^{(2)i} = y^{(2)i} + \frac{1}{2}\gamma^i_{jk}(x)y^{(1)j}y^{(1)k}$ is a d–vector field on $E = \mathrm{Osc}^2 M$.
2. $L(x, y^{(1)}, y^{(2)}) = \gamma_{ij}(x)z^{(2)i}z^{(2)j}$ is a scalar field on E.
3. $3G^i = \frac{1}{2}\gamma^{ij}\left\{\Gamma\frac{\partial L}{\partial y^{(2)j}} - \frac{\partial L}{\partial y^{(1)j}}\right\}$ are the coefficients of a 2–spray S, Γ being the operator (2.5).
4. Determine the dual coefficients $\left(\underset{(1)}{M^i}_j, \underset{(2)}{M^i}_j\right)$ and direct coefficients $\left(\underset{(1)}{N^i}_j, \underset{(2)}{N^i}_j\right)$ of the nonlinear connection N defined by the previous 2–spray S.
5. Write the differential equations of the autoparallel curve of the nonlinear connection N of the point 4.

Chapter 3

N–Linear Connections Structure Equations

The theory presented in the previous chapter is very useful in the study of the geometry of total space of an osculator bundle $E = \mathrm{Osc}^2 M$. Thus, the direct decomposition (5.1), Ch.2, allows to decompose the tensor fields on E or on $\widetilde{E}$ in the components, with respect to the distributions N_0, N_1, V_2. But these components are special tensor fields, called *distinguished* tensor fields. In the geometry of the manifold E we have determined the so–called distinguished vector fields or covector fields. The coordinates of d–vectors or d–covectors have the same rules of transformations as those of the vector or covector fields on the base manifold M. Similar considerations can be done for the notion of linear connection on E.

Therefore, in this chapter we shall study the algebra of distinguished tensor fields, the concept of N–linear connection, structure equations, curvatures and torsions.

3.1 The Algebra of d–Tensor Fields

Let N be a nonlinear connection, apriori given, on the total space $E = \mathrm{Osc}^2 M$ and the distributions $N_0 = N$, $N_1 = J(N_0)$, $V_2 = J(N_1)$ determined by N. We have, cf. §5, Ch.1, the direct decomposition:

(1.1) $$T_u E = N_0(u) \oplus N_1(u) \oplus V_2(u), \quad \forall u \in E.$$

According to it, for any $X \in \mathcal{X}(E), \omega \in \mathcal{X}^*(E)$, there are

$$X = X^H + X^{V_1} + X^{V_2}, \quad \omega = \omega^H + \omega^{V_1} + \omega^{V_2}. \tag{1.2}$$

Each term of these sums is a distinguished vector field or 1–form field. Evidently, we can write

$$\omega^H(X) = \omega(X^H), \quad \omega^{V_1}(X) = \omega(X^{V_1}), \quad \omega^{V_2}(X) = \omega(X^{V_2}). \tag{1.3}$$

Definition 3.1.1 A tensor field T on E of type (r, s) is called distinguished, briefly *d–tensor field*, if it has the following property:

$$\begin{aligned} &T(\overset{1}{\omega}, ..., \overset{r}{\omega}, \underset{1}{X}, ..., \underset{s}{X}) = T(\overset{1}{\omega}{}^H, ..., \overset{r}{\omega}{}^{V_2}, \underset{1}{X}{}^H, ..., \underset{s}{X}{}^{V_2}), \\ &\forall\, \underset{1}{X}, ..., \underset{s}{X} \in \mathcal{X}(E),\ \forall\, \overset{1}{\omega}, ..., \overset{r}{\omega} \in \mathcal{X}^*(E). \end{aligned} \tag{1.4}$$

If $T \in \mathcal{T}^r_s(E)$ is not a d–tensor, then using (1.2) in $T(\overset{1}{\omega}, ..., \overset{r}{\omega}, \underset{1}{X}, ..., \underset{s}{X})$ we get

$$\begin{aligned} T(\overset{1}{\omega}{}^H + \overset{1}{\omega}{}^{V_1} + \overset{1}{\omega}{}^{V_2}, ..., \underset{s}{X}{}^H + \underset{s}{X}{}^{V_1} + \underset{s}{X}{}^{V_2}) = T(\overset{1}{\omega}{}^H, ..., \underset{s}{X}{}^H) + \\ + T(\overset{1}{\omega}{}^H, ..., \underset{s}{X}{}^{V_1}) + \cdots + T(\overset{1}{\omega}{}^{V_2}, ..., \underset{s}{X}{}^{V_2}). \end{aligned}$$

Then, every term in the second member is a d–tensor field.

Let us consider the coordinates of a d–tensor field with respect to the adapted basis (5.5), Ch.2 and adapted cobasis (6.1), Ch.2:

$$T^{i_1 \dots i_r}_{j_1 \dots j_s}(x, y^{(1)}, y^{(2)}) = T\left(dx^{i_1}, ..., \delta y^{(2)i_r}, \frac{\delta}{\delta x^{j_1}}, ..., \frac{\partial}{\partial y^{(2)j_s}}\right). \tag{1.5}$$

It follows that T can be locally written in the form

$$T = T^{i_1 \dots i_r}_{j_1 \dots j_s}(x, y^{(1)}, y^{(2)}) \frac{\delta}{\delta x^{i_1}} \otimes \cdots \otimes \frac{\partial}{\partial y^{(2)i_r}} \otimes dx^{j_1} \otimes \cdots \otimes \delta y^{(2)j_s}. \tag{1.6}$$

Using $(5.5)'$ and (6.5), Ch.2, we deduce that, with respect to a transformation of coordinates (1.7), Ch.2, on E, the local coordinates $T^{i_1 \dots i_r}_{j_1 \dots j_s}(x, y^{(1)}, y^{(2)})$ are transformed by the classical rule

$$\begin{aligned} &\widetilde{T}^{i_1 \dots i_r}_{j_1 \dots j_s}(\tilde{x}, \tilde{y}^{(1)}, \tilde{y}^{(2)}) = \\ &= \frac{\partial \tilde{x}^{i_1}}{\partial x^{h_s}} \cdots \frac{\partial \tilde{x}^{i_r}}{\partial x^{h_r}} \cdot \frac{\partial x^{k_1}}{\partial \tilde{x}^{j_1}} \cdots \frac{\partial x^{k_s}}{\partial \tilde{x}^{j_s}} T^{h_1 \dots h_r}_{k_1 \dots k_s}(x, y^{(1)}, y^{(2)}). \end{aligned} \tag{1.7}$$

But this is possible only for coordinates of a d–tensor field in adapted basis. In the natural basis $\left(\frac{\partial}{\partial x^i}, \frac{\partial}{\partial y^{(1)i}}, \frac{\partial}{\partial y^{(2)i}}\right)$ and natural cobasis $(dx^i, dy^{(1)i}, dy^{(2)i})$ the rule of transformation of the coordinates of a tensor field T on E are very complicated. This is the reason for which we systematically use the d–tensor fields T on E represented locally in the adapted basis $\left(\frac{\delta}{\delta x^i}, \frac{\delta}{\delta y^{(1)i}}, \frac{\partial}{\partial y^{(2)i}}\right)$ and adapted cobasis $(dx^i, \delta y^{(1)i}, \delta y^{(2)i})$.

Taking into account (1.5) and (1.7), we can prove that the sum and tensor product of d–tensor fields give us d–tensor fields. It follows that $\left\{1, \frac{\delta}{\delta x^i}, \frac{\delta}{\delta y^{(1)i}}, \frac{\partial}{\partial y^{(2)i}}\right\}$ span the tensor algebra of d–tensor fields over the ring of functions on E. The elements of this algebra is a system of generators of the tensorial algebra of tensor fields on E.

For instance, let us consider a Riemannian metric G on E and assume that the distributions N_0, N_1, V_2 are orthogonal, in pairs, with respect to G:

$$(1.8)\quad G(X^H, Y^{V_1}) = G(X^H, Y^{V_2}) = G(X^{V_1}, Y^{V_2}) = 0,\ \forall X, Y \in \mathcal{X}(E).$$

In this case G can be uniquely written as a sum of d–tensors:

$$(1.8)'\qquad G = G^H + G^{V_1} + G^{V_2},$$

where, for any $X, Y \in \mathcal{X}(E)$, we have

$$(1.8)''\qquad \begin{aligned} &G^H(X,Y) = G(X^H, Y^H),\\ &G^{V_1}(X,Y) = G(X^{V_1}, Y^{V_2}),\quad G^{V_2}(X,Y) = G(X^{V_2}, Y^{V_2}). \end{aligned}$$

Consequently, in the adapted cobasis, G can be uniquely written as

$$(1.9)\quad G = g^{(0)}{}_{ij}\, dx^i \otimes dx^j + g^{(1)}{}_{ij}\, \delta y^{(1)i} \otimes \delta y^{(1)j} + g^{(2)}{}_{ij}\, \delta y^{(2)i} \otimes \delta y^{(2)j},$$

where

$$(1.9)'\qquad g^{(\alpha)}{}_{ij}(x, y^{(1)i}, y^{(2)i}) = g^{(\alpha)}{}_{ji}(x, y^{(1)i}, y^{(2)i}),\quad (i = 0, 1, 2),$$

$$(1.9)''\qquad \operatorname{rank} \|g^{(\alpha)}{}_{ji}\| = n,\quad (\alpha = 0, 1, 2).$$

One can make similar considerations for the tensor fields of type $(1,1)$ on E. For instance for $\mathbb{P}$ and $\mathbb{F}$, the structures given in the last section of Chapter 2.

3.2 N–Linear Connection on $\mathrm{Osc}^2 M$

On the total space $E = \mathrm{Osc}^2 M$ of the 2–osculator bundle there are linear connections compatible with the direct decomposition (1.1). The advantage of considering these linear connections is that in the adapted basis they have as coefficients some simple geometrical objects, easily to find in common cases.

Definition 3.2.1 A linear connection D on E is called an N–linear connection if:
(1) D preserves by parallelism the horizontal distribution N.
(2) The 2–tangent structure J is absolute parallel with respect to D, i.e. $D_X J = 0,\ \forall X \in \mathcal{X}(E)$.

A theorem of characterization of the N–linear connections is the following

Theorem 3.2.1 *A linear connection D on the manifold E is an N–linear connection if, and only if, the following properties are verified:*

$$(2.1)\qquad \begin{aligned} &(D_X Y^H)^{V_1} = (D_X Y^H)^{V_2} = 0,\\ &(D_X Y^{V_1})^H = (D_X Y^{V_1})^{V_2} = 0,\\ &(D_X Y^{V_2})^H = (D_X Y^{V_2})^{V_1} = 0, \end{aligned}$$

$$(2.1)'\quad D_X(JY^H) = J(D_X Y^H),\ D_X(JY^{V_\alpha}) = J(D_X Y^{V_\alpha}),\ (\alpha = 1,2)$$

for any vector fields $X, Y \in \mathcal{X}(E)$.

Proof. If D is an N–linear connection, then we have

$$(*)\qquad (D_X J)(Y) = D_X(JY) - J(D_X Y) = 0,\ \ \forall X, Y \in \mathcal{X}(E).$$

Taking $Y = Y^H$ or $Y = Y^{V_\alpha}$, $(\alpha = 1,2)$ in the last equation we obtain (2.1)$'$. Also, from $D_X Y^H \in N_0$ for any $X, Y \in \mathcal{X}(E)$ it follows $(D_X Y^H)^{V_\alpha} = 0$, $(\alpha = 1,2)$. Setting $Y^{V_1} = J(Y^H)$, we deduce $D_X Y^{V_1} = J(D_X Y^H) \in N_1$. This implies $D_X Y^{V_1})^H = 0$ and $(D_X Y^{V_1})^{V_2} = 0$. Similarly, we get $(D_X Y^{V_2})^H = (D_X Y^{V_2})^{V_1} = 0$. Conversely, the equations (2.1)$'$ give $(D_X J)(Y) = 0, \forall X, Y \in \mathcal{X}(E)$ and the first line of (2.1) implies $D_X Y^H \in N_0$ for any $X, Y \in \mathcal{X}(E)$. **q.e.d.**

Theorem 3.2.2 *For any N–linear connection D we have*

$$D_X h = 0, \quad D_X v_1 = D_X v_2 = 0 \tag{2.2}$$

$$D_X I\!P = 0. \tag{2.3}$$

Proof. For any vector fields X, Y on E, $(D_X h)(Y) = D_X(hY) - h(D_X Y)$. By means of (2.1) it follows $D_X h = 0$. Similarly, we deduce $D_X v_\alpha = 0$, $(\alpha = 1, 2)$.

Now, taking into account $I\!P = I - 2v$, it follows $D_X I\!P = 0$. **q.e.d.**

Remark. Later, we shall prove $D_X I\!F = 0$.

Now, let us consider a vector field X on E, written in the form (1.2) and D an N–linear connection. Since $D_X Y$ is $\mathcal{F}(E)$–linear with respect to X, we have

$$D_X Y = D_{X^H} Y + D_{X^{V_1}} Y + D_{X^{V_2}} Y, \quad \forall X, Y \in \mathcal{X}(E). \tag{2.4}$$

We find here new operators in the algebra of d-tensor fields: $D_{X^H}, D_{X^{V_1}}, D_{X^{V_2}}$ denoted by

$$D^H_X = D_{X^H}, \; D^{V_1}_X = D_{X^{V_1}}, \; D^{V_2}_X = D_{X^{V_2}}. \tag{2.5}$$

These operators are not covariant derivations in the algebra of d–tensor fields, since $D^H_X f = X^H f \neq X f$, etc. But they have similar properties with the covariant derivations.

From (2.5) we can write (2.4) in the form

$$D_X Y = D^H_X Y + D^{V_1}_X Y + D^{V_2}_X Y, \quad \forall X, Y \in \mathcal{X}(E). \tag{2.6}$$

Theorem 3.2.3 *The operators D^H, D^{V_1}, D^{V_2} have the properties:*

1° *All equalities on the first line of (2.1) are verified for $X = X^H$, $X = X^{V_1}, X = X^{V_2}$ and $D^H_X f = X^H f, D^{V_1}_X f = X^{V_1} f, D^{V_2}_X f = X^{V_2} f$.*

2° $D^H_X(fY) = X^H f Y + f D^H_X Y, D^{V_\alpha}_X(fY) = X^{V_\alpha} f Y + f D^{V_\alpha}_X Y, (\alpha = 1, 2)$.

3° $(D^H_X Y)_{|U} = D^H_{X|U} Y_{|U}, (D^{V_\alpha}_X Y)_{|U} = D^{V_\alpha}_{X|U} Y_{|U}$ *for any open set $U \subset E$.*

4° $D^H_{X+Y} = D^H_X + D^H_Y, \; D^{V_\alpha}_{X+Y} = D^{V_\alpha}_X + D^{V_\alpha}_Y, \; (\alpha = 1, 2)$.

5° $D^H_{fX} = fD^H_X, \; D^{V_\alpha}_{fX} = fD^{V_\alpha}_X$.

6° $D^H_X(JY) = JD^H_X Y, \; D^{V_\alpha}_X(JY) = JD^{V_\alpha}_X Y, \quad (\alpha = 1, 2)$
for any $f \in \mathcal{F}(E)$ and any vector fields X, Y on E.

These statements can be proved starting from the properties of the N–linear connection D and of the defining properties (2.5) of the operators D^H, D^{V_1}, D^{V_2}.

We also can prove that if these operators, having the properties 1° – 6°, are given, then D from (2.6) is an N–linear connection on E.

D^H, D^{V_1}, D^{V_2} are called $h-$, v_1- and v_2- *covariant derivatives,* respectively.

According to the last theorem, we can extend the action of the $h-$, v_1- and v_2- derivatives to any tensor field on E, particularly to d–tensor fields. So, for any $\omega \in \mathcal{X}^*(E)$ and for any $X, Y \in \mathcal{X}(E)$, we have

$$\begin{aligned} &(D^H_X \omega)(Y) = X^H \omega(Y) - \omega(D^H_X Y) \\ &(D^{V_\alpha}_X \omega)(Y) = X^{V_\alpha}\omega(Y) - \omega(D^{V_\alpha}_X Y), \; (\alpha = 1, 2). \end{aligned} \tag{2.7}$$

If $T \in \mathcal{T}^r_s(E)$, taking in $D_X T$, $X = X^H$ or $X = X^{V_\alpha}$, we have

$$\begin{aligned} &(D^H_X T)(\overset{1}{\omega}, ..., \overset{r}{\omega}, \underset{1}{X}, ..., \underset{s}{X}) = X^H T(\overset{1}{\omega}, ..., \overset{r}{\omega}, \underset{1}{X}, ..., \underset{s}{X}) - \\ &\quad -T(D^H_X \overset{1}{\omega}, ..., \overset{r}{\omega}, \underset{1}{X}, ..., \underset{s}{X}) - \cdots - T(\overset{1}{\omega}, ..., \overset{r}{\omega}, \underset{1}{X}, ..., D^H_X \underset{s}{X}), \\ &(D^{V_\alpha}_X T)(\overset{1}{\omega}, ..., \overset{r}{\omega}, \underset{1}{X}, ..., \underset{s}{X}) = X^{V_\alpha} T(\overset{1}{\omega}, ..., \overset{r}{\omega}, \underset{1}{X}, ..., \underset{s}{X}) - \\ &\quad -T(D^{V_\alpha}_X \overset{1}{\omega}, ..., \overset{r}{\omega}, \underset{1}{X}, ..., \underset{s}{X}) - \cdots - T(\overset{1}{\omega}, ..., \overset{r}{\omega}, \underset{1}{X}, ..., D^{V_\alpha}_X \underset{s}{X}), \\ &\qquad\qquad (\alpha = 1, 2). \end{aligned} \tag{2.8}$$

If γ is a parametrized curve, as in the §6, Ch.2, then its tangent vector field $\dfrac{d\gamma}{dt} = \dot{\gamma}$ can be represented as $\dot{\gamma} = \dot{\gamma}^H + \dot{\gamma}^{V_1} + \dot{\gamma}^{V_2}$.

A vector field Y on E is parallel along γ, with respect to an N–linear connection, if $D_{\dot{\gamma}} Y = 0$, or $D_{\dot{\gamma}} Y^H + D_{\dot{\gamma}} Y^{V_1} + D_{\dot{\gamma}} Y^{V_2} = 0$.

The parametrized curve γ is an autoparallel curve, with respect to D if:

$$D_{\dot{\gamma}}\dot{\gamma} = D_{\dot{\gamma}}\dot{\gamma}^H + D_{\dot{\gamma}}\dot{\gamma}^{V_1} + D_{\dot{\gamma}}\dot{\gamma}^{V_2} = 0. \tag{2.9}$$

These considerations can be developed in the adapted basis.

3.3 The Coefficients of the N–Linear Connections

Let

$$(3.1) \qquad \left\{\frac{\delta}{\delta x^i}, \frac{\delta}{\delta y^{(1))i}}, \frac{\partial}{\partial y^{(2))i}}\right\} \quad (i = 1, ..., n)$$

be the adapted basis to the decomposition (1.1) and

$$(3.1)' \qquad \left\{dx^i, \delta y^{(1))i}, \delta y^{(2))i}\right\} \quad (i = 1, ..., n)$$

the corresponding adapted cobasis.

An N–linear connection D, in the basis (3.1), can be represented in a very simple form.

Theorem 3.3.1 *In the adapted basis* (3.1) *an N-linear connection D can be uniquely written in the following form:*

$$(3.2) \qquad D_{\frac{\delta}{\delta x^j}} \frac{\delta}{\delta y^{(\alpha)i}} = L^m{}_{ij} \frac{\delta}{\delta y^{(\alpha)m}}, D_{\frac{\delta}{\delta y^{(\beta)j}}} \frac{\delta}{\delta y^{(\alpha)i}} = \underset{(\beta)}{C}{}^m{}_{ij} \frac{\delta}{\delta y^{(\alpha)m}},$$
$$(\beta = 1, 2;\ \alpha = 0, 1, 2;\ y^{(0)i} = x^i).$$

Proof. Indeed, we can uniquely write

$$D_{\frac{\delta}{\delta x^j}} \frac{\delta}{\delta x^j} = L^m{}_{jk} \frac{\delta}{\delta x^m} + \overset{(1)}{L}{}^m{}_{jk} \frac{\delta}{\delta y^{(1)m}} + \overset{(2)}{L}{}^m{}_{jk} \frac{\partial}{\partial y^{(2)m}}.$$

Now, applying Theorem 3.2.1 and taking into account that $D_X \dfrac{\delta}{\delta x^j}$ belongs to the horizontal distribution N_0, we get $\overset{(1)}{L}{}^m{}_{jk} = \overset{(2)}{L}{}^m{}_{jk} = 0$. Hence, we have the first equality (3.2) for $\alpha = 0$. After this, applying the tangent structure J and looking at (2.1)$'$ we have $D_{\frac{\delta}{\delta x^j}} \dfrac{\delta}{\delta y^{(\alpha)i}} =$ $= L^m{}_{ij} \dfrac{\delta}{\delta y^{(\alpha)m}}$, $(\alpha = 1, 2)$. Therefore, the first equality in (3.2) is proved. We can similarly prove the second one. **q.e.d.**

If an N–linear connection D is given, then the system of functions $(L^i{}_{jk}(x, y^{(1)}, y^{(2)}),\ \underset{(1)}{C}{}^i{}_{jk}(x, y^{(1)}, y^{(2)}),\ \underset{(2)}{C}{}^i{}_{jk}(x, y^{(1)}, y^{(2)}))$ is well determined from (3.2). It is called the system of coefficients of D in the adapted basis (3.1).

The last Theorem has some consequences, given by the following proposition:

Proposition 3.3.1 *With respect to* (1.7), *Ch.2, the coefficients* $(L^i{}_{jk},$ $\underset{(1)}{C}{}^i{}_{jk}, \underset{(2)}{C}{}^i{}_{jk})$ *of* D *are transformed as follows:*

$$
\begin{aligned}
&\tilde{L}^i{}_{rs}\frac{\partial \tilde{x}^r}{\partial x^j}\frac{\partial \tilde{x}^s}{\partial x^k} = \frac{\partial \tilde{x}^i}{\partial x^r}L^r{}_{jk} - \frac{\partial^2 \tilde{x}^i}{\partial x^j \partial x^k}\\
&\underset{(\alpha)}{\tilde{C}}{}^i{}_{rs}\frac{\partial \tilde{x}^r}{\partial x^j}\frac{\partial \tilde{x}^s}{\partial x^k} = \frac{\partial \tilde{x}^i}{\partial x^r}\underset{(\alpha)}{C}{}^r{}_{jk}, \quad (\alpha = 1, 2).
\end{aligned} \tag{3.3}
$$

It is important to remark that the coefficients $\underset{(1)}{C}{}^i{}_{jk}(x, y^{(1)}, y^{(2)})$ and $\underset{(2)}{C}{}^i{}_{jk}(x, y^{(1)}, y^{(2)})$ are d–tensor fields of type $(1, 2)$.

Theorem 3.3.2 *If on every domain of local chart* $\pi^{-1}(U)$ *on* E, *a system of functions* $(L^i{}_{jk}(x, y^{(1)}, y^{(2)}), \underset{(1)}{C}{}^i{}_{jk}(x, y^{(1)}, y^{(2)}), \underset{(2)}{C}{}^i{}_{jk}(x, y^{(1)}, y^{(2)}))$ *is given so that* (3.3) *holds, then there exists a unique* N*–linear connection* D *on* E, *which satisfies* (3.2).

Indeed, if $(L^i{}_{jk}\ \underset{(\beta)}{C}{}^i{}_{jk})$, $(\alpha = 1, 2)$, are given on every domain of local chart on E, and the equations (3.3) are verified, then it is easy to see that (3.2) determines every term of an N–connection $D_X = D^H_X +$ $+D^{V_1}_X + D^{V_2}_X$. By contradiction, we prove that this is unique, having as coefficients exactly the given functions.

Example. Let N be a nonlinear connection on E, with the coefficients $\underset{(1)}{N}{}^i{}_j, \underset{(2)}{N}{}^i{}_j$. With respect to (1.7), Ch.2, we have the rules of transformations (4.11), Ch.2. Hence, we get:

Proposition 3.3.2 *The system of functions* $B\Gamma(N) = (L^i{}_{jk}, \underset{(1)}{C}{}^i{}_{jk}, \underset{(2)}{C}{}^i{}_{jk})$ *given by*

$$(3.4) \qquad L^i{}_{jk} = \frac{\delta \underset{(1)}{N}{}^i{}_j}{\delta y^{(1)k}}, \quad \underset{(\alpha)}{C}{}^i{}_{jk} = 0, \quad (\alpha = 1, 2)$$

determines an N-linear connection, which depends only on the nonlinear connection N.

The N–linear connection D with the coefficients (3.4) will be called the *Berwald connection* determined by the nonlinear connection N.

This example, together with the Theorems 2.3.4 and 2.3.5, justify the following :

Theorem 3.3.3 *If the base manifold M of the 2–osculator bundle $(\mathrm{Osc}^2 M, \pi, M)$ is paracompact, then on $E = \mathrm{Osc}^2 M$ there exist N–linear connections.*

Let T be the d–tensor field given in (1.6) and $X = X^i \dfrac{\delta}{\delta x^i}$ a horizontal vector field. Taking into account (3.2) the h–covariant derivation of T is expressed by

$$(3.5) \quad D^H_X T = X^m T^{i_1 \dots i_r}_{j_1 \dots j_s | m} \frac{\delta}{\delta x^{i_1}} \otimes \cdots \otimes \frac{\partial}{\partial y^{(2) i_r}} \otimes dx^{j_1} \otimes \cdots \otimes \delta y^{(2) j_s},$$

where

$$(3.6) \quad \begin{aligned} T^{i_1 \dots i_r}_{j_1 \dots j_s | m} &= \frac{\delta T^{i_1 \dots i_r}_{j_1 \dots j_s}}{\delta x^m} + L^{i_1}{}_{hm} T^{h i_2 \dots i_r}_{j_1 \dots j_s} + \cdots + \\ &+ L^{i_r}{}_{hm} T^{i_1 \dots h}_{j_1 \dots j_s} - L^h{}_{j_1 m} T^{i_1 \dots i_r}_{h j_2 \dots j_s} - \cdots - L^h{}_{j_s m} T^{i_1 \dots i_r}_{j_1 \dots h}. \end{aligned}$$

Therefore, the operator $|$ from (3.6) is called the *h–covariant derivation.* It has the same properties as in the Proposition 1.3.4, or as an ordinary operator of covariant derivative in the tensorial algebra on the base manifold M. We shall not repeat them.

Let us consider the operators $D_X^{V_\alpha}$, for the vector fields $X^{V_\alpha} = \overset{(\alpha)}{X}{}^i \frac{\delta}{\delta y^{(\alpha)i}}$, $(\alpha = 1, 2)$. Then, from (1.6), we derive:

$$(3.7)\quad \begin{aligned} D_X^{V_\alpha} T = \overset{(\alpha)}{X}{}^m T^{i_1 \dots i_r}_{j_1 \dots j_s} \overset{(\alpha)}{|}_m \frac{\delta}{\delta x^{i_1}} \otimes \cdots \otimes \\ \otimes \frac{\partial}{\partial y^{(2)i_r}} \otimes dx^{j_1} \otimes \cdots \otimes \delta y^{(2)j_s}, \ (\alpha = 1, 2) \end{aligned}$$

where

$$(3.8)\quad \begin{aligned} T^{i_1 \dots i_r}_{j_1 \dots j_s} \overset{(\alpha)}{|}_m = \frac{\delta T^{i_1 \dots i_r}_{j_1 \dots j_s}}{\delta y^{(\alpha)m}} + \underset{(\alpha)}{C}{}^{i_1}{}_{hm} T^{h i_2 \dots i_r}_{j_1 \dots j_s} + \cdots + \underset{(\alpha)}{C}{}^{i_r}{}_{hm} T^{i_1 \dots h}_{j_1 \dots j_s} - \\ - \underset{(\alpha)}{C}{}^{h}{}_{j_1 m} T^{i_1 \dots i_r}_{h j_2 \dots j_s} - \cdots - \underset{(\alpha)}{C}{}^{h}{}_{j_s m} T^{i_1 \dots i_r}_{j_1 \dots h}, \ (\alpha = 1, 2). \end{aligned}$$

The operators $\overset{(\alpha)}{|}$ are called *v_α–covariant derivations.*

The properties of v_α–covariant derivations are similar to that of the h–covariant derivation. For instance, we have

$$(3.8)'\quad (f T^{\dots}_{\dots}) \overset{(\alpha)}{|}_m = \frac{\delta f}{\delta y^{(\alpha)m}} T^{\dots}_{\dots} + f T^{\dots}_{\dots} \overset{(\alpha)}{|}_m, \ (\alpha = 1, 2).$$

As application, we determine the h– and v_α– covariant derivatives of the Liouville d–vector fields $z^{(1)i}, z^{(2)i}$. We obtain the so–called *deflection* tensor fields of the N–linear connection D. They are as follows:

$$(3.9)\quad \begin{aligned} &\overset{(1)}{D}{}^i{}_j = z^{(1)i}{}_{|j}, \quad \overset{(11)}{d}{}^i{}_j = z^{(1)i} \overset{(1)}{|}_j, \quad \overset{(12)}{d}{}^i{}_j = z^{(1)i} \overset{(2)}{|}_j, \\ &\overset{(2)}{D}{}^i{}_j = z^{(2)i}{}_{|j}, \quad \overset{(21)}{d}{}^i{}_j = z^{(2)i} \overset{(1)}{|}_j, \quad \overset{(22)}{d}{}^i{}_j = z^{(2)i} \overset{(2)}{|}_j. \end{aligned}$$

By a straightforward calculus, using (3.6) and (3.8), we have

$$(3.9)'\quad \begin{aligned} &\overset{(\alpha)}{D}{}^i{}_j = \frac{\delta z^{(\alpha)i}}{\delta x^j} + z^{(\alpha)m} L^i{}_{mj}, \quad \overset{(1\alpha)}{d}{}^i{}_j = \frac{\delta z^{(1)i}}{\delta y^{(\alpha)j}} + z^{(1)m} \underset{(\alpha)}{C}{}^i{}_{mj}, \\ &\overset{(2\alpha)}{d}{}^i{}_j = \frac{\delta z^{(2)i}}{\delta y^{(\alpha)j}} + z^{(2)m} \underset{(\alpha)}{C}{}^i{}_{mj}, \ (\alpha = 1, 2). \end{aligned}$$

We shall use these d–tensors in the construction of the electromagnetic tensors of the connections D on the manifold E.

Also, as an application of the previous theory we prove an important property of the almost 2–contact structure $\mathbb{F}$ determined by the nonlinear connection N (see (8.6), Ch.2).

Theorem 3.3.4 *Any N–linear connection d is compatible with the almost n–contact structure* $\mathbb{F}$ *associated to the nonlinear connection N, i.e.,*

$$D_X \mathbb{F} = 0, \quad \forall X \in \mathcal{X}(E). \tag{3.10}$$

Proof. For any vector field $Y \in \mathcal{X}(E)$ we have

$$(D_X \mathbb{F})(Y) = D_X(\mathbb{F}Y) - \mathbb{F}(D_X Y). \tag{$*$}$$

Thus, it is sufficiently to prove that (3.10) holds on the adapted basis (3.1). One obtains from $(*)$

$$\begin{aligned} &D_X\left(\mathbb{F}\,\frac{\delta}{\delta x^j}\right) - \mathbb{F}\left(D_X\,\frac{\delta}{\delta x^j}\right) = 0,\\ &D_X\left(\mathbb{F}\,\frac{\delta}{\delta y^{(1)j}}\right) - \mathbb{F}\left(D_X\,\frac{\delta}{\delta y^{(1)j}}\right) = 0,\\ &D_X\left(\mathbb{F}\,\frac{\partial}{\partial y^{(2)j}}\right) - \mathbb{F}\left(D_X\,\frac{\partial}{\partial y^{(2)j}}\right) = 0. \end{aligned} \tag{$**$}$$

Hence we have, for $X = \dfrac{\delta}{\delta x^k}$ or $X = \dfrac{\delta}{\delta y^{(1)k}}$ or $X = \dfrac{\partial}{\partial y^{(2)k}}$ in the first equality $(**)$,

$$-L^i{}_{jk}\,\frac{\partial}{\partial y^{(2)i}} + L^i{}_{jk}\,\frac{\partial}{\partial y^{(2)i}} = 0,\ \underset{(\alpha)}{C}{}^i{}_{jk}\,\frac{\partial}{\partial y^{(2)i}} - \underset{(\alpha)}{C}{}^i{}_{jk}\,\frac{\partial}{\partial y^{(2)i}} = 0.$$

In the same way, we prove the second or the third of these equations $(**)$. It follows that the equation (3.10) holds. **q.e.d.**

The last Theorem states that: *The almost n–contact structure* $\mathbb{F}$, *determined by the nonlinear connection N is absolutely parallel with respect to any N–linear connection D.*

Remark. The Berwald connection $B\Gamma(N) = (B^i{}_{jk}, 0, 0)$ has very simple expressions for its $h-$ and $v_\alpha-$ $(\alpha = 1,2)$ covariant derivatives. Indeed, denoting by $_{||}$ and $\overset{(\alpha)}{||}$ $(\alpha = 1,2)$ the operators of $h-$ and $v_\alpha-$ covariant derivatives, with respect to $B\Gamma(N)$, for any d-tensor field $T^{i_1\dots i_r}_{j_1\dots j_s}$ we have

$$\text{(3.11)}\quad \begin{aligned} &T^{i_1\dots i_r}_{j_1\dots j_s||m} = \frac{\delta T^{i_1\dots i_r}_{j_1\dots j_s}}{\delta x^m} + B^{i_1}{}_{hm}\, T^{h\dots i_r}_{j_1\dots j_s} + \dots - B^h{}_{j_s m}\, T^{i_1\dots i_r}_{j_1\dots h},\\ &T^{i_1\dots i_r}_{j_1\dots j_s}\overset{(1)}{||}_m = \frac{\delta T^{i_1\dots i_r}_{j_1\dots j_s}}{\delta y^{(1)m}};\quad T^{i_1\dots i_r}_{j_1\dots j_s}\overset{(2)}{||}_m = \frac{\partial T^{i_1\dots i_r}_{j_1\dots j_s}}{\partial y^{(2)m}}. \end{aligned}$$

3.4 d–Tensors of Torsion

On the total space E of the 2–osculator bundle $(\mathrm{Osc}^2 M, \pi, M)$ endowed with a nonlinear connection N and an N–linear connection D we can consider the torsion and curvature of D and report them to the direct decomposition (1.1).

Let $\mathbb{T}$ be the tensor of torsion of D. We can write, for any vector fields $X, Y \in \mathcal{X}(E)$:

$$\text{(4.1)}\qquad \mathbb{T}(X,Y) = D_X Y - D_Y X - [X,Y].$$

The tensor $\mathbb{T}$ from (4.1) can be evaluated for the pairs of d–tensor fields $(X^H, Y^H), (X^H, Y^{V_\alpha}), (X^{V_\alpha}, Y^{V_\beta})$. We obtain the vector fields

$$\text{(4.2)}\qquad \mathbb{T}(X^H,Y^H), \mathbb{T}(X^H,Y^{V_\alpha}), \mathbb{T}(X^{V_\alpha},Y^{V_\beta}),\quad (\alpha,\beta = 1,2).$$

Proposition 3.4.1 *The tensor of torsion $\mathbb{T}$ of an N–linear connection D is well determined by the following components, where in the right hand we have d–tensor fields of type* $(1,2)$:

$$\text{(4.3)}\quad \begin{aligned} &\mathbb{T}(X^H,Y^H) = h\mathbb{T}(X^H,Y^H) + v_1\mathbb{T}(X^H,Y^H) + v_2\mathbb{T}(X^H,Y^H),\\ &\mathbb{T}(X^H,Y^{V_\alpha}) = h\mathbb{T}(X^H,Y^{V_\alpha}) + v_1\mathbb{T}(X^H,Y^{V_\alpha}) + v_2\mathbb{T}(X^H,Y^{V_\alpha}),\\ &\hfill (\alpha = 1,2)\\ &\mathbb{T}(X^{V_1},Y^{V_\alpha}) = v_1\mathbb{T}(X^{V_1},Y^{V_\alpha}) + v_2\mathbb{T}(X^{V_1},Y^{V_\alpha}),\\ &\hfill (\alpha = 1,2)\\ &\mathbb{T}(X^{V_2},Y^{V_2}) = v_1\mathbb{T}(X^{V_2},Y^{V_2}) + v_2\mathbb{T}(X^{V_2},Y^{V_2}). \end{aligned}$$

Indeed, the first two equalities are obvious, since the set (4.2) is formed by the vector fields. We shall prove only the equations $h\mathbb{T}(X^{V_1}, Y^{V_\alpha}) = h\mathbb{T}(X^{V_2}, Y^{V_2}) = 0$. Using (4.1) we have

$$\begin{aligned} h\mathbb{T}(X^{V_1}, Y^{V_\alpha}) &= h\left\{D_X^{V_1} Y^{V_\alpha} - D_Y^{V_\alpha} X^{V_1} - [X^{V_1}, Y^{V_\alpha}]\right\} = \\ &= -h[X^{V_1}, Y^{V_\alpha}] = 0, \end{aligned}$$

since the vertical distribution V is integrable. Analogously for the second equation. **q.e.d.**

The d–tensors from the right hand of (4.3) are called the *d–tensors of torsion* of the N–linear connection D.

Theorem 3.4.1 *The N-linear connection D is torsion free if, and only if, all d–tensors of torsion of D vanish.*

Indeed, the vector fields (4.2) vanish if, and only if, $\mathbb{T}(X, Y) = 0$, $\forall X, Y \in \mathcal{X}(E)$. But these vector fields vanish if, and only if, all the terms in the right hand of (4.3) vanish. **q.e.d.**

Theorem 3.4.2 *The following properties hold:*

(a) $v_\alpha\mathbb{T}(X^H, Y^H) = 0$, $(\alpha = 1, 2)$ *if, and only if, the horizontal distribution N is integrable.*

(b) $v_\alpha\mathbb{T}(X^{V_1}, Y^{V_2}) = 0$, $(\alpha = 1, 2)$ *if, and only if, the J–vertical distribution N_1 is integrable.*

(c) $v_\alpha\mathbb{T}(X^H, Y^{V_\alpha}) = 0$, *if, and only if,* $D_X^H Y^{V_\alpha} = [X^H, Y^{V_\alpha}]^{V_\alpha}$.

(d) $v_2\mathbb{T}(X^{V_1}, Y^{V_2}) = 0$, *if, and only if,* $D_X^{V_1} Y^{V_2} = [X^{V_1}, Y^{V_2}]^{V_2}$.

Proof. (a) $v_\alpha\mathbb{T}(X^H, Y^H) = -[X^H, Y^H]^{V_\alpha}$, $(\alpha = 1, 2)$. Applying Theorem 2.5.2, the property (a) is immediate. In the same manner we prove (b), (c) and (d). **q.e.d.**

Now let us express the d–tensors of torsion of the N–linear connection D in the adapted basis (3.1). In order to do this we can prove by a direct calculus and using (6.10), Ch.2, the following theorem:

Theorem 3.4.3 *The Lie brackets of the vector fields of the adapted basis are given by*

$$
\begin{aligned}
&\left[\frac{\delta}{\delta x^j}, \frac{\delta}{\delta x^k}\right] = \underset{(01)}{R}{}^i{}_{jk}\,\frac{\delta}{\delta y^{(1)i}} + \underset{(02)}{R}{}^i{}_{jk}\,\frac{\partial}{\partial y^{(2)i}},\\
&\left[\frac{\delta}{\delta x^j}, \frac{\delta}{\delta y^{(1)k}}\right] = \underset{(11)}{B}{}^i{}_{jk}\,\frac{\partial}{\partial y^{(1)i}} + \underset{(12)}{B}{}^i{}_{jk}\,\frac{\partial}{\partial y^{(2)i}},\\
(4.4)\qquad &\left[\frac{\delta}{\delta x^j}, \frac{\partial}{\partial y^{(2)k}}\right] = \underset{(21)}{B}{}^i{}_{jk}\,\frac{\delta}{\delta y^{(1)i}} + \underset{(22)}{B}{}^i{}_{jk}\,\frac{\partial}{\partial y^{(2)i}},\\
&\left[\frac{\delta}{\delta y^{(1)j}}, \frac{\delta}{\delta y^{(1)k}}\right] = \underset{(12)}{R}{}^i{}_{jk}\,\frac{\partial}{\partial y^{(2)i}},\\
&\left[\frac{\delta}{\delta y^{(1)j}}, \frac{\partial}{\partial y^{(2)k}}\right] = \underset{(21)}{B}{}^i{}_{jk}\,\frac{\partial}{\partial y^{(2)i}},
\end{aligned}
$$

where

$$
\begin{aligned}
&\underset{(01)}{R}{}^i{}_{jk} = \frac{\delta \underset{(1)}{N}{}^i{}_j}{\delta x^k} - \frac{\delta \underset{(1)}{N}{}^i{}_k}{\delta x^j};\\
&\underset{(02)}{R}{}^i{}_{jk} = \underset{(1)}{N}{}^i{}_m\,\underset{(01)}{R}{}^m{}_{jk} + \frac{\delta \underset{(2)}{N}{}^i{}_j}{\delta x^k} - \frac{\delta \underset{(2)}{N}{}^i{}_k}{\delta x^j},\\
(4.5)\qquad &\underset{(11)}{B}{}^i{}_{jk} = \frac{\delta \underset{(1)}{N}{}^i{}_j}{\delta y^{(1)k}},\quad \underset{(12)}{B}{}^i{}_{jk} = \underset{(1)}{N}{}^i{}_m\,\underset{(11)}{B}{}^m{}_{jk} + \frac{\delta \underset{(2)}{N}{}^i{}_j}{\delta y^{(1)k}} - \frac{\delta \underset{(2)}{N}{}^i{}_k}{\delta y^{(1)j}},\\
&\underset{(21)}{B}{}^i{}_{jk} = \frac{\partial \underset{(1)}{N}{}^i{}_j}{\delta y^{(2)k}},\quad \underset{(22)}{B}{}^i{}_{jk} = \underset{(1)}{N}{}^i{}_m\,\underset{(21)}{B}{}^m{}_{jk} + \frac{\partial \underset{(2)}{N}{}^i{}_j}{\partial y^{(2)k}};\\
&\underset{(12)}{R}{}^i{}_{jk} = \frac{\delta \underset{(1)}{N}{}^i{}_j}{\delta y^{(1)k}} - \frac{\delta \underset{(1)}{N}{}^i{}_k}{\delta y^{(1)j}}.
\end{aligned}
$$

The functions $\underset{(11)}{B}{}^i{}_{jk}$ are the coefficients of the Berwald connection associated to the nonlinear connection N.

It is not difficult to prove that $\underset{(01)}{R}{}^i{}_{jk}$, $\underset{(02)}{R}{}^i{}_{jk}$ and $\underset{(12)}{R}{}^i{}_{jk}$ are d–tensor fields.

Theorem 3.4.4

(a) *The horizontal distribution N is integrable if, and only if, the following d–tensor fields vanish:*

$$\underset{(01)}{R}{}^{i}{}_{jk} = \underset{(02)}{R}{}^{i}{}_{jk} = 0. \tag{4.6}$$

(b) *The J–vertical distribution N_1 is integrable if, and only if, we have:*

$$\underset{(12)}{R}{}^{i}{}_{jk} = 0. \tag{4.6$'$}$$

Now, we can give the local expressions of the d–tensors of torsion of an N–linear connection D.

Theorem 3.4.5 *The d–tensors of torsion of an N–linear connection with the coefficients $D\Gamma(N) = (L^{i}{}_{jk}, \underset{(1)}{C}{}^{i}{}_{jk}, \underset{(2)}{C}{}^{i}{}_{jk})$ in the adapted basis (3.1) have the following expressions:*

$$\begin{aligned}
&h\mathbb{T}\left(\frac{\delta}{\delta x^k}, \frac{\delta}{\delta x^j}\right) = \underset{(0)}{T}{}^{i}{}_{jk}\frac{\delta}{\delta x^i}; v_\alpha\mathbb{T}\left(\frac{\delta}{\delta x^k}, \frac{\delta}{\delta x^j}\right) = \underset{(0\alpha)}{R}{}^{i}{}_{jk}\frac{\delta}{\delta y^{(\alpha)i}}, \quad (\alpha=1,2)\\
&h\mathbb{T}\left(\frac{\delta}{\delta y^{(\alpha)k}}, \frac{\delta}{\delta x^j}\right) = \underset{(\alpha)}{C}{}^{i}{}_{jk}\frac{\delta}{\delta x^i}; v_\beta\mathbb{T}\left(\frac{\delta}{\delta y^{(\alpha)k}}, \frac{\delta}{\delta x^j}\right) = \underset{(\alpha\beta)}{P}{}^{i}{}_{jk}\frac{\delta}{\delta y^{(\beta)i}}, \quad (\alpha,\beta=1,2)\\
&v_\alpha\mathbb{T}\left(\frac{\partial}{\partial y^{(2)k}}, \frac{\delta}{\delta y^{(1)j}}\right) = \underset{(\alpha 2)}{Q}{}^{i}{}_{jk}\frac{\delta}{\delta y^{(\alpha)i}};\\
&v_\alpha\mathbb{T}\left(\frac{\delta}{\delta y^{(1)k}}, \frac{\delta}{\delta y^{(1)j}}\right) = \underset{(\alpha 1)}{Q}{}^{i}{}_{jk}\frac{\delta}{\delta y^{(\alpha)i}}, \quad (\alpha=1,2)\\
&v_2\mathbb{T}\left(\frac{\partial}{\partial y^{(2)k}}, \frac{\partial}{\partial y^{(2)j}}\right) = \underset{(2)}{S}{}^{i}{}_{jk}\frac{\partial}{\partial y^{(2)i}};\ v_1\mathbb{T}\left(\frac{\partial}{\partial y^{(2)k}}, \frac{\partial}{\partial y^{(2)j}}\right) = 0,
\end{aligned} \tag{4.7}$$

where

$$\begin{aligned}
&\underset{(0)}{T}{}^{i}{}_{jk} = L^{i}{}_{jk} - L^{i}{}_{kj};\ \underset{(11)}{P}{}^{i}{}_{jk} = \underset{(11)}{B}{}^{i}{}_{jk} - L^{i}{}_{kj};\ \underset{(12)}{P}{}^{i}{}_{jk} = \underset{(12)}{B}{}^{i}{}_{jk}\\
&\underset{(21)}{P}{}^{i}{}_{jk} = \underset{(21)}{B}{}^{i}{}_{jk};\ \underset{(22)}{P}{}^{i}{}_{jk} = \underset{(22)}{B}{}^{i}{}_{jk} - L^{i}{}_{kj};\ \underset{(12)}{Q}{}^{i}{}_{jk} = \underset{(2)}{C}{}^{i}{}_{jk},\\
&\underset{(22)}{Q}{}^{i}{}_{jk} = \underset{(21)}{B}{}^{i}{}_{jk} - \underset{(1)}{C}{}^{i}{}_{kj},\ \underset{(11)}{Q}{}^{i}{}_{jk} = \underset{(1)}{S}{}^{i}{}_{jk} = \underset{(1)}{C}{}^{i}{}_{jk} - \underset{(1)}{C}{}^{i}{}_{kj};\\
&\underset{(21)}{Q}{}^{i}{}_{jk} = \underset{(12)}{R}{}^{i}{}_{jk};\ \underset{(2)}{S}{}^{i}{}_{jk} = \underset{(2)}{C}{}^{i}{}_{jk} - \underset{(2)}{C}{}^{i}{}_{kj}.
\end{aligned} \tag{4.8}$$

Indeed, (4.1), (4.3) and (4.4) imply (4.7) and (4.8).

We pay a special attention to the d–tensors of torsion

$$(4.9)\quad \underset{(0)}{T}{}^{i}{}_{jk} = L^{i}{}_{jk} - L^{i}{}_{kj},\ \underset{(1)}{S}{}^{i}{}_{jk} = \underset{(1)}{C}{}^{i}{}_{jk} - \underset{(1)}{C}{}^{i}{}_{kj},\ \underset{(2)}{S}{}^{i}{}_{jk} = \underset{(2)}{C}{}^{i}{}_{jk} - \underset{(2)}{C}{}^{i}{}_{kj}.$$

Therefore,

Proposition 3.4.2 *The following statements are equivalent*

(a) $\underset{(0)}{T}{}^{i}{}_{jk} = \underset{(1)}{S}{}^{i}{}_{jk} = \underset{(2)}{S}{}^{i}{}_{jk} = 0$

(b) $L^{i}{}_{jk} = L^{i}{}_{kj},\ \underset{(1)}{C}{}^{i}{}_{jk} = \underset{(1)}{C}{}^{i}{}_{kj},\ \underset{(2)}{C}{}^{i}{}_{jk} = \underset{(2)}{C}{}^{i}{}_{kj}.$

Proposition 3.4.3 *The d–tensors of torsion of the Berwald connection $B\Gamma(N) = (B^{i}{}_{jk}, 0, 0)$ are given by*

$$(4.10)\quad \begin{aligned} &\underset{(0)}{T}{}^{i}{}_{jk} = B^{i}{}_{jk} - B^{i}{}_{kj};\ \underset{(11)}{P}{}^{i}{}_{jk} = \underset{(0)}{T}{}^{i}{}_{kj};\ \underset{(12)}{P}{}^{i}{}_{jk};\ \underset{(22)}{P}{}^{i}{}_{jk},\\ &\underset{(12)}{Q}{}^{i}{}_{jk} = 0,\ \underset{(22)}{Q}{}^{i}{}_{jk} = \underset{(21)}{P}{}^{i}{}_{jk},\ \underset{(1)}{S}{}^{i}{}_{jk} = \underset{(2)}{S}{}^{i}{}_{jk} = 0,\\ &\underset{(21)}{Q}{}^{i}{}_{jk} = \underset{(12)}{R}{}^{i}{}_{jk} = \underset{(0)}{T}{}^{i}{}_{jk}. \end{aligned}$$

3.5 d–Tensors of Curvature

The curvature tensor $I\!R$ of an N–linear connection D is expressed by

$$(5.1)\quad I\!R(X,Y) = [D_X, D_Y]Z - D_{[X,Y]}Z,\ \forall X,Y,Z \in \mathcal{X}(E).$$

We know that the 2–tangent structure J has the property $D_X(JY) = = J(D_XY)$, for any vector fields $X, Y \in \mathcal{X}(E)$.

Therefore, we can assert:

Proposition 3.5.1 *For any N–linear connection D the curvature tensor field $I\!R$ has the properties*

$$(5.2)\qquad \begin{aligned} J[I\!R(X,Y)Z] &= I\!R(X,Y)(JZ);\\ J^2[I\!R(X,Y)Z] &= I\!R(X,Y)(J^2Z). \end{aligned}$$

Indeed, using (5.1), it follows

$$J[I\!R(X,Y)Z] = J\{D_XD_YZ - D_YD_XZ - D_{[X,Y]}Z\} = D_XD_Y(JZ)- \\ -D_YD_X(JZ) - D_{[X,Y]}(JZ) = I\!R(X,Y)(JZ), \text{ etc.}$$

q.e.d.

Let us consider a horizontal vector field $Z = Z^H$ and its transformations by the 2–tangent mapping J, i.e. $JZ^H = Z^{V_1}$, $J^2Z^H = Z^{V_2}$.

Thus, the components $I\!R(X,Y)Z^{V_1}$ and $I\!R(X,Y)Z^{V_2}$ of $I\!R$ are obtained as follows:

$$(5.3)\qquad \begin{aligned} I\!R(X,Y)Z^{V_1} &= J(I\!R(X,Y)Z^H),\\ I\!R(X,Y)Z^{V_2} &= J^2(I\!R(X,Y)Z^H). \end{aligned}$$

Hence: *The essential components of the curvature tensor field $I\!R$ are $I\!R(X,Y)Z^H$ for any vector fields $X, Y, Z \in \mathcal{X}(E)$.* Another important remark is given by the following assertion: *The vector field $I\!R(X,Y)Z^H$ is horizontal.*

Indeed, Z^H being a horizontal vector field, D_XZ^H has the same property. Hence (4.1) implies that $I\!R(X,Y)Z^H$ is a horizontal vector field.

Consequently, we can formulate:

Proposition 3.5.2 *The curvature tensor field $I\!R$ of an N–linear connection D has the properties:*

$$(5.4)\qquad \begin{aligned} &v_\beta[I\!R(X,Y)Z^H]=0,\ h[I\!R(X,Y)Z^{V_\beta}]=0,\ (\beta=1,2)\\ &v_1[I\!R(X,Y)Z^{V_2}]=0,\ v_2[I\!R(X,Y)Z^{V_1}]=0,\ \forall X,Y,Z\in\mathcal{X}(E). \end{aligned}$$

Thus, the curvature tensor field $I\!R$ gives rise to the d–vector fields

$$(5.5)\qquad \begin{aligned} I\!R(X^H,Y^H)Z^H,\ I\!R(X^{V_\alpha},Y^H)Z^H,\ I\!R(X^{V_\gamma},Y^{V_\alpha})Z^H,\\ (\alpha=1,2;\ \gamma\le\alpha). \end{aligned}$$

and to its transformations by the 2–tangent mapping J:

$$(5.5)' \qquad \begin{aligned} J\{\mathbb{R}(X^H, Y^H)Z^H\} &= \mathbb{R}(X^H, Y^H)Z^{V_1};\\ J\{\mathbb{R}(X^{V_\alpha}, Y^H)Z^H\} &= \mathbb{R}(X^{V_\alpha}, Y^H)Z^{V_1};\\ J\{\mathbb{R}(X^{V_\gamma}, Y^{V_\alpha})Z^H\} &= \mathbb{R}(X^{V_\gamma}, Y^{V_\alpha})Z^{V_1}, \end{aligned}$$

as well as to its transformations by J^2:

$$(5.5)'' \qquad \begin{aligned} J^2\{\mathbb{R}(X^H, Y^H)Z^H\} &= \mathbb{R}(X^H, Y^H)Z^{V_2};\\ J^2\{\mathbb{R}(X^{V_\alpha}, Y^H)Z^H\} &= \mathbb{R}(X^{V_\alpha}, Y^H)Z^{V_2};\\ J^2\{\mathbb{R}(X^{V_\gamma}, Y^{V_\alpha})Z^H\} &= \mathbb{R}(X^{V_\gamma}, Y^{V_\alpha})Z^{V_2},\ \{\alpha=1,2,\gamma\leq\alpha\} \end{aligned}$$

The d–tensors (5.5)–(5.5)″ are called *d–tensors of curvature* of the N–linear connection D.

We have:

Theorem 3.5.1 *The following properties hold:*

(a) *The d–tensors of curvature* (5.5) *have the expressions:*

$$(5.6) \qquad \begin{aligned} \mathbb{R}(X^H, Y^H)Z^H &= [D^H_X, D^H_Y]Z^H - D^H_{[X^H,Y^H]}Z^H -\\ &\quad -D^{V_1}_{[X^H,Y^H]}Z^H - D^{V_2}_{[X^H,Y^H]}Z^H,\\ \mathbb{R}(X^{V_\alpha}, Y^H)Z^H &= [D^{V_\alpha}_X, D^H_Y]Z^H - D^H_{[X^{V_\alpha},Y^H]}Z^H -\\ &\quad -D^{V_1}_{[X^{V_\alpha},Y^H]}Z^H - D^{V_2}_{[X^{V_\alpha},Y^H]}Z^H,\\ \mathbb{R}(X^{V_\gamma}, Y^{V_\alpha})Z^H &= [D^{V_\gamma}_X, D^{V_\alpha}_Y]Z^H - D^{V_1}_{[X^{V_\gamma},Y^{V_\alpha}]}Z^H -\\ &\quad -D^{V_2}_{[X^{V_\gamma},Y^{V_\alpha}]}Z^H,\ (\alpha=1,2;\ \gamma\leq\alpha). \end{aligned}$$

(b) *The d–tensors of curvature* (5.5)′ *and* (5.5)″ *are obtained from the previous d–tensors applying the mappings J and J^2 and setting $Z^{V_1} = JZ^H$, $Z^{V_2} = J^2Z^H$.*

Proof. (a) Taking into account (5.1) we get

$$\mathbb{R}(X^H, Y^H)Z^H = [D^H_X, D^H_Y]Z^H - D_{[X^H,Y^H]}Z^H.$$

From this expression, remarking that the vector field $[X^H, Y^H]$ can be written as follows

$$[X^H, Y^H] = h[X^H, Y^H] + v_1[X^H, Y^H] + v_2[X^H, Y^H],$$

we obtain the first formula (5.6). By the same technique, we deduce the other formulae of (5.6).

(b) The d–tensors (5.5)′ and (5.5)″ are obtained from (5.6) using the properties (5.3). **q.e.d.**

The conclusion of the last theorem is as follows: *The essential components of the d–tensors of curvature* $\mathbb{R}$ *of an N–linear connection D are those of* (5.6).

Remark. Using the previous results we can express the Bianchi identities, verified by the d–tensors of torsion and d–tensors of curvature of D. In this respect we need the following form of the Bianchi identities of D:

$$(5.7)\quad \begin{aligned} &\sum_{XYZ}\{(D_X\mathbb{T})(Y,Z)-\mathbb{R}(X,Y)Z+\mathbb{T}(\mathbb{T}(X,Y),Z)\}=0,\\ &\sum_{XYZ}\{(D_X\mathbb{R})(U,Y,Z)+\mathbb{R}(\mathbb{T}(X,Y),Z)U\}=0\end{aligned}$$

In the applications it is suitable to consider the equalities (5.6) as being the Ricci identities.

Theorem 3.5.2 *For any N–linear connection D the following Ricci identities hold:*

$$(5.8)\quad \begin{aligned} &D^H_XD^H_YZ^H-D^H_YD^H_XZ^H=\mathbb{R}(X^H,Y^H)Z^H+D^H_{[X^H,Y^H]}Z^H+\\ &\qquad\qquad +D^{V_1}_{[X^H,Y^H]}Z^H+D^{V_2}_{[X^H,Y^H]}Z^H,\\ &D^{V_\alpha}_XD^H_YZ^H-D^H_YD^{V_\alpha}_XZ^H=\mathbb{R}(X^{V_\alpha},Y^H)Z^H+D^H_{[X^{V_\alpha},Y^H]}Z^H+\\ &\qquad\qquad +D^{V_1}_{[X^{V_\alpha},Y^H]}Z^H+D^{V_2}_{[X^{V_\alpha},Y^H]}Z^H,\\ &D^{V_\gamma}_XD^{V_\alpha}_YZ^H-D^{V_\alpha}_YD^{V_\gamma}_XZ^H=\mathbb{R}(X^{V_\gamma},Y^{V_\alpha})Z^H+D^{V_1}_{[X^{V_\gamma},Y^{V_\alpha}]}Z^H+\\ &\qquad\qquad +D^{V_2}_{[X^{V_\gamma},Y^{V_\alpha}]}Z^H.\end{aligned}$$

The identities obtained from (5.8), *applying the mappings J and* J^2 *and setting* $Z^{V_1}=J(Z^H)$, $Z^{V_2}=J^2(Z^H)$, *also hold.*

The local expressions of the d–tensors of curvature of an N–linear connection D with the coefficients $D\Gamma(N) = (L^i{}_{jk}, \underset{(1)}{C}{}^i{}_{jk}, \underset{(2)}{C}{}^i{}_{jk})$, in the adapted basis (3.1) can be found from the formulae (5.6).

Therefore, by setting

$$
(5.9)\qquad
\begin{aligned}
&I\!R\left(\frac{\delta}{\delta x^k}, \frac{\delta}{\delta x^j}\right)\frac{\delta}{\delta x^h} = R_h{}^i{}_{jk}\,\frac{\delta}{\delta x^i},\\
&I\!R\left(\frac{\delta}{\delta y^{(\alpha)k}}, \frac{\delta}{\delta x^j}\right)\frac{\delta}{\delta x^h} = \underset{(\alpha)}{P}{}_h{}^i{}_{jk}\,\frac{\delta}{\delta x^i}, \qquad (\alpha = 1,2)\\
&I\!R\left(\frac{\delta}{\delta y^{(2)k}}, \frac{\delta}{\delta y^{(1)j}}\right)\frac{\delta}{\delta x^h} = \underset{(21)}{S}{}_h{}^i{}_{jk}\,\frac{\delta}{\delta x^i},\\
&I\!R\left(\frac{\delta}{\delta y^{(\alpha)k}}, \frac{\delta}{\delta y^{(\alpha)j}}\right)\frac{\delta}{\delta x^h} = \underset{(\alpha\alpha)}{S}{}_h{}^i{}_{jk}\,\frac{\delta}{\delta x^i}, \qquad (\alpha = 1,2)
\end{aligned}
$$

and noting Theorem 3.5.1, we deduce what other d–tensors of curvature one obtains from (5.9) by appling the mappings J and J^2. But they have the same coefficients as in the equalities (5.9):

$$
(5.10)\qquad R_h{}^i{}_{jk},\ \underset{(\alpha)}{P}{}_h{}^i{}_{jk},\ \underset{(21)}{S}{}_h{}^i{}_{jk},\ \underset{(\alpha\alpha)}{S}{}_h{}^i{}_{jk},\ (\alpha = 1,2).
$$

Thus, the set of d–tensors of curvature is characterized by the d–tensor fields (5.10).

Theorem 3.5.3 *An N–linear connection D, with the coefficients $D\Gamma(N) = (L^i{}_{jk}, \underset{(1)}{C}{}^i{}_{jk}, \underset{(2)}{C}{}^i{}_{jk})$ has the d–tensors of curvature* (5.10) *expressed by the following formulae:*

(5.11)

$$R_{h}{}^{i}{}_{jk} = \frac{\delta L^{i}{}_{hj}}{\delta x^{k}} - \frac{\delta L^{i}{}_{hk}}{\delta x^{j}} + L^{s}{}_{hj} L^{i}{}_{sk} - L^{s}{}_{hk} L^{i}{}_{sj} + $$
$$+ \underset{(1)}{C}{}^{i}{}_{hm} \underset{(01)}{R}{}^{m}{}_{jk} + \underset{(2)}{C}{}^{i}{}_{hm} \underset{(02)}{R}{}^{m}{}_{jk},$$

$$\underset{(\alpha)}{P}{}_{h}{}^{i}{}_{jk} = \frac{\delta L^{i}{}_{hj}}{\delta y^{(\alpha)k}} - \frac{\delta \underset{(\alpha)}{C}{}^{i}{}_{hk}}{\delta x^{j}} + L^{s}{}_{hj} \underset{(\alpha)}{C}{}^{i}{}_{sk} - \underset{(\alpha)}{C}{}^{s}{}_{hk} L^{i}{}_{sj} +$$
$$+ \underset{(1)}{C}{}^{i}{}_{hm} \underset{(\alpha 1)}{B}{}^{m}{}_{jk} + \underset{(2)}{C}{}^{i}{}_{hm} \underset{(\alpha 2)}{B}{}^{m}{}_{jk},$$

$$\underset{(21)}{S}{}_{h}{}^{i}{}_{jk} = \frac{\partial \underset{(1)}{C}{}^{i}{}_{hj}}{\delta y^{(2)k}} - \frac{\delta \underset{(2)}{C}{}^{i}{}_{hk}}{\delta y^{(1)j}} + \underset{(1)}{C}{}^{s}{}_{hj} \underset{(2)}{C}{}^{i}{}_{sk} -$$
$$- \underset{(2)}{C}{}^{s}{}_{hk} \underset{(1)}{C}{}^{i}{}_{sj} + \underset{(2)}{C}{}^{i}{}_{hm} \underset{(21)}{B}{}^{m}{}_{jk},$$

$$\underset{(\alpha\alpha)}{S}{}_{h}{}^{i}{}_{jk} = \frac{\delta \underset{(\alpha)}{C}{}^{i}{}_{hj}}{\delta y^{(\alpha)k}} - \frac{\delta \underset{(\alpha)}{C}{}^{i}{}_{hjk}}{\delta y^{(\alpha)j}} + \underset{(\alpha)}{C}{}^{s}{}_{hj} \underset{(\alpha)}{C}{}^{i}{}_{sk} -$$
$$- \underset{(\alpha)}{C}{}^{s}{}_{hk} \underset{(\alpha)}{C}{}^{i}{}_{sj} + \underset{(2)}{C}{}^{i}{}_{hm} \underset{(\alpha 1)}{R}{}^{m}{}_{jk},$$
$$\left(\alpha = 1, 2; \ \frac{\delta}{\delta y^{(2)i}} = \frac{\partial}{\partial y^{(2)i}}; \ \underset{(22)}{R}{}^{i}{}_{jk} = 0\right).$$

Proof. These formulae are obtained from (5.6) and (5.9) taking into account the Lie brackets (4.4). **q.e.d.**

The Ricci identities (5.8) in the adapted basis (3.1) have many applications.

Theorem 3.5.4 *For any N–linear connection D, with the coefficients* $D\Gamma(N) = \left(L^i{}_{jk}, \underset{(1)}{C}{}^i{}_{jk}, \underset{(2)}{C}{}^i{}_{jk}\right)$ *the following Ricci identities hold:*

$$
\begin{aligned}
&X^i{}_{|j|k} - X^i{}_{|k|j} = X^m R_m{}^i{}_{jk} - \underset{(0)}{T}{}^m{}_{jk} X^i{}_{|m} - \underset{(01)}{R}{}^m{}_{jk} X^i \overset{(1)}{|}_m - \\
&\qquad\qquad - \underset{(02)}{R}{}^m{}_{jk} X^i \overset{(2)}{|}_m \\
&X^i{}_{|j} \overset{(\alpha)}{|}_k - X^i \overset{(\alpha)}{|}_{k|j} = X^m \underset{(\alpha)}{P}{}_m{}^i{}_{jk} - \underset{(\alpha)}{C}{}^m{}_{jk} X^i{}_{|m} - \\
&\qquad\qquad - \underset{(\alpha 1)}{P}{}^m{}_{jk} X^i \overset{(1)}{|}_m - \underset{(\alpha 2)}{P}{}^m{}_{jk} X^i \overset{(2)}{|}_m \\
&X^i \overset{(1)}{|}_j \overset{(2)}{|}_k - X^i \overset{(2)}{|}_k \overset{(1)}{|}_j = X^m \underset{(21)}{S}{}_m{}^i{}_{jk} - \underset{(2)}{C}{}^m{}_{jk} X^i \overset{(1)}{|}_m - \\
&\qquad\qquad - \underset{(22)}{Q}{}^m{}_{jk} X^i \overset{(2)}{|}_m, \\
&X^i \overset{(\alpha)}{|}_j \overset{(\alpha)}{|}_k - X^i \overset{(\alpha)}{|}_k \overset{(\alpha)}{|}_j = X^m \underset{(\alpha\alpha)}{S}{}_m{}^i{}_{jk} - \underset{(\alpha)}{S}{}^m{}_{jk} X^i \overset{(\alpha)}{|}_m - \\
&\qquad\qquad - \underset{(\alpha 2)}{R}{}^m{}_{jk} X^i \overset{(2)}{|}_m, \ (\alpha = 1, 2; \ \underset{(22)}{R}{}^i{}_{jk} = 0).
\end{aligned}
\tag{5.12}
$$

These identities can be established directly by using the operators of h– and v_α– covariant derivations.

In the case of Berwald connection (3.11) we get:

Proposition 3.5.3 *The d–tensor of curvature of the Berwald connection* $B\Gamma(N) = (B^i{}_{jk}, 0, 0)$ *associated to the nonlinear connection* N *has the following expressions:*

$$(5.11)' \qquad \begin{aligned} &R_h{}^i{}_{jk} = \frac{\delta B^i{}_{hj}}{\delta x^k} - \frac{\delta B^i{}_{hk}}{\delta x^j} + B^s{}_{hj} B^i{}_{sk} - B^s{}_{hk} B^i{}_{sj}, \\ &\underset{(\alpha)}{P}{}_h{}^i{}_{jk} = \frac{\delta B^i{}_{hj}}{\delta y^{(\alpha)k}}, \ (\alpha = 1, 2); \\ &\underset{(21)}{S}{}_h{}^i{}_{jk} = 0, \ \underset{(\alpha\alpha)}{S}{}_h{}^i{}_{jk} = 0, \ (\alpha = 1, 2). \end{aligned}$$

Of course, the Ricci identities (5.12) corresponding to the Berwald connection $B\Gamma(N)$ can be easily written.

Let us consider the system of partial differential equations

$$(5.13) \qquad X^i{}_{|j} = 0, \ X^i \overset{(1)}{|}_j = 0, \ X^i \overset{(2)}{|}_j = 0,$$

where $X^i(x, y^{(1)}, y^{(2)})$ is a d–vector field.

The conditions of integrability of this system, by (5.12), are given by

$$(5.13)' \qquad \begin{aligned} &X^m R_m{}^i{}_{jk} = 0, \ X^m \underset{(\alpha)}{P}{}_m{}^i{}_{jk} = 0, \ X^m \underset{(21)}{S}{}_m{}^i{}_{jk} = 0, \\ &X^m \underset{(\alpha\alpha)}{S}{}_m{}^i{}_{jk} = 0, \qquad (\alpha = 1, 2). \end{aligned}$$

So: *If there exists a system of n–independent d–vector fields, solutions of the equations* (5.13), *then all d–tensors of curvature of the N-linear connection D vanish.*

Of course, the Ricci identities can be extended to any d–tensor field. For instance, in the case of a d–tensor field $K^i{}_j$ we have the Ricci identities:

$$K^i{}_{j|h|k} - K^i{}_{j|k|h} = K^m{}_j R_m{}^i{}_{hk} - K^i{}_m R_j{}^m{}_{hk} - \underset{(0)}{T}{}^m{}_{hk} K^i{}_{j|m} -$$

$$- \underset{(01)}{R}{}^m{}_{hk} K^i{}_j \overset{(1)}{|}_m - \underset{(02)}{R}{}^m{}_{hk} K^i{}_j \overset{(2)}{|}_m,$$

$$K^i{}_{j|h} \overset{(\alpha)}{|}_k - K^i{}_j \overset{(\alpha)}{|}_{k|h} = K^m{}_j \underset{(\alpha)}{P}{}_m{}^i{}_{hk} - K^i{}_m \underset{(\alpha)}{P}{}_j{}^m{}_{hk} -$$

$$- \underset{(\alpha)}{C}{}^m{}_{hk} K^i{}_{j|m} - \underset{(\alpha 1)}{P}{}^m{}_{hk} K^i{}_j \overset{(1)}{|}_m - \underset{(\alpha 2)}{P}{}^m{}_{hk} K^i{}_j \overset{(2)}{|}_m,$$

(5.14)

$$K^i{}_j \overset{(1)}{|}_h \overset{(2)}{|}_k - K^i{}_j \overset{(2)}{|}_k \overset{(1)}{|}_h = K^m{}_j \underset{(21)}{S}{}_m{}^i{}_{hk} - K^i{}_m \underset{(21)}{S}{}_j{}^m{}_{hk} -$$

$$- \underset{(2)}{C}{}^m{}_{hk} K^i{}_j \overset{(1)}{|}_m - \underset{(22)}{Q}{}^m{}_{hk} K^i{}_j \overset{(2)}{|}_m,$$

$$K^i{}_j \overset{(\alpha)}{|}_h \overset{(\alpha)}{|}_k - K^i{}_j \overset{(\alpha)}{|}_k \overset{(\alpha)}{|}_h = K^m{}_j \underset{(\alpha\alpha)}{S}{}_m{}^i{}_{hk} - K^i{}_m \underset{(\alpha\alpha)}{S}{}_j{}^m{}_{hk} -$$

$$- \underset{(\alpha)}{S}{}^m{}_{hk} K^i{}_j \overset{(\alpha)}{|}_m - \underset{(\alpha 2)}{Q}{}^m{}_{hk} K^i{}_j \overset{(2)}{|}_m,$$

for $\alpha = 1, 2$ and $\underset{(22)}{R}{}^i{}_{jk} = 0$.

Examples

1. Let us consider a d–tensor field $g_{ij}(x, y^{(1)}, y^{(2)})$ having the properties

$$g_{ij|k} = 0, \quad g_{ij} \overset{(1)}{|}_k = 0, \quad g_{ij} \overset{(2)}{|}_k = 0, \tag{5.15}$$

with respect to an N–linear connection D. In these conditions, the following identities are satisfied, by virtue of Ricci identities:

$$g_{im}R_j{}^m{}_{kh}+g_{mj}R_i{}^m{}_{kh}=0,\qquad g_{im}\underset{(\alpha)}{P}{}_j{}^m{}_{kh}+g_{mj}\underset{(\alpha)}{P}{}_i{}^m{}_{kh}=0$$

$$(5.16)\quad g_{im}\underset{(21)}{S}{}_j{}^m{}_{kh}+g_{mj}\underset{(21)}{S}{}_i{}^m{}_{kh}=0,\quad g_{im}\underset{(\alpha\alpha)}{S}{}_j{}^m{}_{kh}+g_{mj}\underset{(\alpha\alpha)}{S}{}_i{}^m{}_{kh}=0$$

$$(\alpha=1,2).$$

2. If $L(x,y^{(1)},y^{(2)})$ is a scalar field, then applying the Ricci identities we obtain

$$(5.17)\quad \begin{aligned}
&L_{|j|k}-L_{|k|j}=-\underset{(0)}{T}{}^m{}_{jk}L_{|m}-\underset{(01)}{R}{}^m{}_{jk}L\overset{(1)}{|}_m-\underset{(02)}{R}{}^m{}_{jk}L\overset{(2)}{|}_m,\\
&L_{|j}\overset{(\alpha)}{|}_k-L\overset{(\alpha)}{|}_{k|j}=\underset{(\alpha)}{C}{}^m{}_{jk}L_{|m}-\underset{(\alpha1)}{P}{}^m{}_{jk}L\overset{(1)}{|}_m-\underset{(\alpha2)}{P}{}^m{}_{jk}L\overset{(2)}{|}_m,\\
&L\overset{(1)}{|}_j\overset{(2)}{|}_k-L\overset{(2)}{|}_k\overset{(1)}{|}_j=-\underset{(2)}{C}{}^m{}_{jk}L\overset{(1)}{|}_m-\underset{(22)}{Q}{}^m{}_{jk}L\overset{(2)}{|}_m,\\
&L\overset{(\alpha)}{|}_j\overset{(\alpha)}{|}_k-L\overset{(\alpha)}{|}_k\overset{(\alpha)}{|}_j=-\underset{(\alpha)}{S}{}^m{}_{jk}L\overset{(\alpha)}{|}_m-\underset{(\alpha2)}{R}{}^m{}_{jk}L\overset{(2)}{|}_m,
\end{aligned}$$

$$(\alpha=1,2;\ \underset{(22)}{R}{}^i{}_{jk}=0).$$

For instance, if the scalar $L(x,y^{(1)},y^{(2)})$ is a solution of the partial differential equation

$$(5.18)\qquad L_{|i}=\frac{\delta L}{\delta x^i}=\frac{\partial L}{\partial x^i}-\underset{(1)}{N}{}^m{}_i\frac{\partial L}{\partial y^{(1)m}}-\underset{(2)}{N}{}^m{}_i\frac{\partial L}{\partial y^{(2)m}}=0,$$

then L is a solution of the equation

$$(5.18)'\qquad \underset{(01)}{R}{}^m{}_{jk}L\overset{(1)}{|}_m+\underset{(02)}{R}{}^m{}_{jk}L\overset{(2)}{|}_m=0.$$

Finally, by means of (5.12), applied to the Liouville d–vector fields $z^{(1)i}, z^{(2)i}$, we can deduce some fundamental identities for the deflection tensors (3.9).

Theorem 3.5.5 *For any N–linear connection D, with the coefficients* $D\Gamma(N) = \left(L^i{}_{jk}, \underset{(1)}{C}{}^i{}_{jk}, \underset{(2)}{C}{}^i{}_{jk}\right)$, *the deflection tensor fields satisfy the following identities:*

$$\begin{aligned}
&\overset{(\beta)}{D}{}^i{}_{j|k} - \overset{(\beta)}{D}{}^i{}_{k|j} = z^{(\beta)m} R_m{}^i{}_{jk} - \underset{(0)}{T}{}^m{}_{jk}\, \overset{(\beta)}{D}{}^i{}_m - \\
&\qquad - \underset{(01)}{R}{}^m{}_{jk}\, \overset{(\beta1)}{d}{}^i{}_m - \underset{(02)}{R}{}^m{}_{jk}\, \overset{(\beta2)}{d}{}^i{}_m, \\
&\overset{(\beta)}{D}{}^i{}_j \overset{(\alpha)}{|}_k - \overset{(\beta\alpha)}{d}{}^i{}_{k|j} = z^{(\beta)m} \underset{(\alpha)}{P}{}_m{}^i{}_{jk} - \underset{(\alpha)}{C}{}^m{}_{jk}\, \overset{(\beta)}{D}{}^i{}_m - \\
&\qquad - \underset{(\alpha1)}{P}{}^m{}_{jk}\, \overset{(\beta1)}{d}{}^i{}_m - \underset{(\alpha2)}{P}{}^m{}_{jk}\, \overset{(\beta2)}{d}{}^i{}_m, \\
&\overset{(\beta1)}{d}{}^i{}_j \overset{(2)}{|}_k - \overset{(\beta2)}{d}{}^i{}_k \overset{(1)}{|}_j = z^{(\beta)m} \underset{(21)}{S}{}_m{}^i{}_{jk} - \underset{(2)}{C}{}^m{}_{jk}\, \overset{(\beta1)}{d}{}^i{}_m - \\
&\qquad - \underset{(22)}{Q}{}^m{}_{jk}\, \overset{(\beta2)}{d}{}^i{}_m, \\
&\overset{(\beta\alpha)}{d}{}^i{}_j \overset{(\alpha)}{|}_k - \overset{(\beta\alpha)}{d}{}^i{}_k \overset{(\alpha)}{|}_j = z^{(\beta)m} \underset{(\alpha\alpha)}{S}{}_m{}^i{}_{jk} - \underset{(\alpha)}{S}{}^m{}_{jk}\, \overset{(\beta1)}{d}{}^i{}_m - \\
&\qquad - \underset{(\alpha)}{R}{}^m{}_{jk}\, \overset{(\beta2)}{d}{}^i{}_m, \ (\alpha = 1,2;\ \underset{(22)}{R}{}^i{}_{jk} = 0;\ \beta = 1,2).
\end{aligned} \tag{5.19}$$

An important particular case is given by $\overset{(1)}{D}{}^i{}_j = \overset{(2)}{D}{}^i{}_j = 0$.

We shall apply these identities for studying the electromagnetic fields in the Lagrange spaces of second order.

3.6 Structure Equations of an N–Linear Connection

We shall establish the structure equations of an N–linear connection D, having its coefficients $D\Gamma(N) = \left(L^i{}_{jk}, \underset{(1)}{C}{}^i{}_{jk}, \underset{(2)}{C}{}^i{}_{jk}\right)$ in the adapted basis (3.1).

Let $\gamma : I \longrightarrow E$ be a parametrized curve, locally given by

$$x^i = x^i(t),\ y^{(1)i} = y^{(1)i}(t),\ y^{(2)i} = y^{(2)i}(t),\quad t \in I. \tag{6.1}$$

Thus, the tangent vector field $\dot{\gamma}$ can be written in the form (cf. (6.14), Ch.2):

$$\dot{\gamma} = \frac{dx^i}{dt}\frac{\delta}{\delta x^i} + \frac{\delta y^{(1)i}}{dt}\frac{\delta}{\delta y^{(1)i}} + \frac{\delta y^{(2)i}}{dt}\frac{\partial}{\partial y^{(2)i}}. \tag{6.2}$$

We denote

$$D_{\dot{\gamma}}X = \frac{DX}{dt},\quad DX = \frac{DX}{dt}\,dt,\qquad \forall X \in \mathcal{X}(E). \tag{6.3}$$

DX is *the covariant differential* of the vector field X and $\dfrac{DX}{dt}$ is the covariant differential of X along the curve γ.

If a vector field $X \in \mathcal{X}(E)$ is written in the form:

$$X = X^H + X^{V_1} + X^{V_2} = \overset{(0)}{X}{}^i \frac{\delta}{\delta x^i} + \overset{(1)}{X}{}^i \frac{\delta}{\delta y^{(1)i}} + \overset{(2)}{X}{}^i \frac{\partial}{\partial y^{(2)i}}, \tag{6.4}$$

then its covariant differential DX can be set in the form:

$$\begin{aligned} DX &= DX^H + DX^{V_1} + DX^{V_2} = (D_{\dot{\gamma}}X^H + D_{\dot{\gamma}}X^{V_1} + D_{\dot{\gamma}}X^{V_2})dt = \\ &= (D^H_{\dot{\gamma}} + D^{V_1}_{\dot{\gamma}} + D^{V_2}_{\dot{\gamma}})X^H\,dt + (D^H_{\dot{\gamma}} + D^{V_1}_{\dot{\gamma}} + D^{V_2}_{\dot{\gamma}})X^{V_1}\,dt + \\ &+ (D^H_{\dot{\gamma}} + D^{V_1}_{\dot{\gamma}} + D^{V_2}_{\dot{\gamma}})X^{V_2}\,dt. \end{aligned}$$

By means of (6.2) we have

$$\begin{aligned} DX &= \left\{ \overset{(0)}{X}{}^i{}_{|k}dx^k + \overset{(0)}{X}{}^i \overset{(1)}{|}_k \delta y^{(1)k} + \overset{(0)}{X}{}^i \overset{(2)}{|}_k \delta y^{(2)k} \right\} \frac{\delta}{\delta x^i} + \\ &+ \left\{ \overset{(1)}{X}{}^i{}_{|k}dx^k + \overset{(1)}{X}{}^i \overset{(1)}{|}_k \delta y^{(1)k} + \overset{(1)}{X}{}^i \overset{(2)}{|}_k \delta y^{(2)k} \right\} \frac{\delta}{\delta y^{(1)k}} + \\ &+ \left\{ \overset{(2)}{X}{}^i{}_{|k}dx^k + \overset{(2)}{X}{}^i \overset{(1)}{|}_k \delta y^{(1)k} + \overset{(2)}{X}{}^i \overset{(2)}{|}_k \delta y^{(2)k} \right\} \frac{\partial}{\partial y^{(2)i}}. \end{aligned}$$

Taking into account the expressions of $h-$ and $v_\alpha-$ covariant derivations of the d–vector fields $\overset{(0)}{X}{}^i, \overset{(1)}{X}{}^i$ and $\overset{(2)}{X}{}^i$, and denoting

$$\omega^i{}_j = L^i{}_{jk}\, dx^k + \underset{(1)}{C}{}^i{}_{jk}\, \delta y^{(1)k} + \underset{(2)}{C}{}^i{}_{jk}\, \delta y^{(2)k}, \tag{6.5}$$

the previous expression of the differential covariant DX can be written in a very convenient form:

$$\begin{aligned} DX &= \left\{ d\overset{(0)}{X}{}^i + \omega^i{}_m \overset{(0)}{X}{}^m \right\} \frac{\delta}{\delta x^i} + \left\{ d\overset{(1)}{X}{}^i + \omega^i{}_m \overset{(1)}{X}{}^m \right\} \frac{\delta}{\delta y^{(1)i}} + \\ &+ \left\{ d\overset{(2)}{X}{}^i + \omega^i{}_m \overset{(2)}{X}{}^m \right\} \frac{\partial}{\partial y^{(2)i}}. \end{aligned} \tag{6.6}$$

The 1–forms $\omega^i{}_j$ from (6.5) are called *connection 1–forms* of the N–linear connection D.

As an application we can study the parallelism of the vector fields along the curve $\gamma : I \longrightarrow E$, defined in the section 2, Ch.2.

Proposition 3.6.1 *The vector field* (6.4) *is parallel along the parametrized curve* $\gamma : I \to E$, *if, and only if, its components* $\overset{(\alpha)}{X}{}^i(x, y^{(1)}, y^{(2)})$, $(\alpha = 0, 1, 2)$ *are solutions of the differential equations:*

$$\frac{dZ^i}{dt} + \frac{\omega^i{}_m}{dt} Z^m \left(x(t), y^{(1)}(t), y^{(2)}(t)\right) = 0.$$

A theorem of existence and uniqueness for the parallel vector fields along of a given parametrized curve in E can be formulated.

A *horizontal path* of an N–linear connection D is a horizontal parametrized curve $\gamma : I \longrightarrow E$ with the property $D_{\dot\gamma}\dot\gamma = 0$.

Theorem 3.6.1 *The horizontal paths of an N–linear connection D are characterized by the system of differential equations*

$$\frac{d^2x^i}{dt^2} + L^i{}_{jk}(x, y^{(1)}, y^{(2)}) \frac{dx^j}{dt} \frac{dx^k}{dt} = 0, \quad \frac{\delta y^{(1)i}}{dt} = \frac{\delta y^{(2)i}}{dt} = 0. \tag{6.7}$$

A parametrized curve $\gamma : I \to E$ is vertical in the point x_0, $x_0 \in M$ if the vector field $\dot{\gamma}$ belongs to the vertical distribution V.

A vertical curve γ_{x_0} in the point $x_0 \in M$ is represented by the equations $x^i = x^i_0$, $y^{(\alpha)i} = y^{(\alpha)i}(t)$, $t \in I$, $(\alpha = 1, 2)$. It is called a *v–path*, with respect to D if $D_{\dot{\gamma}_{x_0}} \dot{\gamma}_{x_0} = 0$.

Theorem 3.6.2 *The vertical paths at the point $x_0 \in M$, with respect to the N-linear connection D are characterized by the system of differential equations*

$$(6.7)' \qquad x^i = x^i_0,\ \frac{dx^i}{dt} = 0,\ \frac{d}{dt}\frac{\delta y^{(\alpha)i}}{dt} + \frac{\omega^i{}_m}{dt}\frac{\delta y^{(\alpha)i}}{dt} = 0,\ (\alpha{=}1,2).$$

Remarks

1° If the curve γ_{x_0} belongs to the distribution V_2, then $x^i = x^i_0$, $\dfrac{dx^i}{dt} = 0$, $\dfrac{\delta y^{(1)i}}{dt} = 0$. A v_2–path is characterized by

$$(6.7)'' \qquad \begin{aligned} & x^i = x^i_0,\ \frac{dx^i}{dt} = 0,\ \frac{dy^{(1)i}}{dt} = 0,\\ & \frac{d^2 y^{(2)i}}{dt^2} + \underset{(2)}{C}{}^i{}_{jk}\frac{dy^{(2)j}}{dt}\frac{dy^{(2)k}}{dt} = 0. \end{aligned}$$

2° The local existence and uniqueness of v–paths or v_2– paths, when the initial conditions are given, can be studied by means of the system of equations (6.7)′ or (6.7)″.

Now, we establish the so–called structure equations of an N–linear connection. Firstly, it is necessary to calculate the exterior differentials of the 1–forms from the adapted cobasis (3.1)′:

$$(6.8) \qquad \begin{aligned} & dx^i,\ \delta y^{(1)i} = dy^{(1)i} + \underset{(1)}{M}{}^i{}_j\, dx^j,\\ & \delta y^{(2)i} = dy^{(2)i} + \underset{(1)}{M}{}^i{}_j\, dy^{(1)j} + \underset{(2)}{M}{}^i{}_j\, dx^j, \end{aligned}$$

using the relations between the coefficients $\left(\underset{(1)}{N}{}^i{}_j, \underset{(2)}{N}{}^i{}_j\right)$ and the dual coefficients $\left(\underset{(1)}{M}{}^i{}_j, \underset{(2)}{M}{}^i{}_j\right)$ expressed in (6.4), Ch.2.

Lemma 3.6.1 *The exterior differentials of the 1–forms (6.8) are given by the following formulae*

$$
(6.9)\quad
\begin{aligned}
d(dx^i) &= 0,\\
d(\delta y^{(1)i}) &= \frac{1}{2}\underset{(01)}{R}{}^{i}{}_{jm}\, dx^m \wedge dx^j + \underset{(11)}{B}{}^{i}{}_{jm}\, \delta y^{(1)m} \wedge dx^j +\\
&+ \underset{(21)}{B}{}^{i}{}_{jm}\, \delta y^{(2)m} \wedge dx^j,\\
d(\delta y^{(2)i}) &= \frac{1}{2}\underset{(02)}{R}{}^{i}{}_{jm}\, dx^m \wedge dx^j + \underset{(12)}{B}{}^{i}{}_{jm}\, \delta y^{(1)m} \wedge dx^j +\\
&+ \underset{(22)}{B}{}^{i}{}_{jm}\, \delta y^{(2)m} \wedge dx^j + \frac{1}{2}\underset{(12)}{R}{}^{i}{}_{jm}\, \delta y^{(1)m} \wedge \delta y^{(1)j} +\\
&+ \underset{(21)}{B}{}^{i}{}_{jm}\, \delta y^{(2)m} \wedge \delta y^{(1)j}.
\end{aligned}
$$

Indeed, we have

$$
d(\delta y^{(1)i}) = d\,\underset{(1)}{M}{}^{i}{}_{j} \wedge dx^j,\ \delta y^{(2)i} = d\,\underset{(1)}{M}{}^{i}{}_{j} \wedge dy^{(1)j} + d\,\underset{(2)}{M}{}^{i}{}_{j} \wedge dx^j,
$$

where we substitute $dy^{(1)i}$ and $dy^{(2)i}$ from (6.8) and take into account the relations $\underset{(2)}{M}{}^{i}{}_{j} = \underset{(2)}{N}{}^{i}{}_{j} + \underset{(1)}{N}{}^{i}{}_{m}\ \underset{(1)}{N}{}^{m}{}_{j}$ and the formulae (4.5). **q.e.d.**

Considering the connection 1–forms $\omega^i{}_j$ from (6.5) we can prove the following important theorem.

Theorem 3.6.3 *The structure equations of an N-linear connection D on the total space of 2–osculator bundle E are given by*

$$
(6.9)\quad
\begin{aligned}
d(dx^i) - dx^m \wedge \omega^i{}_m &= -\overset{(0)}{\Omega}{}^{i},\\
d(\delta y^{(1)i}) - \delta y^{(1)m} \wedge \omega^i{}_m &= -\overset{(1)}{\Omega}{}^{i},\\
d(\delta y^{(2)i}) - \delta y^{(2)m} \wedge \omega^i{}_m &= -\overset{(2)}{\Omega}{}^{i},\\
d\omega^i{}_j - \omega^m{}_j \wedge \omega^i{}_m &= -\Omega^i{}_j,
\end{aligned}
$$

where $\overset{(\alpha)}{\Omega}{}^{i}$, $\alpha = 1, 2, 3)$, *are the 2–forms of torsion*

$$
(6.10)\quad
\begin{aligned}
\overset{(0)}{\Omega}{}^{i} &= \frac{1}{2}\underset{(0)}{T}{}^{i}{}_{jk}\, dx^{j}\wedge dx^{k} + \underset{(1)}{C}{}^{i}{}_{jk}\, dx^{j}\wedge \delta y^{(1)k} + \\
&+ \underset{(2)}{C}{}^{i}{}_{jk}\, dx^{j}\wedge \delta y^{(2)k},\\
\overset{(1)}{\Omega}{}^{i} &= \frac{1}{2}\underset{(01)}{R}{}^{i}{}_{jk}\, dx^{j}\wedge dx^{k} + \underset{(11)}{P}{}^{i}{}_{jk}\, dx^{j}\wedge \delta y^{(1)k} + \\
&+ \underset{(21)}{P}{}^{i}{}_{jk}\, dx^{j}\wedge \delta y^{(2)k} + \frac{1}{2}\underset{(1)}{S}{}^{i}{}_{jk}\, \delta y^{(1)j}\wedge \delta y^{(1)k} + \\
&+ \underset{(2)}{C}{}^{i}{}_{jk}\, \delta y^{(1)j}\wedge \delta y^{(2)k},\\
\overset{(2)}{\Omega}{}^{i} &= \frac{1}{2}\underset{(02)}{R}{}^{i}{}_{jk}\, dx^{j}\wedge dx^{k} + \underset{(12)}{P}{}^{i}{}_{jk}\, dx^{j}\wedge \delta y^{(1)k} + \\
&+ \underset{(22)}{P}{}^{i}{}_{jk}\, dx^{j}\wedge \delta y^{(2)k} \frac{1}{2}\underset{(12)}{R}{}^{i}{}_{jk}\, \delta y^{(1)j}\wedge \delta y^{(1)k} + \\
&+ \underset{(22)}{Q}{}^{i}{}_{jk}\, \delta y^{(1)j}\wedge \delta y^{(2)k} + \frac{1}{2}\underset{(2)}{S}{}^{i}{}_{jk}\, \delta y^{(2)j}\wedge \delta y^{(2)k},
\end{aligned}
$$

and where $\Omega^{i}{}_{j}$ are the 2–forms of curvature:

$$
(6.11)\quad
\begin{aligned}
\Omega^{i}{}_{j} &= \frac{1}{2} R_{j}{}^{i}{}_{km}\, dx^{k}\wedge dx^{m} + \underset{(1)}{P}{}_{j}{}^{i}{}_{km}\, dx^{k}\wedge \delta y^{(1)m} + \\
&+ \underset{(2)}{P}{}_{j}{}^{i}{}_{km}\, dx^{k}\wedge \delta y^{(2)m} + \frac{1}{2}\underset{(11)}{S}{}_{j}{}^{i}{}_{km}\, \delta y^{(1)k}\wedge \delta y^{(1)m} + \\
&+ \underset{(21)}{S}{}_{j}{}^{i}{}_{km}\, \delta y^{(1)k}\wedge \delta y^{(2)m} + \tfrac{1}{2}\underset{(22)}{S}{}_{j}{}^{i}{}_{km}\, \delta y^{(2)k}\wedge \delta y^{(2)m}.
\end{aligned}
$$

Proof. Using the differential covariant DX, from (6.6), we can see that the connection 1–forms $\omega^{i}{}_{j}$ are the same in every coordinate of DX in the adapted basis (3.1). So, it is not difficult to prove that the structure equations of an N–linear connection D on E have the form (6.9). The calculation of the 2–forms of torsion $\overset{(\alpha)}{\Omega}{}^{i}$, $(\alpha = 1, 2, 3)$, and of the 2–forms of curvature $\Omega^{i}{}_{j}$ can be made directly, developing the first member of (6.9) and taking into account Lemma 3.6.1.

Now Bianchi identities of an N–linear connection D with the coefficients $D\Gamma(N) = (L^{i}{}_{jk}, \underset{(1)}{C}{}^{i}{}_{jk}, \underset{(2)}{C}{}^{i}{}_{jk})$, adapted basis (3.1), can be obtained from (6.9) applying the operator of exterior differentiation and calculating the exterior differentials of the 2–forms of torsion and curvatures, by virtue of the system (6.9).

All the results from this chapter serve to study the notion of the second order Lagrange spaces, as well as that of the Lagrangian Mechanics of second order.

3.7 Problems

1. Assuming that the d–tensor field $g_{ij}(x, y^{(1)}, y^{(2)})$ is symmetric and nonsingular, determine the N–linear connection $D\Gamma(N) = (L^i{}_{jk}, \underset{(1)}{C}{}^i{}_{jk}, \underset{(2)}{C}{}^i{}_{jk})$ for which

   (a) $g_{ij|k} = 0,\ g_{ij} \overset{(1)}{\big|}_k = 0,\ g_{ij} \overset{(2)}{\big|}_k = 0$, (b) $\underset{(0)}{T}{}^i{}_{jk} = 0,\ \underset{(1)}{S}{}^i{}_{jk} = 0,\ \underset{(2)}{S}{}^i{}_{jk} = 0.$

2. The same problem in the case when (b) is substituted with

   (b)′ The d–tensors $\underset{(0)}{T}{}^i{}_{jk}$, $\underset{(1)}{S}{}^i{}_{jk}$ and $\underset{(2)}{S}{}^i{}_{jk}$, skew–symmetric in the bottom indices are given apriori. Look for the N–linear connection D with the property that they are exactly $h-$ and v_α–torsions of D.

3. In the conditions of the problem 1, determine the covariant d–tensors of curvature

$$R_{ijkh} = g_{jm} R_i{}^m{}_{kh},\quad \underset{(\alpha)}{P}{}_{ijkh} = g_{jm} \underset{(\alpha)}{P}{}_i{}^m{}_{kh},\quad (\alpha = 1, 2),\ \text{etc.}$$

   Show that these d–tensors are skew–symmetric with respect to the first two indices.

4. Write the Ricci identities for the Berwald connection (3.1) associated to the nonlinear connection N. Write the Bianchi identities for the Berwald connection (3.1).

5. Determine the tensors of deflection of the Berwald connection (3.1).

Chapter 4

Lagrangians of Second Order. Variational Problem. Nöther Type Theorems

The Analytical Mechanics based on the Lagrangians defined on the higher order osculator bundles has been studied, with remarkable results by many people: M. Crampin et al., M. de Léon et al., D. Krupka, D. Grigore, K. Kondo, R. Miron and many others [see References].

Here we shall present some basic concepts of the Mechanics of the Lagrangians of second order as a natural extension of that studied in some sections of the first chapter, using the geometrical methods described in the geometry of the 2–osculator bundle. The theory which follows derives from the variational problem on the Lagrangians of second order.

Therefore, we define the notion of differential Lagrangians of order 2 by introducing the main invariants, which allow us to express the so–called Zermello conditions. We study the variational problem for the integral of action and determine the Euler–Lagrange equations for the Lagrangians of second order. Consequently, we study the notions of higher order energies and their laws of conservations. Based on the notion of symmetry we prove two Nöther theorems. The end of the chapter is devoted to the case of regular Lagrangians of second order. In this case we determine some canonical geometrical objects fields, such as 2–sprays or nonlinear connections determined only by the regular Lagrangians of second order. We finish with the problem of prolongation of the Riemannian, Finslerian and Lagrangian structures.

4.1 Lagrangians of Second Order. Zermelo Conditions

A Lagrangian of second order is a map $L : E = \mathrm{Osc}^2 M \longrightarrow R$. It is called a *differentiable Lagrangian* if L is of the class C^∞ on the manifolds $\widetilde{E}$ (cf.(2.1), Ch.2) and L is continuous in the points $(x, 0, y^{(2)})$ of E.

Let us consider the Hessian of a differentiable Lagrangian $L(x, y^{(1)}, y^{(2)})$ with respect to $y^{(2)}$. Its elements are given by $2g_{ij}(x, y^{(1)}, y^{(2)})$, where

$$g_{ij}(x, y^{(1)}, y^{(2)}) = \frac{1}{2} \frac{\partial^2 L}{\partial y^{(2)i} \partial y^{(2)j}}. \tag{1.1}$$

By a transformation of a local coordinate on E, (1.7), Ch.2, g_{ij} is transformed by the following rule

$$\tilde{g}_{ij} = \frac{\partial x^r}{\partial \tilde{x}^i} \frac{\partial x^s}{\partial \tilde{x}^j} g_{rs}. \tag{1.2}$$

We can say that g_{ij} is a differentiable d–tensor field on $\tilde{E}$, symmetric, covariant of order two.

Example 1.1 The function (3.7), Ch.2, i.e.,

$$L(x, y^{(1)}, y^{(2)}) = \gamma_{ij}(x) z^{(2)i} z^{(2)j} \tag{1.3}$$

where $z^{(2)i}$ is the d–vector field from (3.6), Ch.2, is a differentiable Lagrangian on $\tilde{E}$.

One can notice that the Lagrangian of second order L from (1.3) is determined only by the Riemannian structure $\gamma_{ij}(x)$. By virtue of Theorem 2.3.4, it follows:

Theorem 4.1.1 *If the base manifold M is paracompact, then on E there exists the differentiable Lagrangian $L(x, y^{(1)}, y^{(2)})$.*

For any differentiable Lagrangian $L(x, y^{(1)}, y^{(2)})$ we consider its Lie derivative with respect to a vector field X on $\tilde{E}$, denoted by

$$X(L(x, y^{(1)}, y^{(2)})) = \pounds_X L(x, y^{(1)}, y^{(2)}). \tag{1.4}$$

Therefore, with respect to the Liouville vector fields $\overset{1}{\Gamma}$ and $\overset{2}{\Gamma}$ we get the scalar fields on $\tilde{E}$

$$\overset{1}{I}(L) = \pounds_{\overset{1}{\Gamma}} L, \quad \overset{2}{I}(L) = \pounds_{\overset{2}{\Gamma}} L. \tag{1.5}$$

They will be called the *main invariants* of the Lagrangian L.

The expanded expressions of the main invariants are as follows:

$$\overset{1}{I}(L) = y^{(1)} \frac{\partial L}{\partial y^{(2)i}}, \quad \overset{2}{I}(L) = y^{(1)} \frac{\partial L}{\partial y^{(1)i}} + 2y^{(2)i} \frac{\partial L}{\partial y^{(2)i}}. \tag{1.6}$$

Let us consider $c : [0,1] \longrightarrow M$ a smooth parametrized curve, represented in a domain U of a local chart by $c(t) = (x^i(t))$, $t \in [0,1]$. Its extension (cf. §3, Ch.2), $\tilde{c} : [0,1] \longrightarrow \pi^{-1}(U) \subset E$ is represented by

$$x^i = x^i(t), \; y^{(1)i} = \frac{dx^i}{dt}(t), \; y^{(2)i} = \frac{1}{2}\frac{d^2x^i}{dt^2}(t), \quad t \in [0,1]. \tag{1.7}$$

The integral of action of the Lagrangian L along c is defined by

$$I(c) = \int_0^1 L\left(x(t), \frac{dx}{dt}, \frac{1}{2}\frac{d^2x}{dt^2}\right) dt. \tag{1.8}$$

Now, we can prove:

Theorem 4.1.2 *The necessary conditions that the integral of action $I(c)$ does not depend on the parametrization of the curve c are the following ones:*

$$\overset{1}{I}(L) = 0, \quad \overset{2}{I}(L) = L. \tag{1.9}$$

Proof. Let $\tilde{t} = \tilde{t}(t)$, $t \in [0,1]$ be a differentiable diffeomorphism. In order that the integral of action $I(c)$ did not depend on the parametrization of the curve c it is necessary that:

$$\tilde{L}\left(\tilde{x}, \frac{d\tilde{x}}{d\tilde{t}}, \frac{1}{2}\frac{d^2\tilde{x}^i}{d\tilde{t}^2}\right)\frac{d\tilde{t}}{dt} = L\left(x, \frac{dx}{d\tilde{t}}\frac{d\tilde{t}}{dt}, \frac{1}{2}\frac{d}{dt}\left(\frac{dx}{d\tilde{t}}\frac{d\tilde{t}}{dt}\right)\right). \tag{$*$}$$

The previous equality holds for any diffeomorphism $\tilde{t} = \tilde{t}(t)$. Deriving $(*)$ with respect to $\frac{d\tilde{t}}{dt}$ and taking $\tilde{t} = t$ we get $L = y^{(1)i} \frac{\partial L}{\partial y^{(1)i}} + +2y^{(2)i} \frac{\partial L}{\partial y^{(2)i}}$ or $L = \overset{2}{I}(L)$. Deriving $(*)$ again with respect to $\frac{d^2\tilde{t}}{dt^2}$ and taking $\tilde{t} = t$ we obtain $\overset{1}{I}(L) = 0$. **q.e.d.**

A.Kawaguchi and K.Kondo named the equations (1.9) *Zermelo conditions* [150], [152].

An important result is given by the following theorem:

Theorem 4.1.3 *If the differentiable Lagrangian $L(x, y^{(1)}, y^{(2)})$ satisfies the Zermelo conditions* (1.9), *then the following property holds:*

$$rank\, \|g_{ij}(x, y^{(1)}, y^{(2)})\| < n. \tag{1.10}$$

Proof. $\overset{1}{I}(L) = 0$ implies $y^{(1)i} \frac{\partial^2 L}{\partial y^{(2)i} \partial y^{(2)j}} = 0$ or $y^{(1)i} g_{ij} = 0$. The last equation, on $\tilde{E}$, has (1.10) as consequence. **q.e.d.**

4.2 Variational Problem

Following H.V.Craig and J.L.Synge's ideas, [62], [285], we present now the study of the variational problem for the integral of action $I(c)$, adding some new considerations about Euler–Lagrange equations and new important operators, useful in the proof of a Nöther theorem.

Let $c : [0,1] \longrightarrow M$ be the curve considered in previous section, whose image belongs to the domain U of a local chart, $U \subset M$. And let $\tilde{c} : [0,1] \longrightarrow \pi^{-1}(U) \subset E$ its extension, given by (1.7).

On the open set U we consider the curves

$$c_\varepsilon : t \in [0,1] \longrightarrow (x^i(t) + \varepsilon V^i(t)) \in M, \tag{2.1}$$

where ε is a real number, sufficiently small in absolute value so that $Im\, c_\varepsilon \subset U$, $V^i(x(t))$, denoted as $V^i(t)$, being a regular vector field on U, restricted to the curve c. We assume that the curves c_ε have the same end points $c(0)$ and $c(1)$, with the curve c and at these points they have the same tangents. Therefore, the vector field $V^i(t)$ satisfies the conditions

$$V^i(0) = V^i(1) = 0, \quad \frac{dV^i}{dt}(0) = \frac{dV^i}{dt}(1) = 0. \tag{2.2}$$

The extension to $\tilde{E}$ of the curves c_ε is

$$\tilde{c}_\varepsilon : t \in [0,1] \to \left(x^i(t) + \varepsilon V^i(t), \frac{dx^i}{dt} + \varepsilon \frac{dV^i}{dt}, \frac{1}{2}\left(\frac{d^2x^i}{dt^2} + \varepsilon \frac{d^2V^i}{dt^2} \right) \right) \in \pi^{-1}(U).$$

The integral of action of the differentiable Lagrangian $L(x, y^{(1)}, y^{(2)})$ on the curves c_ε is

$$I(c_\varepsilon) = \int_0^1 L\left(x + \varepsilon V, \frac{dx}{dt} + \varepsilon \frac{dV}{dt}, \frac{1}{2}\left(\frac{d^2x}{dt^2} + \varepsilon \frac{d^2V}{dt^2} \right) \right) dt. \tag{2.3}$$

A necessary condition that $I(c)$ be an extremal value for $I(c_\varepsilon)$ is

$$\left. \frac{dI(c_\varepsilon)}{d\varepsilon} \right|_{\varepsilon=0} = 0. \tag{2.4}$$

In our conditions of differentiability, the operator $\frac{d}{d\varepsilon}$ is permuting with the operator of integration. From (2.3) we get:

$$\frac{dI(c_\varepsilon)}{d\varepsilon} = \int_0^1 \frac{d}{d\varepsilon} L\left(x + \varepsilon V, \frac{dx}{dt} + \varepsilon \frac{dV}{dt}, \frac{1}{2}\left(\frac{d^2x}{dt^2} + \varepsilon \frac{d^2V}{dt^2} \right) \right) dt. \tag{2.5}$$

A straightforward calculation leads to

$$\left. \frac{dI(c_\varepsilon)}{d\varepsilon} \right|_{\varepsilon=0} = \int_0^1 \left[\frac{\partial L}{\partial x^i} V^i + \frac{\partial L}{\partial y^{(1)i}} \frac{dV^i}{dt} + \frac{1}{2} \frac{\partial L}{\partial y^{(2)i}} \frac{d^2V^i}{dt^2} \right] dt. \tag{2.5$'$}$$

Setting

$$\overset{1}{I}_V(L) = V^i \frac{\partial L}{\partial y^{(2)i}}, \quad \overset{2}{I}_V(L) = V^i \frac{\partial L}{\partial y^{(1)i}} + \frac{dV^i}{dt} \frac{\partial L}{\partial y^{(2)i}} \tag{2.6}$$

and

$$\overset{0}{E}_i(L) = \frac{\partial L}{\partial x^i} - \frac{d}{dt}\frac{\partial L}{\partial y^{(1)i}} + \frac{1}{2}\frac{d^2}{dt^2}\frac{\partial L}{\partial y^{(2)i}}, \tag{2.7}$$

one deduces an important identity:

$$\begin{aligned} &V^i\frac{\partial L}{\partial x^i} + \frac{dV^i}{dt}\frac{\partial L}{\partial y^{(1)i}} + \frac{1}{2}\frac{d^2V^i}{dt^2}\frac{\partial L}{\partial y^{(2)i}} = \\ &= \overset{0}{E}_i(L)V^i + \frac{d}{dt}\overset{2}{I}_V(L) - \frac{1}{2}\frac{d^2}{dt^2}\overset{1}{I}_V(L). \end{aligned} \tag{2.8}$$

This identity can also be verified by a direct calculus. Obviously, the conditions (2.2) imply, for $\overset{1}{I}_V(L)$ and $\overset{2}{I}_V(L)$ the properties

$$\overset{\alpha}{I}_V(L)(c(0)) = \overset{\alpha}{I}_V(L)(c(1)) = 0, \ (\alpha = 1, 2). \tag{2.2$'$}$$

These considerations allow to write (2.5)$'$ in the following form:

$$\left.\frac{dI(c_\varepsilon)}{d\varepsilon}\right|_{\varepsilon=0} = \int_0^1 \overset{0}{E}_i(L)V^i dt + \int_0^1 \frac{d}{dt}\left\{\overset{2}{I}_V(L) - \frac{1}{2}\frac{d}{dt}\overset{1}{I}_V(L)\right\}dt. \tag{2.5$''$}$$

The second term of the previous equality disappears, by virtue of (2.2)$'$ and (2.2). It follows

$$\left.\frac{dI(c_\varepsilon)}{d\varepsilon}\right|_{\varepsilon=0} = \int_0^1 \overset{0}{E}_i(L)V^i\, dt. \tag{2.5$'''$}$$

Taking into account the fact that the vector field V^i on the curve c is an arbitrary one, the condition of extremum (2.4) and the equality (2.5)$'''$ have as consequence the following theorem:

Theorem 4.2.1 *In order that the integral of action $I(c)$ was an extremal value for the functionals $I(c_\varepsilon)$, it is necessary that the following Euler–Lagrange equations hold:*

$$\begin{aligned} &\overset{0}{E}_i(L) \overset{def}{=} \frac{\partial L}{\partial x^i} - \frac{d}{dt}\frac{\partial L}{\partial y^{(1)i}} + \frac{1}{2}\frac{d^2}{dt^2}\frac{\partial L}{\partial y^{(2)i}} = 0, \\ &y^{(1)i} = \frac{dx^i}{dt}, \quad y^{(2)i} = \frac{1}{2}\frac{d^2x^i}{dt^2}. \end{aligned} \tag{2.9}$$

The curves $c : [0,1] \longrightarrow M$, solutions of the equations (2.9) are called *extremal curves of the integral of action* $I(c)$.

The formula $(2.5)'''$ implies:

Theorem 4.2.2 *$E_i(L)$, from (2.7), is a d–covector field.*

Proof. A transformation of coordinates (1.7), Ch.2, on E and $(2.5)'''$ lead to the equation

$$\int_0^1 \left[\overset{0}{\tilde{E}_i} (\tilde{L})\tilde{V}^i - \overset{0}{E_i} (L)V^i \right] dt = \int_0^1 \left[\overset{0}{\tilde{E}_i} (\tilde{L})\frac{\partial \tilde{x}^i}{\partial x^j} - \overset{0}{E_j} (L) \right] V^j \, dt = 0.$$

But V^i is an arbitrary vector field.

Hence the last equation implies $\overset{0}{\tilde{E}_i} (\tilde{L})\dfrac{\partial \tilde{x}^i}{\partial x^j} = \overset{0}{E_j} (L)$. **q.e.d.**

Remarks.

1° Starting from (2.7) and using the rules of transformation of the natural basis of $\mathcal{X}(E)$, we can prove directly the last theorem.

2° The equation $\overset{0}{E_i} (L) = 0$ has a geometrical meaning.

4.3 Operators $\overset{1}{I}_V$, $\overset{2}{I}_V$, $\dfrac{d_V}{dt}$

On further examination of the identity (2.8) we can introduce some new operators, frequently used in the theory of higher order Lagrangians.

Let $c : t \in [0,1] \longrightarrow (x^i(t)) \in M$ be a smooth curve, $\tilde{c}$ from (1.7) its extension to E and $V^i(x(t))$ a differentiable vector field along c.

We have

Lemma 4.3.1 *The mapping $S_V : c \longrightarrow \mathrm{Osc}^2 M$, defined by*

$$(3.1) \qquad \begin{aligned} & x^i = x^i(t) \\ & y^{(1)i} = V^i(x(t)), \quad 2y^{(2)i} = \frac{dV^i(x(t))}{dt}, \quad t \in [0,1] \end{aligned}$$

is a section of the projection $\pi : \mathrm{Osc}^2 M \longrightarrow M$ along the curve c.

Indeed, using (1.7), Ch.2, we get

$$\tilde{y}^{(1)i} = \frac{\partial \tilde{x}^i}{\partial x^j} y^{(1)j} = \frac{\partial \tilde{x}^i}{\partial x^j} V^j = \tilde{V}^i, \; 2\tilde{y}^{(2)i} = \frac{d}{dt}\left(\frac{\partial \tilde{x}^i}{\partial x^j} V^j\right) = \frac{d\tilde{V}^i}{dt}.$$

q.e.d.

Clearly, if $V^i = \dfrac{dx^i}{dt}$, then $S_{\frac{dx}{dt}}(c) = \tilde{c}$.

The identity (2.8) suggests us to introduce the following operator along the curve c

$$\frac{d_V}{dt} = V^i \frac{\partial}{\partial x^i} + \frac{dV^i}{dt}\frac{\partial}{\partial y^{(1)i}} + \frac{1}{2}\frac{d^2V^i}{dt^2}\frac{\partial}{\partial y^{(2)i}}. \tag{3.2}$$

The importance of this operator results from:

Theorem 4.3.1 *The operator* $\dfrac{d_V}{dt}$ *has the properties:*

1° *It is invariant with respect to the coordinate transformations on* E.

2° *For any differentiable Lagrangian* $L(x, y^{(1)}, y^{(2)})$, $\dfrac{d_V L}{dt}$ *is a scalar field along* c.

3° *It is an operator of derivation, i.e.,*

$$\begin{aligned}
&\frac{d_V}{dt}(L + L') = \frac{d_V L}{dt} + \frac{d_V L'}{dt},\\
&\frac{d_V}{dt}(aL) = a\,\frac{d_V L}{dt}, \quad a \in R,\\
&\frac{d_V}{dt}(L \cdot L') = \frac{d_V L}{dt}\cdot L' + L \cdot \frac{d_V L'}{dt}.
\end{aligned} \tag{3.3}$$

4° $\dfrac{d_V L}{dt} = \dfrac{dL}{dt}$, *for* $V^i = \dfrac{dx^i}{dt}$.

Proof. 1° Using (1.7), Ch.2, along of section (3.1), we have $\dfrac{d_V}{dt} =$
$= y^{(1)i}\dfrac{\partial}{\partial x^i} + 2y^{(2)i}\dfrac{\partial}{\partial y^{(1)i}} + \dfrac{dy^{(2)i}}{dt}\dfrac{\partial}{\partial y^{(2)i}} = \tilde{y}^{(1)i}\dfrac{\partial}{\partial \tilde{x}^i} + 2\tilde{y}^{(1)i}\dfrac{\partial}{\partial \tilde{y}^{(1)i}} +$

$+\frac{d\tilde{y}^{(2)i}}{dt}\frac{\partial}{\partial\tilde{y}^{(2)i}}$. Hence, $\frac{d_V}{dt}$ is invariant with respect to the transformations of coordinates on E.

2° From 1° we deduce $\frac{d_V\tilde{L}}{dt}=\frac{d_V L}{dt}$, for any differentiable Lagrangian of second order.

3° The particular form (3.2) of $\frac{d_V}{dt}$ implies (3.3).

4° If $V^i=\frac{dx^i}{dt}$ and remarking that along c we have

$$(3.4)\qquad \begin{aligned} &y^{(1)i}=\frac{dx^i}{dt},\quad y^{(2)i}=\frac{1}{2}\frac{d^2x^i}{dt^2},\\ &\frac{dL}{dt}=\frac{\partial L}{\partial x^i}y^{(1)i}+2\frac{\partial L}{\partial y^{(1)i}}y^{(2)i}+\frac{\partial L}{\partial y^{(2)i}}\frac{dy^{(2)i}}{dt}, \end{aligned}$$

it follows $\frac{d_V L}{dt}=\frac{dL}{dt}$, for $V^i=\frac{dx^i}{dt}$. **q.e.d.**

For reasons derived from the previous theorem, the operator $\frac{d_V}{dt}$ is called the *total derivative* in the direction of the vector V^i.

Now, let us consider the following operators:

$$(3.2)'\qquad \overset{1}{I}_V=V^i\frac{\partial}{\partial y^{(2)i}},\quad \overset{2}{I}_V=V^i\frac{\partial}{\partial y^{(1)i}}+\frac{dV^i}{dt}\frac{\partial}{\partial y^{(2)i}}.$$

Similarly with the previous theorem, we can prove:

Theorem 4.3.2 *The operators* $(3.2)'$ *have the following properties:*

1° $\overset{1}{I}_V,\overset{2}{I}_V$ *are vector fields along the curve* c.

2° $\overset{1}{I}_V(L),\overset{2}{I}_V(L)$ *are the scalar fields* (2.6).

3° $\overset{2}{I}_V=J\left(\frac{d_V}{dt}\right),\ \overset{1}{I}_V=J^2\left(\frac{d_V}{dt}\right)$.

4° *If* $V^i=\frac{dx^i}{dt}$, *then* $\overset{1}{I}_V,\overset{2}{I}_V$ *are the Liouville vector fields* $\overset{1}{\Gamma}$ *and* $\overset{2}{\Gamma}$, *respectively, along the curve* c.

Finally, the identity (2.8) leads to the theorem:

Theorem 4.3.3 *Along any smooth curve c in M, we have*

$$\frac{d_V L}{dt} = V^i \overset{0}{E}_i (L) + \frac{d}{dt} \overset{2}{I}_V (L) - \frac{1}{2} \frac{d^2}{dt^2} \overset{1}{I}_V (L). \tag{3.5}$$

Indeed, (2.8) and (3.2) imply the last equality.

Theorem 4.3.4 *The following property $\frac{d_V L}{dt} = 0$ holds, along an extremal curves of the integral of action $I(c)$, if, and only if, $\overset{2}{I}_V (L) - \frac{1}{2} \frac{d}{dt} \overset{1}{I}_V (L) = const.$ along the curve c.*

Using (3.5), the last statement is immediate.

Corollary 4.3.1 *For any Lagrangian $L(x, y^{(1)}, y^{(2)})$, along a smooth curve c, we have*

$$\frac{dL}{dt} = \frac{dx^i}{dt} \overset{0}{E}_i (L) + \frac{d}{dt} \overset{2}{I} (L) - \frac{1}{2} \frac{d^2}{dt^2} \overset{1}{I} (L). \tag{3.6}$$

Corollary 4.3.2 *If c is a solution curve of the Euler–Lagrange equation $\overset{0}{E}_i (L) = 0$, then L is constant along the curve c if, and only if, $\overset{2}{I} (L) - \frac{1}{2} \frac{d}{dt} \overset{1}{I} (L) = const.$*

4.4 Craig–Synge Covectors

Instead of the covector field $\overset{0}{E}_i (L)$ there are other two covectors $\overset{1}{E}_i (L), \overset{2}{E}_i (L)$ associated to a differentiable Lagrangian of second order L. They were introduced by H.Craig and J.Synge [62], [285]. These covector fields are useful in the geometry of the regular Lagrangians of order 2 which give rise to the Lagrange spaces of second order.

Let us consider a smooth curve $c : [0,1] \longrightarrow M$ and along its extension $\tilde{c}$ the following operators

$$\text{(4.1)} \qquad \begin{array}{l} \overset{0}{E}_i = \dfrac{\partial}{\partial x^i} - \dfrac{d}{dt}\dfrac{\partial}{\partial y^{(1)i}} + \dfrac{1}{2}\dfrac{d^2}{dt^2}\dfrac{\partial}{\partial y^{(2)i}}; \\ \overset{1}{E}_i = -\dfrac{\partial}{\partial x^i} + \dfrac{d}{dt}\dfrac{\partial}{\partial y^{(2)i}}; \ \overset{2}{E}_i = \dfrac{1}{2}\dfrac{\partial}{\partial y^{(2)i}}. \end{array}$$

These operators act R–linearly over the R–linear space of Lagrangians of second order. The main result concerning the previous operators is given by the theorem:

Theorem 4.4.1 *$\overset{\alpha}{E}_i(L)$, $(\alpha = 0,1,2)$ are d–covector fields.*

Indeed, according to Theorem 4.2.2, $\overset{0}{E}_i(L)$ is a d–covector field. Evidently, $\overset{2}{E}_i(L)$ has the same property. For $\overset{1}{E}_i(L)$, on the curve c, we successively have

$$\overset{1}{E}_i(L) = -\frac{\partial L}{\partial y^{(1)i}} + \frac{d}{dt}\frac{\partial L}{\partial y^{(2)i}} = -\frac{\partial \tilde{y}^{(1)j}}{\partial y^{(1)i}}\frac{\partial \tilde{L}}{\partial \tilde{y}^{(1)j}} - \frac{\partial \tilde{y}^{(2)j}}{\partial y^{(1)i}}\frac{\partial \tilde{L}}{\partial \tilde{y}^{(2)j}} +$$

$$+\frac{d}{dt}\left(\frac{\partial \tilde{x}^j}{\partial x^i}\frac{\partial \tilde{L}}{\partial y^{(2)j}}\right) = -\frac{\partial \tilde{x}^j}{\partial x^i}\frac{\partial \tilde{L}}{\partial \tilde{y}^{(1)j}} - \frac{\partial \tilde{y}^{(1)j}}{\partial x^i}\frac{\partial \tilde{L}}{\partial \tilde{y}^{(2)j}} + \frac{\partial \tilde{y}^{(1)j}}{\partial x^i}\frac{\partial \tilde{L}}{\partial \tilde{y}^{(2)j}} +$$

$$+\frac{\partial \tilde{x}^j}{\partial x^i}\frac{d}{dt}\frac{\partial \tilde{L}}{\partial \tilde{y}^{(2)j}} = \frac{\partial \tilde{x}^j}{\partial x^i}\left(-\frac{\partial \tilde{L}}{\partial \tilde{y}^{(1)j}} + \frac{d}{dt}\frac{\partial \tilde{L}}{\partial \tilde{y}^{(2)j}}\right) = \frac{\partial \tilde{x}^j}{\partial x^i}\overset{1}{\tilde{E}}_j(\tilde{L}).$$

q.e.d.

In the general case of the differentiable Lagrangians of higher order, the previous theorem can be easier proved, by means of the following lemma:

Lemma 4.4.1 *For any differentiable Lagrangian $L(x, y^{(1)}, y^{(2)})$ and any differentiable function $\phi(t), t \in [0,1]$, along the curve c we have*

$$\text{(4.2)} \qquad \overset{0}{E}_i(\phi L) = \phi \overset{0}{E}_i(L) + \frac{d\phi}{dt}\overset{1}{E}_i(L) + \frac{d^2\phi}{dt^2}\overset{2}{E}_i(L).$$

The proof is immediate if we remark that

$$\frac{\partial(\phi L)}{\partial x^i} = \phi\frac{\partial L}{\partial x^i}, \quad \frac{\partial(\phi L)}{\partial y^{(\alpha)i}} = \phi\frac{\partial(L)}{\partial y^{(\alpha)i}}, \quad (\alpha = 1, 2),$$

and then we apply the Leibniz rule.

Another auxiliary result is given by:

Lemma 4.4.2 *If F is a differentiable Lagrangian of second order with the property $\frac{\partial F}{\partial y^{(2)i}} = 0$, then the following equations hold, along the curve c:*

$$(4.3)\qquad \begin{aligned} \frac{\partial}{\partial x^i}\frac{dF}{dt} &= \frac{d}{dt}\frac{\partial F}{\partial x^i}, \\ \frac{\partial}{\partial y^{(1)i}}\frac{dF}{dt} &= \frac{\partial F}{\partial x^i} + \frac{d}{dt}\frac{\partial F}{\partial y^{(1)i}}, \\ \frac{\partial}{\partial y^{(2)i}}\frac{dF}{dt} &= 2\frac{\partial F}{\partial y^{(1)i}}. \end{aligned}$$

Indeed, by means of (3.4) we obtain

$$\frac{dF}{dt} = \frac{\partial F}{\partial x^i}y^{(1)i} + 2\frac{\partial F}{\partial y^{(1)i}}y^{(2)i}, \quad \frac{d}{dt}\frac{\partial F}{\partial x^i} = \frac{\partial^2 F}{\partial x^i \partial x^j}y^{(1)j} + 2\frac{\partial^2 F}{\partial x^i \partial y^{(1)j}}y^{(2)j}.$$

Hence, the first equality (4.3) follows immediately. The other formulae (4.3) can be proved using the same ideas.

Now we are able to prove an extension of the Caratheodory result.

Theorem 4.4.2 *For any differentiable Lagrangian $L(x, y^{(1)}, y^{(2)})$ and any function $F(x, y^{(1)})$, along the curve c we have*

$$(4.4)\qquad \overset{0}{E}_i\left(L + \frac{dF}{dt}\right) = \overset{0}{E}_i(L).$$

Proof. Taking into account the property $\overset{0}{E}_i\left(L + \frac{dF}{dt}\right) = \overset{0}{E}_i(L) + \overset{0}{E}_i\left(\frac{dF}{dt}\right)$ and using the previous lemma, we get $\overset{0}{E}_i\left(\frac{dF}{dt}\right) = 0.$

Remark. The formula (4.4) is true, only if $\frac{\partial F}{\partial y^{(2)i}} = 0$.

A new version of the last theorem is given in the following

Theorem 4.4.3 *The integrals of action* $I(c) = \int_0^1 L\,dt$ *and* $I'(c) = = \int_0^1 \left(L + \frac{dF}{dt}\right) dt$ *have the same extremal curves, for any differentiable Lagrangian F with the property* $\frac{\partial F}{\partial y^{(2)i}} = 0$.

Based on the Lemma 4.4.2, we can prove, without any difficulties:

Theorem 4.4.4 *For any differentiable Lagrangian F, having the property* $\frac{\partial F}{\partial y^{(2)i}} = 0$, *the following equations hold:*

$$(4.6)\quad \overset{0}{E}_i \left(\frac{dF}{dt}\right) = 0,\ \overset{1}{E}_i \left(\frac{dF}{dt}\right) = -\overset{0}{E}_i (F),\ \overset{2}{E}_i \left(\frac{dF}{dt}\right) = -\overset{1}{E}_i (F).$$

Therefore, we deduce:

Corollary 4.4.1

a. *If the Lagrangian F has the properties* $\frac{\partial F}{\partial y^{(2)i}} = 0$, $\overset{1}{E}_i \left(\frac{dF}{dt}\right) = 0$, *then it has the following property, too:* $\overset{0}{E}_i (F) = 0$.

b. *If the Lagrangian F has the property* $\overset{2}{E}_i \left(\frac{dF}{dt}\right) = 0$, *then we have* $\frac{\partial F}{\partial y^{(1)i}} = 0$.

4.5 The Energies $\overset{1}{\mathcal{E}}_c (L)$, $\overset{2}{\mathcal{E}}_c (L)$

The notion of energy of a Lagrangian $L(x, y^{(1)}, y^{(2)})$ is not so simple. Indeed, the function

$$\overset{2}{I} (L) - L = y^{(1)i} \frac{\partial L}{\partial y^{(1)i}} + 2y^{(2)i} \frac{\partial L}{\partial y^{(2)i}} - L$$

is a scalar field on the manifold $E = \mathrm{Osc}^2 M$ (not only along a curve c). It extends the classical notion of energy. But, it is not convenient for the Lagrangians $L(x, y^{(1)}, y^{(2)})$ since it does not satisfy the law of conservation.

Studying the variational problem on the functional $I(c)$, from (1.8), we get to consider the notion of energies of higher order along a curve $c : [0,1] \longrightarrow M$. Let us consider again the curve c and its extension $\tilde{c}$ to $\tilde{E}$, given by (1.7).

Definition 4.5.1 Along the curve c, the following function

$$\overset{2}{\mathcal{E}}_c (L) = \overset{2}{I} (L) - \frac{1}{2} \frac{d}{dt} \overset{1}{I} (L) - L \tag{5.1}$$

is called the energy of second order of the differentiable Lagrangian $L(x, y^{(1)}, y^{(2)})$.

This notion was introduced and studied by M.de Leon, D.Krupka et al. [164], [155], [157]. They proved the law of conservation for $\overset{2}{\mathcal{E}}_c (L)$.

From the variational problem on $I(c)$ it follows that it is necessary to introduce also the function

$$\overset{1}{\mathcal{E}}_c (L) = -\frac{1}{2} \overset{1}{I} (L), \tag{5.2}$$

called the energy of order 1 of the Lagrangian L. Of course, it does not depend on the curve c. This property is important in the geometry of Lagrangians of second order.

Theorem 4.5.1 *If the differentiable Lagrangian $L(x, y^{(1)}, y^{(2)})$ satisfies the Zermelo conditions, then the energies $\overset{1}{\mathcal{E}}_c (L)$ and $\overset{2}{\mathcal{E}}_c (L)$ vanish.*

An important result on the energy of second order is given by:

Theorem 4.5.2 *For any differentiable Lagrangian $L(x, y^{(1)}, y^{(2)})$ along a smooth curve $c : [0,1] \to (x^i(t)) \in M$ the variation of the energy of second order $\overset{2}{\mathcal{E}}_c (L)$ is given by*

$$\frac{d \overset{2}{\mathcal{E}}_c (L)}{dt} = -\frac{dx^i}{dt} \overset{0}{E}_i (L). \tag{5.3}$$

Proof. From (5.1) and (3.6), we deduce

$$\frac{d \overset{2}{\mathcal{E}_c} (L)}{dt} = \frac{d \overset{2}{I} (L)}{dt} - \frac{1}{2} \frac{d^2 \overset{1}{I} (L)}{dt^2} - \frac{dL}{dt} = -\frac{dx^i}{dt} \overset{0}{E_i} (L).$$

q.e.d.

The last theorem has a remarkable consequence:

Theorem 4.5.3 *For any differentiable Lagrangian $L(x, y^{(1)}, y^{(2)})$, the energy of second order $\overset{2}{\mathcal{E}_c} (L)$ is conserved along every solution curve c of the Euler–Lagrange equations $\overset{0}{E_i} (L) = 0$.*

A similar result can be proved for the energy $\overset{1}{\mathcal{E}_c} (L)$.

Theorem 4.5.4 *Along a smooth curve c in M we have*

$$(5.3)' \qquad \frac{d \overset{1}{\mathcal{E}_c} (L)}{dt} + \frac{1}{2} \overset{2}{I} (L) = -\frac{1}{2} \frac{dx^i}{dt} \overset{1}{E_i} (L).$$

Proof. The variation on c of the energy of order 1, $\overset{1}{\mathcal{E}_c} (L)$, is as follows:

$$\begin{aligned} \frac{d \overset{1}{\mathcal{E}_c} (L)}{dt} &= -\frac{1}{2} \left\{ 2y^{(2)i} \frac{\partial L}{\partial y^{(2)i}} + y^{(1)i} \frac{d}{dt} \frac{\partial L}{\partial y^{(2)i}} \right\} = \\ &= -\frac{1}{2} \overset{2}{I} (L) - \frac{1}{2} \frac{dx^i}{dt} \overset{1}{E_i} (L). \end{aligned}$$

q.e.d.

Consequently, we get:

Corollary 4.5.1 *Along a solution curve c of the differential equations $\overset{1}{E_i} (L) = 0$, the following equation holds:*

$$(5.3)'' \qquad \frac{d \overset{1}{\mathcal{E}_c} (L)}{dt} = -\frac{1}{2} \overset{2}{I} (L).$$

Example. Let us consider the following Lagrangian

$$L(x,y^{(1)},y^{(2)}) = \gamma_{ij}(x)z^{(2)i}z^{(2)j} + b_i(x,y^{(1)})z^{(2)i} + b(x,y^{(1)}), \tag{5.4}$$

where $\gamma_{ij}(x)$ is a Riemannian metric on the base manifold M, $b_i(x,y^{(1)})$ is a d–covector field on E (assuming that this d–covector field exists) and $b(x,y^{(1)})$ is a scalar field on E, where $z^{(2)i}$ is the Liouville d–vector field (3.6), Ch.2:

$$z^{(2)i} = y^{(2)i} + \frac{1}{2}\gamma^i{}_{jk}(x)y^{(1)j}y^{(1)k}, \tag{5.4'}$$

$\gamma^i{}_{jk}(x)$ being the Christoffel symbols of $\gamma_{ij}(x)$. This Lagrangian is a natural extension of the Lagrangian (12.2), Ch.1, from Electrodynamics.

A straightforward calculus, allows to get the expressions of the main invariants $\overset{1}{I}(L)$ and $\overset{2}{I}(L)$:

$$\begin{aligned} &\overset{1}{I}(L) = \left(2\gamma_{ij}z^{(2)j} + b_i\right)y^{(1)i},\\ &\overset{2}{I}(L) = 2\left(2\gamma_{ij}z^{(2)j} + b_i\right)z^{(1)i} + \left(\frac{\partial b_j}{\partial y^{(1)i}}z^{(2)j} + \frac{\partial b}{\partial y^{(1)i}}\right)y^{(1)i}. \end{aligned} \tag{5.5}$$

Now the energy $\overset{2}{\mathcal{E}_c}(L)$ can be easily written.

We end this section with an auxiliary result:

Lemma 4.5.1 *For any differentiable Lagrangian $L(x,y^{(1)},y^{(2)})$ and any differentiable function $\tau : M\times[0,1]\longrightarrow R$, along a smooth curve $c:[0,1]\rightarrow(x^i(t))\in M$, the functions $\tau(x(t),t)$ and $L\left(x(t),\frac{dx}{dt},\frac{1}{2}\frac{d^2x}{dt^2}\right)$ satisfy the following identities:*

$$\begin{aligned} &\frac{d\tau}{dt}L - \left[\frac{d\tau}{dt}\overset{1}{I}(L)\right] =\\ &= \tau\,\frac{d\overset{2}{\mathcal{E}_c}(L)}{dt} + \frac{d}{dt}\left\{-\tau\overset{2}{\mathcal{E}_c}(L) + \frac{d\tau}{dt}\overset{1}{\mathcal{E}_c}(L)\right\}. \end{aligned} \tag{5.6}$$

Proof. The right hand of this formula is given by

$$\begin{aligned} &-\frac{d\tau}{dt}\left\{\overset{2}{\mathcal{E}_c}(L) - \frac{d\overset{2}{\mathcal{E}_c}(L)}{dt}\right\} + \frac{d^2\tau}{dt^2}\overset{1}{\mathcal{E}_c}(L) =\\ &= -\frac{d\tau}{dt}\left(\overset{2}{I}(L) - L\right) - \frac{1}{2}\overset{1}{I}(L)\frac{d^2\tau}{dt^2}. \end{aligned}$$

But this is exactly the first member of (5.6). **q.e.d.**

We shall use this lemma in the next section.

4.6 Nöther Theorems

By Theorem 4.4.3 the integrals of action $I(c) = \int_0^1 L\,dt$ and $I'(c) = = \int_0^1 \left(L + \frac{dF}{dt}\right) dt$ have the same extremal curves, for any differentiable Lagrangian F with the property $\frac{\partial F}{\partial y^{(2)i}} = 0$.

Therefore, we can formulate:

Definition 4.6.1 A symmetry of the differentiable Lagrangian $L(x,y^{(1)}, y^{(2)})$ is a C^∞–diffeomorphism $\varphi : M \times R \longrightarrow M \times R$, which preserves the variational principle of the integral of action $I(c)$ from (1.8).

Generally, the variational principle is considered on an open set $U \subset M$. So, we can consider the notion of the local symmetry of the Lagrangian L, taking the local diffeomorphism φ.

Therefore, in the following considerations we study the infinitesimal symmetries, given on an open set $U \times (a, b)$ in $M \times R$, in the form

$$\begin{aligned} x'^i &= x^i + \varepsilon V^i(x,t), \quad (i = 1, ..., n) \\ t' &= t + \varepsilon\tau(x,t), \end{aligned} \tag{6.1}$$

where ε is a real number, sufficiently small in absolute value, such that the points (x, t) and (x', t') belong to the same domain of a local chart in $U \times (a, b)$, where the curve $c : t \in [0, 1] \longrightarrow (x^i(t), t) \in U \times (a, b)$ is defined.

Of course, c is a parametrized curve, t being the parameter on c.

In the following considerations terms of order greater than 1 in ε will be neglected, $V^i(x,t)$ being a vector field on the open set $U \times (a, b)$.

The inverse transformation of the local diffeomorphism (6.1) is as follows:

$$x^i = x'^i - \varepsilon V^i(x,t), \quad t = t' - \varepsilon V^i(x,t).$$

In the end points $c(0)$ and $c(1)$ we assume that the vector field $V^i(x(t),t) = V^i(t)$ satisfies the conditions (2.2).

We can prove that $S_V : c \longrightarrow \mathrm{Osc}^2 M \times R$ defined by

$$\text{(6.2)} \qquad \begin{aligned} x^i &= x^i(t) \\ y^{(1)i} &= V^i(x(t),t) \\ 2y^{(2)i} &= \frac{dV^i(x(t),t)}{dt} \\ t &= t \end{aligned}$$

is a section of the mapping $\check{\pi} : (u,t) \in \mathrm{Osc}^2 M \times R \longrightarrow (\pi(u),t) \in M \times R$. The proof is the same as in the Lemma 4.3.1.

Using the Definition 4.5.1, the infinitesimal transformation (6.1) is a symmetry of the Lagrangian $L(x, y^{(1)}, y^{(2)})$ if, and only if, for any C^∞-function $F(x, y^{(1)})$, the following equation holds:

$$\text{(6.3)} \qquad L\left(x', \frac{dx'}{dt'}, \frac{1}{2}\frac{d^2x'}{dt'^2}\right) dt' = \left\{L\left(x, \frac{dx}{dt}, \frac{1}{2}\frac{d^2x}{dt^2}\right) + F\left(x, \frac{dx}{dt}\right)\right\} dt.$$

We shall substitute in (6.3) $x', \dfrac{dx'}{dt'}, \dfrac{1}{2}\dfrac{d^2x'}{dt'^2}$ from (6.1). First, we obtain:

$$\text{(6.4)} \qquad \begin{aligned} \frac{dt'}{dt} &= 1 + \varepsilon \frac{d\tau}{dt}, \\ \frac{dx'^i}{dt'} &= \frac{dx^i}{dt} + \varepsilon\varphi^{(1)i} \\ \frac{1}{2}\frac{d^2x'^i}{dt'^2} &= \frac{1}{2}\left[\frac{d^2x^i}{dt^2} + \varepsilon\varphi^{(2)i}\right] \end{aligned}$$

where

$$\text{(6.4)}' \qquad \begin{aligned} \varphi^{(1)i} &= \frac{dV^i}{dt} - \frac{dx^i}{dt}\frac{d\tau}{dt} \\ \varphi^{(2)i} &= \frac{d^2V^i}{dt^2} - 2\frac{d^2x^i}{dt^2}\frac{d\tau}{dt} - \frac{dx^i}{dt}\frac{d^2\tau}{dt^2}. \end{aligned}$$

Taking into account (6.4), (6.4)$'$, from (6.3), neglecting the terms in $\varepsilon^2, \varepsilon^3, \ldots$ and setting $\phi = \varepsilon F$, we get

$$\text{(6.5)} \qquad L\frac{d\tau}{dt} + \frac{\partial L}{\partial x^i}V^i + \frac{\partial L}{\partial y^{(1)i}}\varphi^{(1)i} + \frac{1}{2}\frac{\partial L}{\partial y^{(2)i}}\varphi^{(2)i} = \frac{d\phi}{dt}.$$

Conversely, if (6.5) holds, for L, V^i, τ and c given apriori, then setting $\varepsilon^{-1}\phi(x, y^{(1)i}) = F(x, y^{(1)i})$ it follows that (6.3) is verified for the infinitesimal transformation (6.1), up to terms of order ≥ 2 in ε. But $\varphi^{(1)i}$, $\varphi^{(2)i}$ are given by (6.4)′. We deduce that the equation (6.5) is equivalent to

$$\text{(6.6)} \qquad \begin{aligned} &V^i \frac{\partial L}{\partial x^i} + \frac{dV^i}{dt}\frac{\partial L}{\partial y^{(1)i}} + \frac{1}{2}\frac{d^2V^i}{dt^2}\frac{\partial L}{\partial y^{(2)i}} + \\ &+ \left\{ L\frac{d\tau}{dt} - \left[\overset{2}{I}(L)\frac{d\tau}{dt} + \frac{1}{2}\overset{1}{I}(L)\frac{d^2\tau}{dt^2} \right] \right\} = \frac{d\phi}{dt}. \end{aligned}$$

Now, using the operator (3.2), we can state:

Theorem 4.6.1 *A necessary and sufficient condition that an infinitesimal transformation (6.1) be a symmetry of the Lagrangian $L(x, y^{(1)}, y^{(2)})$ is that the left hand of the equality*

$$\text{(6.7)} \qquad \frac{d_V L}{dt} + \left\{ L\frac{d\tau}{dt} - \left[\overset{2}{I}(L)\frac{d\tau}{dt} + \frac{1}{2}\overset{1}{I}(L)\frac{d^2\tau}{dt^2} \right] \right\} = \frac{d\phi}{dt}$$

be of the form $\dfrac{d}{dt}\,\phi(x, y^{(1)})$ along the smooth curve c.

The Theorem 4.3.3 and Lemma 4.5.1 imply the fact that (6.7) is equivalent to

$$\text{(6.8)} \qquad \begin{aligned} &V^i \overset{0}{E}_i(L) + \frac{d}{dt}\overset{2}{I}_V(L) - \frac{1}{2}\frac{d^2}{dt^2}\overset{1}{I}_V(L) + \tau\frac{d}{dt}\overset{2}{\mathcal{E}}_c(L) + \\ &+ \frac{d}{dt}\left[-\tau\,\overset{2}{\mathcal{E}}_c(L) + \frac{d\tau}{dt}\overset{1}{\mathcal{E}}_c(L) \right] = \frac{d\phi}{dt}. \end{aligned}$$

Now, applying Theorem 4.5.3 and looking to the previous equation, in which we take $\overset{0}{E}_i(L) = 0$ and $\dfrac{d}{dt}\overset{2}{\mathcal{E}}_c(L) = 0$, the Nöther type theorem follows:

Theorem 4.6.2 *For any infinitesimal symmetry (6.1) (which satisfies (6.7)) for a Lagrangian $L(x, y^{(1)}, y^{(2)})$ and for any C^∞-function $\phi(x, y^{(1)})$, the following function*

$$\text{(6.9)} \qquad \mathcal{F}^2(L, \phi) \overset{def}{=} \overset{2}{I}_V(L) - \frac{1}{2}\frac{d}{dt}\overset{1}{I}_V(L) - \tau\,\overset{2}{\mathcal{E}}_c(L) + \frac{d\tau}{dt}\overset{1}{\mathcal{E}}_c(L) - \phi$$

is conserved along the solution curves of the Euler–Lagrange equation $\overset{0}{E}_i(L) = 0$.

Clearly, the functions $\mathcal{F}^2(L,\phi)$ from (6.9) contain the relative invariants $\overset{1}{I}_V(L), \overset{2}{I}_V(L)$, the energies of order 1 and 2, $\overset{1}{\mathcal{E}}_c(L), \overset{2}{\mathcal{E}}_c(L)$ and the function $\phi(x, y^{(1)})$.

In particular, if the Zermelo conditions (1.9) are verified, then Theorem 4.5.1 and the previous theorem lead to a simpler Nöther theorem:

Theorem 4.6.3 *For any infinitesimal symmetry* (6.1) *of a differentiable Lagrangian* $L(x, y^{(1)}, y^{(2)})$, *which satisfies the Zermelo conditions* (1.9) *and for any* C^∞*-function* $\phi(x, y^{(1)})$, *the following function*

$$\mathcal{F}^2(L,\phi) \overset{def}{=} \overset{2}{I}_V(L) - \frac{1}{2}\frac{d}{dt}\overset{1}{I}_V(L) - \phi \tag{6.10}$$

is conserved along the solution curves of the Euler–Lagrange equation $\overset{0}{E}_i(L) = 0$.

Remark. It is not difficult to extend the Nöther Theorem 4.6.2 to the case when on $M \times R$ there exists a Lie group of transformations, whose infinitesimal transformations are symmetries for a second order differentiable Lagrangian L.

4.7 Jacobi–Ostrogradski Momenta

The classical approach, described in Chapter 1, concerning the notion of moment derived from a Lagrangian can be extended to the Lagrangians of second order. This subject is well treated in the book of Manuel de Leon and P. Rodrigues [164].

In this section, we introduce the notion of Jacobi–Ostrogradski momenta and point out their relations with the energies $\overset{2}{\mathcal{E}}_c(L), \overset{1}{\mathcal{E}}_c(L)$. In a next section we shall use this theory for getting a geometrical model of a second order Lagrange space.

Let us consider the energy of second order $\overset{2}{\mathcal{E}}_c(L) = \overset{2}{I}(L) - \frac{1}{2}\frac{d}{dt}\overset{1}{I}(L) - L$ along a parametrized curve c. Remarking that $\overset{2}{\mathcal{E}}_c(L)$ is a polynomial function of degree one in $\frac{dx^i}{dt}, \frac{d^2x^i}{dt^2}$, we can write:

$$\overset{2}{\mathcal{E}}_c(L) = p_{(1)i}\frac{dx^i}{dt} + p_{(2)i}\frac{d^2x^i}{dt^2} - L \quad \text{where} \tag{7.1}$$

$$p_{(1)i} = \frac{\partial L}{\partial y^{(1)i}} - \frac{1}{2}\frac{d}{dt}\frac{\partial L}{\partial y^{(2)i}},\ p_{(2)i} = \frac{1}{2}\frac{\partial L}{\partial y^{(2)i}}. \tag{7.2}$$

$p_{(1)i}$ and $p_{(2)i}$ are called *Jacobi–Ostrogradski momenta.* Of course, $p_{(2)i}$ does not depend on the curve c.

Proposition 4.7.1 *With respect to a transformation of coordinates (1.7), Ch.2, on E the momenta* $p_{(1)i}$ *and* $p_{(2)i}$ *are transformed by the rules:*

$$p_{(1)i} = \frac{\partial \tilde{y}^{(1)m}}{\partial y^{(1)i}}\tilde{p}_{(1)m} + \frac{\partial \tilde{y}^{(2)m}}{\partial y^{(1)i}}\tilde{p}_{(2)m}, p_{(2)i} = \frac{\partial \tilde{y}^{(2)m}}{\partial y^{(2)i}}\tilde{p}_{(2)m}. \tag{7.3}$$

Indeed:

$$p_{(1)i} = \frac{\partial \tilde{y}^{(1)m}}{\partial y^{(1)i}}\frac{\partial L}{\partial \tilde{y}^{(1)m}} + \frac{\partial \tilde{y}^{(2)m}}{\partial y^{(1)i}}\frac{\partial L}{\partial \tilde{y}^{(2)m}} - \frac{1}{2}\frac{d}{dt}\left(\frac{\partial \tilde{x}^m}{\partial x^i}\frac{\partial L}{\partial \tilde{y}^{(2)m}}\right)$$

and the expressions of $\tilde{p}_{(1)i}, \tilde{p}_{(2)i}$ give us the rule of transformation (7.3).

Theorem 4.7.1 *Along the curve* $c : I \longrightarrow M$ *we have*

$$\frac{d}{dt}p_{(1)i} - \frac{\partial L}{\partial x^i} = -\overset{0}{E}_i(L), \frac{d}{dt}p_{(2)i} - p_{(1)i} = \overset{1}{E}_i(L). \tag{7.4}$$

Proof. The operators (4.1) applied to the Lagrangian L, together with the expressions of momenta (7.2), imply the equations (7.4). **q.e.d.**

Taking into account that both equations $\overset{0}{E}_i(L) = 0$ and $\overset{1}{E}_i(L) = 0$ have a geometrical meaning, the previous theorem has some consequences.

Theorem 4.7.2

1° *Along the solution curves of the Euler–Lagrange equations* $\overset{0}{E}_i(L) = 0$ *we have*

$$\frac{dp_{(1)i}}{dt} = \frac{\partial L}{\partial x^i}, \quad \frac{dp_{(2)i}}{dt} = \frac{\partial L}{\partial y^{(1)i}} - p_{(1)i}. \tag{7.5}$$

2° *Along the solution curves of the Synge equations* $\overset{1}{E}_i(L) = 0$ *the following property holds:* $\dfrac{dp_{(2)i}}{dt} = p_{(1)i}$.

Using the previous theorems we can deduce the Hamilton–Jacobi equations:

Theorem 4.7.3 *Along each solution curve of the Euler–Lagrange equations* $\overset{0}{E}_i(L) = 0$, *the following Hamilton–Jacobi equations hold:*

$$\frac{\partial \overset{2}{\mathcal{E}_c}(L)}{\partial p_{(1)i}} = \frac{dx^i}{dt}, \qquad \frac{\partial \overset{2}{\mathcal{E}_c}(L)}{\partial p_{(2)i}} = \frac{d^2x^i}{dt^2};$$

$$\frac{\partial \overset{2}{\mathcal{E}_c}(L)}{\partial x^i} = -\frac{dp_{(1)i}}{dt}, \quad \frac{\partial \overset{2}{\mathcal{E}_c}(L)}{\partial y^{(1)i}} = -\frac{dp_{(2)i}}{dt}$$

Of course $\overset{2}{\mathcal{E}_c}(L)$ is the Hamiltonian energy of the Lagrangian $L(x, y^{(1)}, y^{(2)})$. We can also establish:

Theorem 4.7.4 *The variation of the energy of order* 1 *along a smooth curve* $c : I \longrightarrow M$ *is given by*

$$\frac{d\,\overset{1}{\mathcal{E}_c}(L)}{dt} = p_{(1)i}\frac{dx^i}{dt} + p_{(2)i}\frac{d^2x^i}{dt^2} - \overset{2}{I}(L).$$

Theorem 4.7.5 *The energy of order* 1 *is conserved along a smooth curve* $c : I \longrightarrow M$ *if, and only if,*

$$\underset{\Gamma}{\pounds_2} L = p_{(1)i}\frac{dx^i}{dt} + p_{(2)i}\frac{d^2x^i}{dt^2}.$$

The Jacobi–Ostrogradski momenta $p_{(1)i}$ and $p_{(2)i}$ allow us to define the Jacobi–Ostrogradski 1–forms:

$$p_{(1)} = p_{(1)i}\, dx^i + p_{(2)i}\, dy^{(1)i}, \quad p_{(2)} = p_{(2)}\, dx^i. \tag{7.6}$$

We observe that $p_{(2)}$ is an 1–form on $\tilde{E}$, while $p_{(1)i}$ is an 1–form along the curve c. The exterior differential of the 1–form $p_{(2)}$ is as follows

$$dp_{(2)} = \frac{1}{2}\left(\frac{\delta p_{(2)i}}{\delta x^j} - \frac{\delta p_{(2)j}}{\delta x^i}\right) dx^j \wedge dx^i + \frac{\delta p_{(2)i}}{\delta y^{(1)j}}\, \delta y^{(1)j} \wedge dx^i + \theta \tag{7.8}$$

where θ is the 2–form

$$\theta = g_{ij} \delta y^{(2)j} \wedge dx^i \tag{7.9}$$

and N is an apriori given nonlinear connection. Generally, N is the canonical nonlinear connection (see the next section) of the Lagrangian L. In this case θ is globally defined on $\tilde{E}$ and depends only on L. Therefore, it is called the *almost presymplectic structure of the regular Lagrangian L.*

Remarking that the coefficients from (7.8) are d–tensors, we have:

Proposition 4.7.2 *The equation $dp_{(2)} = \theta$ holds if, and only if, the tensorial equations:*

$$\frac{\delta p_{(2)i}}{\delta x^j} - \frac{\delta p_{(2)j}}{\delta x^i} = 0, \quad \frac{\delta p_{(2)i}}{\delta y^{(1)j}} = 0 \tag{7.10}$$

are verified.

In the applications it is important to know the exterior differential of the 2–form θ.

Proposition 4.7.3 *The exterior differential of the 2–form θ is as follows:*

$$d\theta = \frac{1}{6}\left\{ g_{im} \underset{(02)}{R}{}^{m}{}_{jk} + g_{jm} \underset{(02)}{R}{}^{m}{}_{ki} + g_{km} \underset{(02)}{R}{}^{m}{}_{ij} \right\} dx^k \wedge$$

$$\wedge dx^j \wedge dx^i + \frac{1}{2}\left\{ g_{im} \underset{(12)}{B}{}^{m}{}_{jk} - g_{jm} \underset{(12)}{B}{}^{m}{}_{ik} \right\} \delta y^{(1)k} \wedge dx^j \wedge dx^i +$$

$$(7.11) \quad + \frac{1}{2}\left\{ \frac{\delta g_{jk}}{\delta x^i} - \frac{\delta g_{ik}}{\delta x^j} + g_{im} \underset{(22)}{B}{}^{m}{}_{jk} - g_{jm} \underset{(22)}{B}{}^{m}{}_{ik} \right\} \delta y^{(2)k} \wedge$$

$$\wedge dx^j \wedge dx^i + \frac{1}{2} g_{jm} \underset{(12)}{R}{}^{m}{}_{jk} \delta y^{(1)k} \wedge \delta y^{(1)j} \wedge dx^i +$$

$$+ \left\{ g_{im} \underset{(21)}{B}{}^{m}{}_{kj} - \frac{\delta g_{ik}}{\delta y^{(1)j}} \right\} \delta y^{(2)k} \wedge \delta y^{(1)j} \wedge dx^i .$$

Corollary 4.7.1 $dp_{(2)} = \theta \Longrightarrow d\theta = 0.$

4.8 Regular Lagrangians. Canonical Nonlinear Connections

Definition 4.8.1 A differentiable Lagrangian of second order $L(x, y^{(1)}, y^{(2)})$ is called *regular* if its Hessian with respect to the variables $y^{(2)i}$ is nonsingular on $\tilde{E}$.

By means of the previous definition, a regular Lagrangian L has the property

$$(8.1) \qquad rank\|g_{ij}(x, y^{(1)}, y^{(2)})\| = n,$$

where g_{ij} is the d–tensor field on $\tilde{E}$:

$$(8.2) \qquad g_{ij}(x, y^{(1)}, y^{(2)}) = \frac{1}{2}\,\frac{\partial^2 L}{\partial y^{(2)i} \partial y^{(2)j}}.$$

From the example 1.9.1 we can deduce that the Lagrangian $L(x, y^{(1)}, y^{(2)}) = \gamma_{ij}(x) z^{(2)i} z^{(2)j}$ is a regular one. It has the property $g_{ij}(x, y^{(1)}, y^{(2)}) = \gamma_{ij}(x)$. So, using the Theorem 4.1.1, we can assert:

Theorem 4.8.1 *If the base manifold M is paracompact, then on the manifold $\widetilde{E}$ there exist regular Lagrangians of second order.*

Example 4.8.1 Let $F^n = (M, F(x, y^{(1)}))$ be a Finsler space and $N^i{}_j(x, y^{(1)})$ its Cartan nonlinear connection (cf. §11, Ch.1). It follows, without difficulties, that $z^{(2)i} = y^{(2)i} + \frac{1}{2} N^i{}_j y^{(1)j}$ is a d-vector field on $\widetilde{E}$.

Let us consider the Lagrangian of second order

$$L(x, y^{(1)}, y^{(2)}) = \gamma_{ij}(x, y^{(1)}) z^{(2)i} z^{(2)j}, \tag{8.3}$$

where $\gamma_{ij}(x, y^{(1)})$ is the fundamental tensor field of the Finsler space F^n. Thus, we have:

Proposition 4.8.1 *The Lagrangian $L(x, y^{(1)}, y^{(2)})$ from* (8.3) *is regular. Its d-tensor g_{ij} is the same with the fundamental tensor of the Finsler space F^n.*

Returning to the general theory, we remark that a regular Lagrangian $L(x, y^{(1)}, y^{(2)})$ determines some geometrical object fields on $\widetilde{E}$ depending only on the Lagrangian L. One of them is a 2–spray.

Let $g^{ij}(x, y^{(1)}, y^{(2)})$ be the contravariant d–tensor corresponding to $g_{ij}(x, y^{(1)}, y^{(2)})$ on $\widetilde{E}$. It satisfies the equations

$$g_{ik}(x, y^{(1)}, y^{(2)}) g^{kj}(x, y^{(1)}, y^{(2)}) = \delta^j{}_i. \tag{8.4}$$

Along a smooth curve $c : I \longrightarrow M$ we consider the *Craig-Synge* covector field

$$\overset{1}{E}_i (L) = -\frac{\partial L}{\partial y^{(1)i}} + \frac{d}{dt} \frac{\partial L}{\partial y^{(2)i}}, \quad \left(y^{(1)i} = \frac{dx^i}{dt},\ y^{(2)i} = \frac{1}{2} \frac{d^2 x^i}{dt^2} \right). \tag{8.5}$$

Its expanding form is as follows:

$$\overset{1}{E}_i (L) = g_{ij} \frac{d^3 x^i}{dt^3} + \left\{ \Gamma \left(\frac{\partial L}{\partial y^{(2)i}} \right) - \frac{\partial L}{\partial y^{(1)i}} \right\}, \tag{8.5$'$}$$

where Γ is the operator

$$\Gamma = y^{(1)i} \frac{\partial}{\partial x^i} + 2 y^{(2)i} \frac{\partial}{\partial y^{(1)i}}. \tag{8.5$''$}$$

Now we can prove:

Theorem 4.8.2 *For any regular Lagrangian $L(x, y^{(1)}, y^{(2)})$ there exist the 2-sprays determined only on the Lagrangian L. One of them has the coefficients*

$$3G^i = \frac{1}{2} g^{ij} \left\{ \Gamma\left(\frac{\partial L}{\partial y^{(2)j}}\right) - \frac{\partial L}{\partial y^{(1)j}} \right\}. \tag{8.6}$$

Proof. $\overset{1}{E}_i (L)$ being a d–covector field, $g^{ij} \overset{1}{E}_j (L)$ is a d-vector field along the curve c. Therefore, the system of differential equations

$$g^{ij} \overset{1}{E}_j (L) = \frac{d^3 x^i}{dt^3} + 3! G^i \left(x, \frac{dx}{dt}, \frac{1}{2} \frac{d^2 x}{dt^2}\right) = 0 \tag{8.7}$$

has a geometrical meaning. Its coefficients $G^i(x, y^{(1)}, y^{(2)})$ are given by (8.6). Applying the Theorem 2.3.3, it follows that the system (8.7) determines a 2–spray S, with the coefficients (8.6). **q.e.d.**

The 2–spray S from the last Theorem is given by

$$S = y^{(1)i} \frac{\partial}{\partial x^i} + 2y^{(2)i} \frac{\partial}{\partial y^{(1)i}} - 3G^i(x, y^{(1)}, y^{(2)}) \frac{\partial}{\partial y^{(2)i}}, \tag{8.8}$$

G^i being the coefficients (8.6).

Now we can prove:

Theorem 4.8.3 *If the Lagrangian $L(x, y^{(1)}, y^{(2)})$ is globally defined on the manifold E, then the 2–spray S from* (8.8), (8.6) *has the same property.*

Proof. By a direct calculus from (8.6), it follows that to a transformation of coordinates (1.7), Ch.2, the coefficients (8.6) obey the transformation (3.4), Ch.2. Applying the Theorem 2.3.2, the previous statement follows.

So, we can call the spray S *canonical*, because it is determined only on the regular Lagrangian L.

Now, we need the canonical 2–spray S to determine a non–linear connection which depends only on the Lagrangian L.

Taking into account the Theorem 2.7.1, we can assert:

Theorem 4.8.4 *If $L(x, y^{(1)}, y^{(2)})$ is a regular differentiable Lagrangian on E, then on $\widetilde{E}$ there exist the nonlinear connections determined only on the Lagrangian L. One of them has the dual coefficients given by*

$$(8.9)\qquad \begin{aligned} \underset{(1)}{M}{}^{i}{}_{j} &= \frac{\partial G^i}{\partial y^{(2)j}} = \frac{1}{3!}\frac{\partial}{\partial y^{(2)j}}\left\{g^{ir}\left[\Gamma\left(\frac{\partial L}{\partial y^{(2)r}}\right) - \frac{\partial L}{\partial y^{(1)r}}\right]\right\} \\ \underset{(2)}{M}{}^{i}{}_{j} &= \frac{1}{2}\left\{S\ \underset{(1)}{M}{}^{i}{}_{j} + \underset{(1)}{M}{}^{i}{}_{r}\ \underset{(1)}{M}{}^{r}{}_{j}\right\} \end{aligned}$$

where S is the canonical 2–spray of the Lagrangian L, with the coefficients (8.6).

This Theorem is important in the building of geometry of second order Lagrange spaces.

The coefficients $\left(\underset{(1)}{N}{}^{i}{}_{j}, \underset{(2)}{N}{}^{i}{}_{j}\right)$ of the previous nonlinear connection are expressed by means of (8.9) in the form

$$(8.9)'\qquad \underset{(1)}{N}{}^{i}{}_{j} = \underset{(1)}{M}{}^{i}{}_{j},\quad \underset{(2)}{N}{}^{i}{}_{j} = \underset{(2)}{M}{}^{i}{}_{j} - \underset{(1)}{M}{}^{i}{}_{r}\ \underset{(1)}{M}{}^{r}{}_{j}.$$

They depend only on the given Lagrangian L. Therefore, the nonlinear connection N with the coefficients (8.9), (8.9)′ is called the *canonical nonlinear connection* determined by the Lagrangian L.

Now we can consider the adapted basis $\left\{\frac{\delta}{\delta x^i}, \frac{\delta}{\delta y^{(1)i}}, \frac{\partial}{\partial y^{(2)i}}\right\}$ built with the coefficients of the canonical nonlinear connection N and its adapted cobasis. We can write, using the theory from the §1.2, Ch. 1, the autoparallel curves of the canonical nonlinear connection N. We can also consider the Liouville d–vectors $z^{(1)i}$ and $z^{(2)i}$:

$$(8.10)\qquad z^{(1)i} = y^{(1)i};\ z^{(2)i} = y^{(2)i} + \frac{1}{2}\underset{(1)}{N}{}^{i}{}_{j} y^{(1)j}$$

and the main invariants $\overset{1}{I}(L)$ and $\overset{2}{I}(L)$ expressed in the adapted basis in the form

$$(8.11)\qquad \overset{1}{I}(L) = z^{(1)i}\frac{\partial L}{\partial y^{(2)i}},\quad \overset{2}{I}(L) = z^{(1)i}\frac{\delta L}{\delta y^{(1)i}} + 2z^{(2)i}\frac{\partial L}{\partial y^{(2)i}}.$$

The previous theory will be applied in order to study, in the next section, two important problem of the higher order geometry.

4.9 Prolongation to $\mathrm{Osc}^2 M$ of the Riemannian Structures

An old problem in differential geometry is that of the prolongation of the Riemannian structures, defined on the base manifold M, to the higher order osculator bundle. This problem belongs to the so–called higher order geometry.

Several remarkable geometers, as E. Bompiani, Ch. Ehresman, A. Morimoto etc. have studied this problem [223].

We present here the solution of the problem in the case of the prolongation of the Riemannian structures to 2–osculator bundle, effectively establishing a Riemannian metric G on $\mathrm{Osc}^2 M$, derived only from a given Riemannian metric g on the base manifold M. The general solution will be described in a next chapter. The problem was solved by R. Miron and Gh. Atanasiu [197]–[202].

Let $R^n = (M, g)$ be a Riemannian space, g being a Riemannian metric defined on the manifold M, having the local coordinates $g_{ij}(x)$, $x \in U \subset M$. We prolonge g_{ij} to $\pi^{-1}(U) \subset E = \mathrm{Osc}^2 M$ setting

$$(g_{ij} \circ \pi)(u) = g_{ij}(x), \quad \forall u \in \pi^{-1}(U), \ \pi(u) = x. \tag{9.1}$$

Of course, $g_{ij} \circ \pi$ is a d–tensor field on E.

The problem of prolongation of the metric structure g to E (or $\widetilde{E}$) can be formulated in the following manner:

The Riemannian structure g, being apriori given on the manifold M, determines a Riemannian structure G on the manifold $\mathrm{Osc}^2 M$ (or $\widetilde{\mathrm{Osc}^2 M}$) so that G is to be determined only by g.

At the begining we give some preliminary results.

Proposition 4.9.1 *The following Liouville d-vector field*

$$z^{(2)i} = y^{(2)i} + \frac{1}{2}\gamma^i{}_{jk}(x) y^{(1)j} y^{(1)k}, \tag{9.2}$$

where $\gamma^i{}_{jk}(x)$ are the Christoffel symbols of the metric g, has the properties:

1° *It is globally defined on $\tilde{E}$.*

2° *It depends only on the metric g.*

Indeed, it is not difficult to prove directly these statements.

Proposition 4.9.2 . *The function*

$$L(x,y^{(1)},y^{(2)}) = g_{ij}(x)z^{(2)i}z^{(2)j} \tag{9.3}$$

is a differentiable Lagrangian having the properties:

1° *L is globally defined on* $\tilde{E}$.

2° *L depends only on the metric g.*

3° *L is a regular Lagrangian.*

Indeed, L is a scalar field on E. According to the previous proposition, it follows that L is determined by the Riemannian structure g, only. A straightforward calculus leads to

$$\frac{1}{2}\frac{\partial^2 L}{\partial y^{(2)i}\partial y^{(2)j}} = g_{ij}(x). \tag{9.4}$$

So that, rank of the Hessian of L, with the respect to $y^{(2)i}$ is $n = \dim M$.

Proposition 4.9.3 . *The canonical* 2*-spray S of the regular Lagrangian L from* (9.3) *has the coefficients*

$$3G^i = g^{ij}(x)\left\{\Gamma\left(g_{jm}z^{(2)m}\right) - g_{mr}z^{(2)m}\frac{\partial z^{(2)r}}{\partial y^{(1)j}}\right\}. \tag{9.5}$$

S is globally defined on the manifold $\tilde{E}$ *and depends only on the Riemannian structure g.*

Proposition 4.9.4 . *The canonical nonlinear connection determined by the Lagrangian* (9.3) *has the dual coefficients:*

$$\begin{aligned} \underset{(1)}{M}{}^i{}_j &= \gamma^i{}_{jk}y^{(1)k}, \\ \underset{(2)}{M}{}^i{}_j &= \frac{1}{2}\left\{\Gamma(\gamma^i{}_{jk}y^{(1)k}) + \underset{(1)}{M}{}^i{}_r \underset{(1)}{M}{}^r{}_j\right\}. \end{aligned} \tag{9.6}$$

It is globally defined on $\tilde{E}$ *and depends only on the structure g.*

According to the previous propositions, the assertions from the last proposition follow.

Let us consider the adapted basis $\left\{\frac{\delta}{\delta x^i}, \frac{\delta}{\delta y^{(1)i}}, \frac{\partial}{\partial y^{(2)i}}\right\}$ and adapted cobasis $\{dx^i, \delta y^{(1)i}, \delta y^{(2)i}\}$ to the canonical nonlinear connection N (with the dual coefficients (9.6)).

We can construct the following covariant tensor on $\tilde{E}$:

$$(9.7)\quad G(u) = g_{ij}(x)dx^i \otimes dx^j + g_{ij}(x)\delta y^{(1)i} \otimes \delta y^{(1)j} + g_{ij}(x)\delta y^{(2)i} \otimes \delta y^{(2)j},$$

for any point $u \in \tilde{E}$, with the property $\pi(u) = x \in M$.

Now we can prove the following theorem:

Theorem 4.9.1 *The following properties hold:*

1° *G is a symmetric tensor field, covariant of type* $(0,2)$, *globally defined on $\tilde{E}$.*

2° *G is a Riemannian structure on $\tilde{E}$.*

3° *G is derived only from the Riemannian structure g.*

Proof. 1° Evidently, every term of (9.7) is a covariant of order 2 tensor field, defined on the manifold E. 2° G is positively defined. Hence, the matrix of G is of rank $3n$. 3° According to the Proposition 4.9.3, it follows that G depends only on g.

The previous theorem solves the enunciated problem.

The Riemannian space $\text{Prol}\,\mathcal{R}^n = (\tilde{E}, G)$ is called the *prolongation of order* 2 *of the Riemannian space* $\mathcal{R}^n$.

We need these Riemannian spaces in order to construct some special regular Lagrangians of order 2.

4.10 Prolongation to $\text{Osc}^2 M$ of the Finslerian and Lagrangian Structures

A natural widening of the problem of extension of the Riemannian structure can be formulated in the same manner for the Finslerian structures or Lagrangian structures. Using the same methods we treat the problem of prolongation of second order for the Finsler spaces.

Let $F^n = (M, F(x, y^{(1)}))$ be a Finsler space. The Finsler function F (or Finsler structure F) is defined on the manifold $\text{Osc}^1 M$, having the properties specified in §11, Ch.1. We assume that its fundamental tensor field:

$$g_{ij}(x, y^{(1)}) = \frac{1}{2} \frac{\partial^2 F^2}{\partial y^{(1)i} \partial y^{(1)j}} \tag{10.1}$$

is positively defined on E.

Let $\pi_1^2 : \text{Osc}^2 M \longrightarrow \text{Osc}^1 M$ be the projection of $\text{Osc}^2 M$ on $\text{Osc}^1 M$, given by $\pi_1^2(x, y^{(1)}, y^{(2)}) = (x, y^{(1)})$.

We prolonge the function F to $\tilde{E} = \widetilde{\text{Osc}^2 M}$, setting

$$F^*(x, y^{(1)}, y^{(2)}) = (F \circ \pi_1^2)(x, y^{(1)}, y^{(2)}) = F(x, y^{(1)}), \quad \text{on } \tilde{E}. \tag{10.2}$$

We also prolonge the fundamental tensor $g_{ij}(x, y^{(1)})$ to $\tilde{E}$ setting

$$g^*_{ij}(x, y^{(1)}, y^{(2)}) = g_{ij} \circ \pi_1^2(x, y^{(1)}, y^{(2)}), \quad \text{on every } \pi^{-1}(U) \subset \tilde{E}. \tag{10.3}$$

We denote F^* and g^*_{ij} with the same letters F and g_{ij}. Clearly, $g_{ij}(x, y^{(1)})$ is a d-tensor field on $\tilde{E}$. Its rank is equal to n.

The problem of prolongation of the Finsler space $F^n = (M, F)$ to $\tilde{E} = \widetilde{\text{Osc}^2 M}$ is the following:

Determine a regular Lagrangian $L(x, y^{(1)}, y^{(2)})$ on $\tilde{E}$ with the following properties:

a. L to be determined on $\tilde{E}$ only by the fundamental function F of the Finsler space F^n.

b. L to determine on $\tilde{E}$ a Riemannian structure G depending only on the fundamental function F of the space F^n.

The solution of this problem is as follows.

Let $N^i{}_j(x, y^{(1)})$ be the Cartan nonlinear connection of the Finsler space $F^n = (M, F)$. We can prove:

Proposition 4.10.1 *The functions*

$$z^{(2)i} = y^{(2)i} + \frac{1}{2} N^i{}_j(x, y^{(1)}) y^{(1)j} \tag{10.4}$$

defined on every domain of local chart in $\tilde{E}$, determine a d–vector field depending only on the fundamental function $F(x, y^{(1)})$.

Indeed, according to the rules of transformation of $y^{(1)i}, y^{(2)i}$ and $N^i{}_j(x, y^{(1)})$, with respect to (1.7), Ch.2, we deduce:

Theorem 4.10.1 *The function $L : \mathrm{Osc}^2 M \longrightarrow R$ given by*

$$L(x, y^{(1)}, y^{(2)}) = g_{ij}(x, y^{(1)}) z^{(2)i} z^{(2)j} \tag{10.5}$$

is a differentiable Lagrangian on the manifold $E = \mathrm{Osc}^2 M$, having the following properties:

1° *L depends only on the fundamental function F of the Finsler space F^n.*

2° *L is regular, its fundamental tensor being coincident to the fundamental tensor of the Finsler space F^n.*

Proof. $g_{ij}(x, y^{(1)})$ being a d–tensor field and $z^{(2)i}$ a d–vector field on $\tilde{E}$ it follows that L is a differentiable Lagrangian on E and depends only on the fundamental function F. Remarking that $z^{(2)i}$ is linear in $y^{(2)i}$, we deduce that the fundamental tensor field of L is just $g_{ij}(x, y^{(1)})$.

q.e.d.

Now, let us consider the canonical nonlinear connection determined by the regular Lagrangian L from (10.5) and $\{dx^i, \delta y^{(1)i}, \delta y^{(2)i}\}$ its local adapted cobasis of the module $\mathcal{X}^*(\tilde{E})$.

Thus, the tensor field

$$\begin{aligned} G(u) &= g_{ij}(x, y^{(1)}) dx^i \otimes dx^j + g_{ij}(x, y^{(1)}) \delta y^{(1)i} \otimes \delta y^{(1)j} + \\ &+ g_{ij}(x, y^{(1)}) \delta y^{(2)i} \otimes \delta y^{(2)j}, \quad \forall u \in \tilde{E}, \ \pi^2_1(u) = (x, y^{(1)}), \end{aligned} \tag{10.6}$$

solves the enunciated problem. Indeed, we have:

Theorem 4.10.2 *G is a covariant symmetric tensor field on $\tilde{E}$. It is a Riemannian structure on $\tilde{E}$, derived only from the fundamental function $F(x, y^{(1)})$ of the Finsler space F^n.*

The problem of the prolongation of the Finsler spaces can be widened to the problem of that of the Lagrange spaces.

We give here, briefly, its solution.

Considering a Lagrange space $L^n = (M, L(x, y^{(1)}))$ and its canonical nonlinear connection $N^i{}_j(x, y^{(1)})$ from (9.4), Ch.2, we can define the d–vector field $z^{(2)i} = y^{(2)i} + \frac{1}{2} N^i{}_j y^{(1)j}$. Denoting by $g_{ij}(x, y^{(1)})$ the fundamental tensor field of the space L^n we can construct the following Lagrangian

$$L^*(x, y^{(1)}, y^{(2)}) = g_{ij}(x, y^{(1)}) z^{(2)i} z^{(2)j}. \tag{10.7}$$

We can state

Theorem 4.10.3 *The function L^* from* (10.7) *has the following properties:*

1° *L^* is a differentiable Lagrangian on $\tilde{E}$ depending only on the fundamental function L of the Lagrange space L^n.*

2° *L^* is regular and has the same fundamental tensor as the fundamental tensor field of the space L^n.*

Taking into account the canonical nonlinear connection N determined by the Lagrangian L^* from (10.7) we construct the Riemannian structure G from (10.6) corresponding to the prolongation of the given Lagrange space L^n. Hence, the existence of this Riemannian structure G on $\tilde{E}$, depending only on the fundamental function L of the Lagrange space L^n, completly solves the problem of prolongation to $\mathrm{Osc}^2 M$ of the Lagrange spaces $L^n = (M, L(x, y^{(1)}))$.

4.11 Problems

Let us consider the Lagrangian of second order of Electrodynamics, from (5.4).

1. Show that it is a regular one and that the pair $L^n = (M, L)$ is a Lagrange space of order two.

2. Write the energies $\overset{1}{\mathcal{E}_c}(L), \overset{2}{\mathcal{E}_c}(L)$.

3. Determine the Euler–Lagrange equations $\overset{0}{E_i}(L) = 0$.

4. Write the coefficients of the canonical 2–spray of L^n.
5. Determine the coefficients of the canonical nonlinear connection.
6. When is the canonical nonlinear connection N integrable?
7. Write the differential equations of the autoparallel curves of the canonical nonlinear connection $N^i{}_j$.
8. Determine the momenta $p_{(1)}$ and $p_{(2)}$ of this space and the Hamilton–Jacobi equations.

Chapter 5

Second Order Lagrange Spaces

The notion of second order Lagrange space is a natural extension of the notion of Lagrange space $L^n = (M, L)$ introduced in the §6, Ch.1. The second order Lagrange space is defined as a pair $L^{(2)n} = (M, L)$ where M is a real n-dimensional manifold and L is a differentiable regular Lagrangian of second order, whose fundamental tensor field has a constant signature on $\tilde{E}$.

L being a regular Lagrangian, we can study $L^{(2)n}$ applying the theory from the previous chapters.

Thus, for a second order Lagrange space we determine: the canonical 2-spray, canonical nonlinear connection and obtain a special theory of the canonical N-linear connection. Some applications will be given in the theory of gravitational and electromagnetic fields.

The notion of Lagrange space of higher order was introduced by the author of the presented monograph together with Gh. Atanasiu. Therefore we follow the papers [196], [201].

5.1 The Notion of Lagrange Space of Order 2

Definition 5.1.1 We call a Lagrange space of order 2 a pair $L^{(2)n} = (M, L)$ formed by a real n-dimensional differentiable manifold M and a regular differentiable Lagrangian on $\mathrm{Osc}^2 M$, whose d–tensor field

$$g_{ij}(x, y^{(1)}, y^{(2)}) = \frac{1}{2} \frac{\partial^2 L}{\partial y^{(2)i} \partial y^{(2)j}} \tag{1.1}$$

has a constant signature on $\widetilde{\mathrm{Osc}^2 M}$.

This definition is a natural extension of that given for Lagrange space in §6, Chapter 1.

We continue to say that L is the *fundamental function* and g_{ij} is the *fundamental tensor field* of the space $L^{(2)n}$. The restriction of $L^{(2)n}$ to an open set $U \subset M$ gives us the notion of local second order Lagrange space.

In the case when M is a paracompact manifold, the existence of second order Lagrange space, with positively defined fundamental tensor field is always assured.

Theorem 5.1.1 *If the manifold M is paracompact, there exist the Lagrangians $L : \mathrm{Osc}^2 M \longrightarrow R$, so that $L^{(2)n} = (M, L)$ is a second order Lagrange space.*

Indeed, since M is paracompact, there exist a Riemannian metric g on M. Then, the d-tensor field (the Liouville d-vector)

$$z^{(2)i} = y^{(2)i} + \frac{1}{2}\gamma^i{}_{jk}(x) y^{(1)j} y^{(1)k} \tag{1.2}$$

is globally defined on $\tilde{E}$, $\gamma^i{}_{jk}(x)$ being the Christoffel symbols of g. So the Lagrangian

$$L(x, y^{(1)}, y^{(2)}) = g_{ij}(x) z^{(2)i} z^{(2)j} \tag{1.3}$$

is a regular one, having the fundamental tensor field $(g_{ij} \circ \pi)(u) = g_{ij}(x)$, $\forall u \in \tilde{E}$, $\pi(u) = x$, positively defined. Consequently, (M, L) is a second order Lagrange space.

Assuming that the geometrical object fields taken into consideration exist, we can give some examples.

Examples of spaces $L^{(2)n}$

1° $L^{(2)n}$ with the fundamental function (1.3).

2° $L^{(2)n}$ with the fundamental function

$$L(x, y^{(1)}, y^{(2)}) = g_{ij}(x) z^{(2)i} z^{(2)j} + A_i(x, y^{(1)}) z^{(2)i} + U(x, y^{(1)}), \tag{1.4}$$

where $z^{(2)i}$ is from (1.2), $A_i(x, y^{(1)})$ is a d–covector field and $U(x, y^{(1)})$ a function on E. The Lagrangian (1.4) is a natural extension of the Lagrangian (6.4), Ch.1, from classical electrodynamics. Therefore the space $L^{(2)n}$ endowed with the Lagrangian (1.4) will be called the *second order Lagrange space of electrodynamics.*

3° The spaces $L^{(2)n}$ of Randers type are defined by the fundamental function

$$L(x, y^{(1)}, y^{(2)}) = \left\{\sqrt{g_{ij}(x)z^{(2)i}z^{(2)j}} + A_i(x, y^{(1)})z^{(2)i}\right\}^2, \tag{1.5}$$

$A_i(x, y^{(1)})$ being a d–covector field and $A_i(x, y^{(1)})z^{(2)i} > 0$.

4° The spaces $L^{(2)n}$ of Kropina type have the Lagrangian

$$L(x, y^{(1)}, y^{(2)}) = \left\{\frac{g_{ij}(x)z^{(2)i}z^{(2)j}}{A_i(x, y^{(1)})z^{(2)i}}\right\}^2, \tag{1.6}$$

where the scalar field $A_i(x, y^{(1)})z^{(2)i}$ is nonvanishing on $\tilde{E}$.

Other examples can be obtained starting from the prolongation to $\mathrm{Osc}^2 M$ of a Finsler space $F^n = (M, F)$. Indeed, taking into account the Cartan nonlinear connection of F^n with the coefficients $N^i{}_j(x, y^{(1)})$ we can consider the Liouville d–vector field

$$u^{(2)i} = y^{(2)i} + \frac{1}{2}\, N^i{}_j(x, y^{(1)})y^{(1)j}. \tag{1.7}$$

Using the d–vector $u^{(2)i}$ and the fundamental tensor $g_{ij}(x, y^{(1)})$ of the Finsler space F^n, we can take back the previous examples. So, we have the following new examples:

5° $L(x, y^{(1)}, y^{(2)}) = g_{ij}(x, y^{(1)})u^{(2)i}u^{(2)j}$

6° $L(x, y^{(1)}, y^{(2)}) = g_{ij}(x, y^{(1)})u^{(2)i}u^{(2)j} + A_i(x, y^{(1)})u^{(2)i} + U(x, y^{(1)})$

7° $L(x, y^{(1)}, y^{(2)}) = \left\{\sqrt{g_{ij}(x, y^{(1)})u^{(2)i}u^{(2)j}} + A_i(x, y^{(1)})u^{(2)i}\right\}^2$

8° $L(x, y^{(1)}, y^{(2)}) = \left\{\dfrac{g_{ij}(x, y^{(1)})u^{(2)i}u^{(2)j}}{A_i(x, y^{(1)})u^{(2)i}}\right\}^2, \quad A_i(x, y^{(1)})u^{(2)i} > 0$

For applications in Biology, we must take into consideration the fundamental function $F(x, y^{(1)})$, introduced by P.L. Antonelli et al. [13], [14], [15].

5.2 Euler–Lagrange Equations of a Lagrange Space $L^{(2)n}$

Let $L^{(2)n} = (M,L)$ be a second order Lagrange space and $c: [0,1] \to M$ a parametrized curve, having the image in a domain of a local chart of the manifold M. The integral of action of the Lagrangian L on c is

$$I(c) = \int_0^1 L\left(x(t), \frac{dx}{dt}(t), \frac{1}{2}\frac{d^2x}{dt^2}(t)\right) dt. \tag{2.1}$$

Let us consider again the main invariants $\overset{1}{I}(L)$ and $\overset{2}{I}(L)$ of the Lagrangian L.

In Chapter 4 we established the necessary conditions that $I(c)$ did not depend on the parametrized curve c. They are given by the Zermelo conditions $\overset{1}{I}(L) = 0$, $\overset{2}{I}(L) = L$ (cf. Th. 12, Ch.4).

Therefore, the following property holds:

Theorem 5.2.1 *For any Lagrange space of second order $L^{(2)n} = (M, L)$ the integral of action $I(c)$ depends essentially on the parametrization of the curve c.*

Proof. Assuming, by contradiction, that $I(c)$ does not depend on the parametrization of the curve c, it follows that the main invariants have the values $\overset{1}{I}(L) = 0$, $\overset{2}{I}(L) = L$ on the manifold $\tilde{E}$. The first one is $y^{(1)i}\dfrac{\partial L}{\partial y^{(2)i}} = 0$. Deriving with respect to $y^{(2)i}$ we get $y^{(1)i}g_{ij} = 0$ on $\tilde{E}$. The last equation, by multiplication by the contravariant tensor g^{ih} and contraction, leads to the equations $y^{(1)i} = 0$ on $\tilde{E}$. This is a contradiction. **q.e.d.**

The variational problem on the functional $I(c)$, from (2.1), leads to the Euler–Lagrange equation

$$\begin{aligned} &\overset{0}{E}_i(L) \overset{def}{=} \frac{\partial L}{\partial x^i} - \frac{d}{dt}\frac{\partial L}{\partial y^{(1)i}} + \frac{1}{2}\frac{d^2}{dt^2}\frac{\partial L}{\partial y^{(2)i}} = 0, \\ &y^{(1)i} = \frac{dx^i}{dt}, \ y^{(2)i} = \frac{1}{2}\frac{dx^2}{dt^2}. \end{aligned} \tag{2.2}$$

Remarking that $\overset{0}{E}_i(L) = 0$ is a d–covector field, it follows that *the equation* $\overset{0}{E}_i(L) = 0$ *holds along any smooth curve of the manifold* M.

Applying the Corollaries 4.3.1 and 4.3.2, we deduce:

Proposition 5.2.1 *If* $L^{(2)n} = (M, L)$ *is a Lagrange space of second order, then on a smooth curve* c *in* M *we have*

$$\frac{dL}{dt} = \frac{dx^i}{dt}\overset{0}{E}_i(L) + \frac{d}{dt}\overset{2}{I}(L) - \frac{1}{2}\frac{d^2\overset{1}{I}(L)}{dt^2}. \tag{2.3}$$

Proposition 5.2.2 *If* c *is a solution curve of the Euler–Lagrange equation* $\overset{0}{E}_i(L) = 0$ *for a second order Lagrange space* $L^{(2)n} = (M, L)$, *then* L *is conserved on the curve* c *if, and only if, the invariant* $\overset{2}{I}(L) - \frac{1}{2}\frac{d}{dt}\overset{1}{I}(L)$ *is conserved on* c, *too.*

Considering the energies $\overset{2}{\mathcal{E}_c}(L)$ and $\overset{1}{\mathcal{E}_c}(L)$ for the fundamental function L of the space $L^{(2)n}$, given by (5.1), (5.2), Ch.4, and taking into account Theorems 4.5.2 and 4.5.3, we obtain:

Theorem 5.2.2 *The energy of second order* $\overset{2}{\mathcal{E}_c}(L)$ *of a Lagrange space* $L^{(2)n}$ *has the following property:*

$$\frac{d\overset{2}{\mathcal{E}_c}(L)}{dt} = -\frac{dx^i}{dt}\overset{0}{E}_i(L), \tag{2.4}$$

on any smooth curve c *of the manifold* M.

Theorem 5.2.3 *For any second order Lagrange space* $L^{(2)n} = (M, L)$ *the energy of second order* $\overset{2}{\mathcal{E}_c}(L)$ *is conserved on every solution curve* c *of the Euler–Lagrange equations* $\overset{0}{E}_i(L) = 0$.

Let us consider an infinitesimal symmetry (6.1), Ch.4, and the invariants $\overset{1}{I}_V(L), \overset{2}{I}_V(L)$ from §6, Ch.4. Applying Theorem 4.6.2, we get a Nöther Theorem for second order Lagrange spaces.

Theorem 5.2.4 *For any infinitesimal symmetry* (6.1)), *Ch.* 4, *of the fundamental function L of a second order Lagrange space* $L^{(2)n} = (M, L)$ *and for any* C^∞*-function* $\phi(x, y^{(1)})$ *the following function*

$$\mathcal{F}^2(L,\phi) = \overset{2}{I}_V(L) - \frac{1}{2}\frac{d}{dt}\overset{1}{I}_V(L) - \tau\,\overset{2}{\mathcal{E}}_c(L) + \frac{d\tau}{dt}\,\overset{1}{\mathcal{E}}_c(L) - \phi$$

is conserved on the solution curve of the Euler–Lagrange equations $\overset{0}{E}_i(L) = 0$.

5.3 Canonical Nonlinear Connections

The fundamental d–tensor field g_{ij} of the second order Lagrange space $L^{(2)n} = (M, L)$ being nonsingular, we can determine the canonical nonlinear connection from the canonical 2–spray of the Lagrangian L. Applying the theory from §8 of the previous chapter, we can first write the Craig–Synge covector field

$$\overset{1}{E}_i(L) = -\frac{\partial L}{\partial y^{(1)i}} + \frac{d}{dt}\frac{\partial L}{\partial y^{(2)i}} \tag{3.1}$$

on a curve $c : I \longrightarrow M$.

This covector field determines the *Synge equation*

$$g^{ij}\,\overset{1}{E}_j(L) = \frac{d^3x^i}{dt^3} + 3!G^i\left(x, \frac{dx}{dt}, \frac{1}{2}\frac{d^2x}{dt^2}\right) = 0, \tag{3.2}$$

where

$$\begin{aligned} 3G^i = \frac{1}{2}g^{ij}\Bigg\{ & y^{(1)m}\frac{\partial}{\partial x^m}\left(\frac{\partial L}{\partial y^{(2)j}}\right) + \\ & + 2y^{(2)m}\frac{\partial}{\partial y^{(1)m}}\left(\frac{\partial L}{\partial y^{(2)j}}\right) - \frac{\partial L}{\partial y^{(1)j}}\Bigg\}. \end{aligned} \tag{3.3}$$

According to the Theorem 4.8.2, we can assert:

Theorem 5.3.1 *For any Lagrange space of second order* $L^{(2)n} = (M, L)$ *there exists a 2–spray S determined only on the fundamental function L. It is given by*

$$S = y^{(1)m}\frac{\partial}{\partial x^m} + 2y^{(2)m}\frac{\partial}{\partial y^{(1)m}} - 3G^m(x, y^{(1)}, y^{(2)})\frac{\partial}{\partial y^{(2)m}} \tag{3.4}$$

where the coefficients G^m are those in (3.3).

S is called the *canonical 2–spray of the space* $L^{(2)n}$. Continuing to apply the theory expounded in Chapter 4, we can determine a canonical nonlinear connection for the space $L^{(2)n}$.

Theorem 5.3.2 *There exist the nonlinear connections N of the second order Lagrange space $L^{(2)n} = (M, L)$ determined only by its fundamental function L. One of them is given by the canonical 2–spray S, having the following dual coefficients*

$$(3.5)\qquad \underset{(1)}{M}{}^{i}{}_{j} = \frac{\partial G^i}{\partial y^{(2)j}},\ \underset{(2)}{M}{}^{i}{}_{j} = \frac{1}{2}\left(S\ \underset{(1)}{M}{}^{i}{}_{j} + \underset{(1)}{M}{}^{i}{}_{m}\ \underset{(1)}{M}{}^{m}{}_{j}\right).$$

This nonlinear connection, depending only on the fundamental function L of the space $L^{(2)n}$ will be called *canonical.* As we know, the coefficients $\underset{(1)}{N}{}^{i}{}_{j}, \underset{(2)}{N}{}^{i}{}_{j}$ of the canonical nonlinear connection N are given by

$$(3.5)'\qquad \begin{aligned} &\underset{(1)}{N}{}^{i}{}_{j} = \underset{(1)}{M}{}^{i}{}_{j},\\ &\underset{(2)}{N}{}^{i}{}_{j} = \underset{(2)}{M}{}^{i}{}_{j} - \underset{(1)}{M}{}^{i}{}_{m}\ \underset{(1)}{M}{}^{m}{}_{j} = \frac{1}{2}\left(S\ \underset{(1)}{N}{}^{i}{}_{j} - \underset{(1)}{N}{}^{i}{}_{m}\ \underset{(1)}{N}{}^{m}{}_{j}\right). \end{aligned}$$

Let us consider the local adapted basis to the direct decomposition

$$(3.6)\qquad T_uE = \underset{(0)}{N}(u) \oplus \underset{(1)}{N}(u) \oplus V_2(u)$$

where $\underset{(0)}{N} = N$ and $\underset{(1)}{N} = J(\underset{(0)}{N})$. It is given by

$$(3.7)\qquad \left(\frac{\delta}{\delta x^i}, \frac{\delta}{\delta y^{(1)i}}, \frac{\delta}{\delta y^{(2)i}}\right),$$

where we are setting

$$(3.7)'\qquad \begin{aligned} &\frac{\delta}{\delta x^i} = \frac{\partial}{\partial x^i} - \underset{(1)}{N}{}^{j}{}_{i}\frac{\partial}{\partial y^{(1)j}} - \underset{(2)}{N}{}^{j}{}_{i}\frac{\partial}{\partial y^{(2)j}},\\ &\frac{\delta}{\delta y^{(1)i}} = \frac{\partial}{\partial y^{(1)i}} - \underset{(1)}{N}{}^{j}{}_{i}\frac{\partial}{\partial y^{(2)j}},\ \frac{\delta}{\delta y^{(2)i}} = \frac{\partial}{\partial y^{(2)i}}. \end{aligned}$$

As one knows, the local adapted cobasis

$$(\delta x^i, \delta y^{(1)i}, \delta y^{(2)i}) \tag{3.8}$$

has the elements

$$\begin{aligned} &\delta x^i = dx^i, \ \delta y^{(1)i} = dy^{(1)i} + \underset{(1)}{M^i}{}_j \, dx^j, \\ &\delta y^{(2)i} = dy^{(2)i} + \underset{(1)}{M^i}{}_j \, dy^{(1)j} + \underset{(2)}{M^i}{}_j \, dy^{(2)j}. \end{aligned} \tag{3.8$'$}$$

The Lie brackets of the vectors from the adapted basis are given by Theorem 3.4.3, and the exterior differentials of the 1–forms (3.8) are expressed in the formulae (6.9), Ch.3. They introduce the d–tensor fields $\underset{(01)}{R^i}{}_{jk}$, $\underset{(02)}{R^i}{}_{jk}$ and $\underset{(12)}{R^i}{}_{jk}$ from (4.5), Ch.3, for the canonical nonlinear connection.

According to Theorem 3.4.4, we have

Theorem 5.3.3 *The following properties hold:*

1° *The canonical nonlinear connection N of a second order Lagrange space $L^{(2)n}$ is integrable if, and only if, we have*

$$\underset{(01)}{R^i}{}_{jk} = \underset{(02)}{R^i}{}_{jk} = 0. \tag{3.9}$$

2° *The J–vertical distribution N_1 of the canonical nonlinear connection N is integrable if, and only if, we have*

$$\underset{(12)}{R^i}{}_{jk} = 0. \tag{3.9$'$}$$

Remembering that the curve $\gamma : I \longrightarrow \tilde{E}$ with the property that the tangent vector field $\dot{\gamma}$ belongs to the horizontal distribution N is a horizontal curve, we obtain:

Theorem 5.3.4 *A smooth curve $\gamma : t \in (a,b) \rightarrow (x^i(t), y^{(1)i}(t), y^{(2)i}(t)) \in$ $\in \tilde{E}$ is horizontal with respect to the canonical nonlinear connection N of the space $L^{(2)n}$ if, and only if, the functions $(x^i(t), y^{(1)i}(t), y^{(2)i}(t))$ are solutions of the system of differential equations*

$$\begin{aligned} &\frac{\delta y^{(1)i}}{dt} \overset{def}{=} \frac{dy^{(1)i}}{dt} + \underset{(1)}{M^i}{}_j \frac{dx^j}{dt} = 0, \\ &\frac{\delta y^{(2)i}}{dt} \overset{def}{=} \frac{dy^{(2)i}}{dt} + \underset{(1)}{M^i}{}_j \frac{dy^{(1)j}}{dt} + \underset{(2)}{M^i}{}_j \frac{dx^j}{dt} = 0. \end{aligned} \tag{3.10}$$

If the curve $c : t \in (a,b) \to (x^i(t)) \in M$ is given, then using (3.10) we can formulate a theorem of existence and uniqueness for the horizontal curves in $\tilde{E}$.

Let $\tilde{c} : t \in (a,b) \to \left(x^i(t), \frac{dx^i}{dt}(t), \frac{1}{2}\frac{d^2x^i}{dt^2}(t)\right)$ be the extension to $\tilde{E}$ of a smooth curve $c : t \in (a,b) \to x^i(t) \in M$. The curve c is called an *autoparallel curve* of the canonical nonlinear connection N if its extension is horizontal with respect to N. Thus, we can state:

Theorem 5.3.5 *The autoparallel curves of the canonical nonlinear connection N of the second order Lagrange space $L^{(2)n}$ are characterized by the following system of differential equations*

$$y^{(1)i} = \frac{dx^i}{dt}, \; y^{(2)i} = \frac{1}{2}\frac{d^2x^i}{dt^2}, \quad \frac{\delta y^{(1)i}}{dt} = \frac{\delta y^{(2)i}}{dt} = 0. \tag{3.11}$$

A theorem of existence and uniqueness for the local solutions of the previous system of differential equations is immediate.

The previous considerations have as consequence:

Theorem 5.3.6 *The fundamental function L of the second order Lagrange space $L^{(2)n}$ is conserved along any autoparallel curve of the canonical nonlinear connection N of $L^{(2)n}$ if, and only if, L is solution of the partial differential equations:*

$$\frac{\delta L}{\delta x^i} = \frac{\partial L}{\partial x^i} - \underset{(1)}{N}{}^m{}_i \frac{\partial L}{\partial y^{(1)m}} - \underset{(2)}{N}{}^m{}_i \frac{\partial L}{\partial y^{(2)m}} = 0. \tag{3.12}$$

Proof. Along a curve $c : I \longrightarrow M$ we get

$$\frac{dL}{dt} = \frac{\delta L}{\delta x^i}\frac{dx^i}{dt} + \frac{\delta L}{\delta y^{(1)i}}\frac{\delta y^{(1)i}}{dt} + \frac{\delta L}{\delta y^{(2)i}}\frac{\delta y^{(2)i}}{dt}.$$

If c is an autoparallel curve of N, according to Theorem 5.3.5, it follows

$$\frac{dL}{dt} = \frac{\delta L}{\delta x^i}\frac{dx^i}{dt} \tag{$*$}$$

Assuming that L is conserved along any autoparallel curve c, we deduce $\frac{\delta L}{\delta x^i} = 0$. Conversely, if (3.12) holds, then along to any autoparallel curve c, we have $\frac{dL}{dt} = 0$. **q.e.d.**

Of course, for the equations (3.12) there are some conditions of integrability, presented in §5, Ch.3.

5.4 Canonical Metrical N–Connections

Let N be the canonical nonlinear connection of a second order Lagrange space $L^{(2)n}$. Applying the theory from Ch.3, we determine an N–linear connection D which depends only on the fundamental function L of the space $L^{(2)n}$. It will be metrical with respect to the fundamental tensor field g_{ij} from (1.1). Therefore, we can establish:

Theorem 5.4.1 *The following properties hold:*
1° *There exists a unique N-linear connection D on $\tilde{E}$ verifying the axioms*

$$g_{ij|k} = 0,\quad g_{ij} \overset{(1)}{|}_k = 0,\quad g_{ij} \overset{(2)}{|}_k = 0 \tag{4.1}$$

$$\underset{(0)}{T}{}^i{}_{jk} = 0,\quad \underset{(1)}{S}{}^i{}_{jk} = 0,\quad \underset{(2)}{S}{}^i{}_{jk} = 0. \tag{4.2}$$

2° *This connection has the coefficients*

$$\begin{aligned} L^i{}_{jk} &= \frac{1}{2} g^{im}\left(\frac{\delta g_{mk}}{\delta x^j} + \frac{\delta g_{jm}}{\delta x^k} - \frac{\delta g_{jk}}{\delta x^m}\right), \\ \underset{(1)}{C}{}^i{}_{jk} &= \frac{1}{2} g^{im}\left(\frac{\delta g_{mk}}{\delta y^{(1)j}} + \frac{\delta g_{jm}}{\delta y^{(1)k}} - \frac{\delta g_{jk}}{\delta y^{(1)m}}\right) \\ \underset{(2)}{C}{}^i{}_{jk} &= \frac{1}{2} g^{im}\left(\frac{\delta g_{mk}}{\delta y^{(2)j}} + \frac{\delta g_{jm}}{\delta y^{(2)k}} - \frac{\delta g_{jk}}{\delta y^{(2)m}}\right) \end{aligned} \tag{4.3}$$

3° *The previous N–linear connection depends only on the fundamental function L.*

Proof. According to (1.1), it is not difficult to prove that the N–connection with the coefficients (4.3) is an N–linear connection. Substituting (4.3) in the first hands of (4.1) and (4.2) we see that these equations are verified. In other words, we have assured the existence of an N–linear connection which satisfies the axioms (4.1) and (4.2). By contradiction, we prove the uniqueness of this connection. Finally, we remark that all coefficients (4.3) depend on the fundamental function L of the space $L^{(2)n}$. **q.e.d.**

The connection D from the previous theorem will be called *canonical metrical N-connection.* Its set of coefficients will be denoted by $C\Gamma(N)$. In the same manner we can establish the following property:

Theorem 5.4.2 *There exists a unique N-linear connection D which verifies the following axioms:*

1° *D is a metric connection with respect to the fundamental tensor field of the second order Lagrange space $L^{(2)n}$ (i.e.(4.1) are verified).*

2° *The h- and v_α- torsions of D are apriori given by the skew-symmetric d-tensors $\underset{(0)}{T}{}^i{}_{jk}, \underset{(1)}{S}{}^i{}_{jk}, \underset{(2)}{S}{}^i{}_{jk}$.*

Theorem 5.4.3 *The coefficients $D\Gamma(N) = \left(L^{*i}{}_{jk}, \underset{(1)}{C}{}^{*i}{}_{jk}, \underset{(2)}{C}{}^{*i}{}_{jk}\right)$ of the N-linear connection from the previous theorem are given by*

$$
(4.4)\qquad
\begin{aligned}
L^{*i}{}_{jk} &= L^i{}_{jk} + \frac{1}{2}g^{im}\left(g_{hk}\underset{(0)}{T}{}^h{}_{mj} + g_{jh}\underset{(0)}{T}{}^h{}_{mk} - g_{mh}\underset{(0)}{T}{}^h{}_{jk}\right)\\
\underset{(1)}{C}{}^{*i}{}_{jk} &= \underset{(1)}{C}{}^i{}_{jk} + \frac{1}{2}g^{im}\left(g_{hk}\underset{(1)}{S}{}^h{}_{mj} + g_{jh}\underset{(1)}{S}{}^h{}_{mk} - g_{mh}\underset{(1)}{S}{}^h{}_{jk}\right)\\
\underset{(2)}{C}{}^{*i}{}_{jk} &= \underset{(2)}{C}{}^i{}_{jk} + \frac{1}{2}g^{im}\left(g_{hk}\underset{(2)}{S}{}^h{}_{mj} + g_{jh}\underset{(2)}{S}{}^h{}_{mk} - g_{mh}\underset{(2)}{S}{}^h{}_{jk}\right)
\end{aligned}
$$

where $\left(L^i{}_{jk}, \underset{(1)}{C}{}^i{}_{jk}, \underset{(2)}{C}{}^i{}_{jk}\right)$ are the coefficients (4.3) *of the canonical metrical N-linear connection of the second order Lagrange space $L^{(2)n}$.*

Let us consider the curvature d-tensors $R_h{}^i{}_{kh}, \underset{(\alpha)}{P}{}_j{}^i{}_{kh}\ldots$ of the canonical metrical N-connection $C\Gamma(N)$ of the space $L^{(2)n}$ and set

$$
(4.5)\qquad R_{jikh} = g_{im}R_j{}^m{}_{kh},\quad \underset{(\alpha)}{P}{}_{jikh} = g_{im}\underset{(\alpha)}{P}{}_j{}^m{}_{kh},\quad \text{etc.}
$$

We deduce:

Theorem 5.4.4 *The covariant d–curvature tensors of the canonical metrical N–connection $C\Gamma(N)$ have the properties*

$$(4.6)\qquad \begin{aligned} &R_{jikh} + R_{ijkh} = 0,\ \underset{(\alpha)}{P}{}_{jikh} + \underset{(\alpha)}{P}{}_{ijkh} = 0,\ \underset{(21)}{P}{}_{jikh} + \underset{(21)}{P}{}_{ijkh} = 0,\\ &\underset{(\alpha)}{S}{}_{jikh} + \underset{(\alpha)}{S}{}_{ijkh} = 0, \qquad (\alpha = 1, 2). \end{aligned}$$

Proof. Applying the Ricci identities to the fundamental tensor $g_{ij}(x, y^{(1)}, y^{(2)})$ and taking into account (4.1) we obtain (4.6).

Remark. The N–linear connection with the coefficients (4.4) satisfies the same property of the previous theorem.

The deflection tensors of the canonical metrical N–connection $C\Gamma(N)$ are given in the formulae (3.9) and (3.9)′, Ch.3.

5.5 Problems

1. Consider the Lagrangian L from (1.4) and determine:
   1° Fundamental tensor field g_{ij} and its contravariant g^{ij}.
   2° Canonical 2–spray.
   3° Canonical nonlinear connection N.
   4° Canonical metrical N–connection D.
   5° Curvatures and torsions of D.
2. Same problems for the Lagrangian (1.5).

Chapter 6

Geometry of the k-Osculator Bundle

The geometrical theory presented in the previous chapters will be extended to the k–osculator bundle and higher order Lagrange spaces. We begin with the geometry of the k–osculator bundle $(\text{Osc}^k M, \pi, M)$ which is the support of the whole theory [197]–[202].

So, we clarify the geometrical structure of the total space $E = = \text{Osc}^k M$, revealing the vertical distributions $V_1, ..., V_k$, the Liouville vector fields $\overset{1}{\Gamma}, ..., \overset{k}{\Gamma}$ and the k–tangent structure J.

The notions of k–spray, nonlinear connection, N–linear connections, structure equations, curvatures and torsions will be introduced.

Only few proofs will be given because most of them are extensions of those from the previous chapters. Throughout the following part of this book, the geometrical objects or mappings will be considered as being of class C^∞.

6.1 The Notion of k–Osculator Bundle

We introduce the concept of k–osculator bundle $(\text{Osc}^k M, \pi, M)$ as a straightforward extension of that of 2–osculator bundle (Ch.2).

Let M be a real n–dimensional manifold. Two curves in M, $\rho, \sigma : I \longrightarrow M$, have in the point $x_0 \in M$, $\rho(0) = \sigma(0) = x_0$, $(0 \in I)$ a "*contact of order k*", $k \in N^*$, if for any function $f \in \mathcal{F}(U)$, $x_0 \in U$ and U an open set in M, we have

$$\frac{d^\alpha(f\circ\rho)(t)}{dt^\alpha}\Big|_{t=0}=\frac{d^\alpha(f\circ\sigma)(t)}{dt^\alpha}\Big|_{t=0},\ (\alpha=1,...,k). \tag{1.1}$$

The relation "*to have a contact of order k*" is an equivalence. Let $[\rho]_{x_0}$ be a class of equivalence. It will be called a "*k–osculator space*" in the point x_0 of the manifold M. We denote by $\mathrm{Osc}^k x_0$ the set of k–osculator spaces in x_0 and put

$$\mathrm{Osc}^k M=\bigcup_{x_0\in M}\mathrm{Osc}^k x_0. \tag{1.2}$$

Defining the mapping

$$\pi:[\rho]_{x_0}\in \mathrm{Osc}^k M\longrightarrow x_0\in M,\ \forall[\rho]_{x_0} \tag{1.3}$$

the set $(\mathrm{Osc}^k M,\pi,M)$ can be endowed with a natural differential structure of fibre bundle.

One can describe this structure exactly as in the §1, Ch.2.

If $U\subset M$ is a coordinate neighbourhood on the manifold M, $x_0\in U$ and the curve $\rho: I\longrightarrow M$, $\rho(0)=x_0$ is represented on U by $x^i=x^i(t)$, $t\in I$ (the indices $i,j,k,...=1,...,n=\dim M$), taking f from (1.1) successively equal to the coordinate functions, the k–osculator space $[\rho]_{x_0}$ will have a representative element given by the curve:

$$x^{*i}=x^i(0)+\frac{t}{1!}\frac{dx^i}{dt}(0)+\cdots+\frac{t^k}{k!}\frac{d^kx^i}{dt^k}(0),\ t\in(-\varepsilon,\varepsilon)\subset I$$

with convenable small $\varepsilon>0$.

This polynomial function is determined by its coefficients:

$$x_0^i=x^i(0),\ y_0^{(1)i}=\frac{1}{1!}\frac{dx^i}{dt}(0),...,y_0^{(k)i}=\frac{1}{k!}\frac{d^kx^i}{dt^k}(0). \tag{1.4}$$

We set

$$\phi:\mathrm{Osc}^k M\longrightarrow R^{(k+1)n},\ \phi([\rho]_{x_0})=(x_0^i,y_0^{(1)i},...,y_0^{(k)i}).$$

It follows that the pair $(\pi^{-1}(U),\phi)$ is a local chart on $\mathrm{Osc}^k M$.

So a differentiable atlas of the manifold M determines a differentiable atlas on $\mathrm{Osc}^k M$ and the triple $(\mathrm{Osc}^k M, \pi, M)$ is a differentiable bundle. Of course the mapping $\pi : \mathrm{Osc}^k M \longrightarrow M$ is a submersion. A transformation of local coordinates $(x^i, y^{(1)i}, ..., y^{(k)i}) \longrightarrow (\tilde{x}^i, \tilde{y}^{(1)i}, ..., \tilde{y}^{(k)i})$ on the manifold $\mathrm{Osc}^k M$ is given by

$$
\begin{aligned}
&\tilde{x}^i = \tilde{x}^i(x^1, ..., x^n), \ rank\|\frac{\partial \tilde{x}^i}{\partial x^j}\| = n,\\
&\tilde{y}^{(1)i} = \frac{\partial \tilde{x}^i}{\partial x^j} y^{(1)j},\\
(1.5)\qquad &2\tilde{y}^{(2)i} = \frac{\partial \tilde{y}^{(1)i}}{\partial x^j} y^{(1)j} + 2\frac{\partial \tilde{y}^{(1)i}}{\partial y^{(1)j}} y^{(2)j},\\
&\dots\dots\dots\dots\dots\dots\dots\dots\dots\dots\dots\dots\\
&ky^{(k)i} = \frac{\partial \tilde{y}^{(k-1)i}}{\partial x^j} y^{(1)j} + 2\frac{\partial \tilde{y}^{(k-1)i}}{\partial y^{(1)j}} y^{(2)j} + \cdots +\\
&\qquad\qquad + k\frac{\partial \tilde{y}^{(k-1)i}}{\partial y^{(k-1)j}} y^{(k)j},
\end{aligned}
$$

An important remark is that we have the identities:

$$
(1.5)' \quad \frac{\partial \tilde{y}^{(\alpha)i}}{\partial x^j} = \frac{\partial \tilde{y}^{(\alpha+1)i}}{\partial y^{(1)j}} = \cdots = \frac{\partial \tilde{y}^{(k)i}}{\partial y^{(k-\alpha)j}}, \ (\alpha = 0, ..., k-1; y^{(0)i} = x^i)
$$

It is notable the simple form of the coordinate transformations (1.5). It follows that $y^{(1)i}$ has a geometrical meaning and the equation $rank\|y^{(1)i}\| = 1$ is a remarkable one.

A point $u \in \mathrm{Osc}^k M$ with the coordinates $(x^i, y^{(1)i}, ..., y^{(k)i})$ will be denoted by $u = (x, y^{(1)}, ..., y^{(k)})$ or $u = (y^{(0)}, y^{(1)}, ..., y^{(k)})$, with $x = y^{(0)}$.

We can consider the projections

$$
(1.6) \quad \pi^k_\ell : (x, y^{(1)}, ..., y^{(k)}) \in \mathrm{Osc}^k M \rightarrow (x, y^{(1)}, ..., y^{(\ell)}) \in \mathrm{Osc}^\ell M, \ (\ell < k)
$$

where $\pi^k_0 = \pi$.

Clearly, the mappings π^k_ℓ $(\ell < k)$ are submersions.

A section $s : M \longrightarrow \mathrm{Osc}^k M$ of the projection $\pi : \mathrm{Osc}^k M \longrightarrow M$ is a mapping with the property $\pi \circ s = 1_M$. It is a local section if $\pi \circ s\big|_U = 1_U$, for U an open sets in M. A section $s : M \longrightarrow \mathrm{Osc}^k M$ along the curve $c : I \longrightarrow M$ has the property $\pi \circ s\big|_c = 1_c$.

Let us consider a curve $c : I \longrightarrow M$, locally represented in a chart (U, φ) by $x^i = x^i(t),\ t \in I$. Thus, the mapping $\tilde{c} : I \longrightarrow \mathrm{Osc}^k M$ given in the chart $(\pi^{-1}(U), \phi)$ by:

$$x^i = x^i(t),\ y^{(1)i} = \frac{1}{1!}\frac{dx^i}{dt}(t), ..., y^{(1)i} = \frac{1}{k!}\frac{d^k x^i}{dt^k}(t),\ t \in I \tag{1.7}$$

is a smooth curve in $\mathrm{Osc}^k M$. It is a section of the projection π along the curve c. $\tilde{c}$ is called the *extension of order* k of the curve c.

More general, if V is a vector field on M and $c : I \longrightarrow M$ is a curve, then the mapping

(1.8) $S_V : c \longrightarrow \mathrm{Osc}^k M$ defined locally by

$$S_V : x^i = x^i(t),\ y^{(1)i} = V^i(x(t)), ..., y^{(k)i} = \frac{1}{k!}\frac{d^{k-1}V^i(x(t))}{dt^{k-1}} \tag{1.8$'$}$$

is a section of the projection π along the curve c. A proof of this property was given in Ch.2, §1.

Of course, such a section could be a local one.

In order to assure the global existence of the various geometrical objects on the manifold $\mathrm{Osc}^k M$ the following theorem is useful. For a proof in the case $k = 2$ see Ch.2, §1.

Theorem 6.1.1 *If the differentiable manifold M is paracompact, then $\mathrm{Osc}^k M$ is a paracompact differentiable manifold, too.*

We can see that $\mathrm{Osc}^k M : Man \longrightarrow Man$ is a covariant functor from the category Man of differentiable manifold to the same category.

Namely,

$$\mathrm{Osc}^k : M \in Ob\, Man \longrightarrow \mathrm{Osc}^k M \in Ob\, Man,$$
$$\mathrm{Osc}^k : \{f : M \longrightarrow N\} \longrightarrow \{\mathrm{Osc}^k f : \mathrm{Osc}^k M \longrightarrow \mathrm{Osc}^k N\}.$$

If $f : x \in M \longrightarrow f(x) \in N$, in local coordinates is represented by

$$x^{i'} = x^{i'}(x^1, ..., x^n),\ (i', j' = 1, ..., m = \dim N) \tag{1.9}$$

then $\mathrm{Osc}^k f : \mathrm{Osc}^k M \longrightarrow \mathrm{Osc}^k N$ is given by

$$(1.9)' \qquad \begin{cases} x^{i'} = x^{i'}(x^1, ..., x^n) \\ y^{(1)i'} = \dfrac{\partial x^{i'}}{\partial x^j} y^{(1)j} \\ \cdots\cdots\cdots\cdots\cdots\cdots\cdots\cdots\cdots\cdots \\ ky^{(k)i'} = \dfrac{\partial y^{(k-1)i'}}{\partial x^j} y^{(1)j} + 2\dfrac{\partial y^{(k-1)i'}}{\partial y^{(1)j}} y^{(2)j} + \cdots + \\ \qquad\qquad + k\dfrac{\partial y^{(k-1)i'}}{\partial y^{(k-1)j}} y^{(k)j} \end{cases}$$

Using $(1.9)'$, one can prove without difficulties that Osc^k is a covariant functor.

In a next chapter, we shall use these considerations for a theory of submanifolds of the manifold $\mathrm{Osc}^k M$.

6.2 Vertical Distributions. Liouville Vector Fields

In Chapter 2, we set $E = \mathrm{Osc}^k M$ and

$$(2.1) \qquad \tilde{E} = \left\{ u = (x, y^{(1)}, ..., y^{(k)}) \in E \mid rank\|y^{(1)i}\| = 1 \right\}.$$

Note that $\tilde{E}$ is an open submanifold of E.

Taking into account the differential $d\pi$ of the mapping $\pi : E \longrightarrow M$, we have $d\pi : TE \longrightarrow TM$ and we can consider $V(E) = \ker d\pi$. Then $V(E)$ is the *vertical bundle.*

It is a subbundle of the tangent bundle (TE, τ_T, E). The fibres of $V(E)$ provide the vertical distribution $V = V_1$ on E:

$$(2.2) \qquad V_1 : u \in E \longrightarrow V_1(u) \subset T_u E.$$

A local basis of the distribution V_1 is as follows

$$(2.2)' \qquad \left\{ \frac{\partial}{\partial y^{(1)i}}, ..., \frac{\partial}{\partial y^{(k)i}} \right\}.$$

It follows that V_1 is an integrable distribution and $\dim V_1 = kn$.

Similarly, the projection $\pi_1^k : E = \mathrm{Osc}^k M \longrightarrow \mathrm{Osc}^1 M$ gives rise to the bundle mapping $d\pi_1^k : T(\mathrm{Osc}^k M) \longrightarrow T(\mathrm{Osc}^1 M)$.

The fibres of $V_2 = \ker d\pi_1^k$ determine a distribution

$$V_2 : u \in E \longrightarrow V_2(u) \subset T_u E. \tag{2.3}$$

A local basis of V_2 is

$$\left\{ \frac{\partial}{\partial y^{(2)i}}, ..., \frac{\partial}{\partial y^{(k)i}} \right\} \tag{2.3)'$$

Consequently, V_2 is an subdistribution of V_1, it is integrable and $\dim V_2 = (k-1)n$.

Inductively, we can continue this process and get the vertical distributions $V_1, V_2, ..., V_k$ with the properties:

a. $V_1, V_2, ..., V_k$ are integrable.

b. $V_1 \supset V_2 \supset \cdots \supset V_k$.

c. $\dim V_1 = kn$, $\dim V_2 = (k-1)n, ..., \dim V_k = n$.

d. A local basis of the distribution V_α is $\left\{ \frac{\partial}{\partial y^{(\alpha)i}}, ..., \frac{\partial}{\partial y^{(k)i}} \right\}$ for $\alpha = 1, ..., k$.

These properties can also be derived from the rules of transformation of the natural basis of the module of vector fields on E.

Indeed, by calculating the Jacobian of the transformations (1.5) we obtain:

Lemma 6.2.1 *A transformation of coordinates* (1.5) *on* E *implies a change of the natural basis according to the following rule:*

$$\begin{aligned}
\frac{\partial}{\partial x^i} &= \frac{\partial \tilde{x}^j}{\partial x^i}\frac{\partial}{\partial \tilde{x}^j} + \frac{\partial \tilde{y}^{(1)j}}{\partial x^i}\frac{\partial}{\partial \tilde{y}^{(1)j}} + \cdots + \frac{\partial \tilde{y}^{(k)j}}{\partial x^i}\frac{\partial}{\partial \tilde{y}^{(k)j}},\\
\frac{\partial}{\partial y^{(1)i}} &= \frac{\partial \tilde{y}^{(1)j}}{\partial y^{(1)i}}\frac{\partial}{\partial \tilde{y}^{(1)j}} + \cdots + \frac{\partial \tilde{y}^{(k)j}}{\partial y^{(1)i}}\frac{\partial}{\partial \tilde{y}^{(k)j}},\\
&\vdots\\
\frac{\partial}{\partial y^{(k)i}} &= \frac{\partial \tilde{y}^{(k)j}}{\partial y^{(k)i}}\frac{\partial}{\partial \tilde{y}^{(k)j}},
\end{aligned} \tag{2.4}$$

Based on (2.4), we can prove without difficulties the following results:

Proposition 6.2.1 *If k is an odd natural number, then* $\mathrm{Osc}^k M$ *is an orientable manifold.*

Theorem 6.2.1

1° *The following operators in the algebra* $\mathcal{F}(E)$

$$(2.5)\qquad \begin{aligned} \overset{1}{\Gamma} &= y^{(1)i}\frac{\partial}{\partial y^{(k)i}} \\ \overset{2}{\Gamma} &= y^{(1)i}\frac{\partial}{\partial y^{(k-1)i}} + 2y^{(2)i}\frac{\partial}{\partial y^{(k)i}} \\ &\cdots\cdots\cdots\cdots\cdots\cdots\cdots\cdots \\ \overset{k}{\Gamma} &= y^{(1)i}\frac{\partial}{\partial y^{(1)i}} + 2y^{(2)i}\frac{\partial}{\partial y^{(2)i}} + \ldots + ky^{(k)i}\frac{\partial}{\partial y^{(k)i}} \end{aligned}$$

are vector fields, globally defined on E.

2° $\overset{1}{\Gamma},\ldots,\overset{k}{\Gamma}$ *are linearly independent on $\tilde{E}$.*

3° $\overset{1}{\Gamma}$ *belongs to the distribution V_k, ..., $\overset{k}{\Gamma}$ belongs to the distribution V_1.*

The vector fields $\overset{1}{\Gamma},\ldots,\overset{k}{\Gamma}$ are called the *Liouville vector fields.* They are very important in the geometry of k–osculator bundle $\mathrm{Osc}^k M$.

In applications we shall use also the operator

$$(2.6)\qquad \Gamma = y^{(1)i}\frac{\partial}{\partial x^i} + 2y^{(2)i}\frac{\partial}{\partial y^{(1)i}} + \cdots + ky^{(k)i}\frac{\partial}{\partial y^{(k-1)i}}.$$

As in the §2, Ch.2, we can prove:

Lemma 6.2.2 *Under a coordinate transformation* (1.5) *the operator* Γ *changes as follows :*

$$(2.7)\qquad \Gamma = \widetilde{\Gamma} + \left\{ y^{(1)i}\frac{\partial \tilde{y}^{(k)j}}{\partial x^i} + \cdots + ky^{(k)i}\frac{\partial \tilde{y}^{(k)j}}{\partial y^{(k-1)i}} \right\}\frac{\partial}{\partial \tilde{y}^{(k)j}}.$$

If $f \in \mathcal{F}(E)$ and $\dfrac{\partial f}{\partial y^{(k)i}} = 0$, then (2.7) *gives:*

$$(2.7)'\qquad \Gamma f = \widetilde{\Gamma}\tilde{f}.$$

6.3 k–Tangent Structure; k–Sprays

The theory from §3, Ch.2, concerning tangent structures or sprays can be extended with several minor modifications to $E = \mathrm{Osc}^k M$. The k–tangent structure is the $\mathcal{F}(E)$–linear mapping $J : \mathcal{X}(E) \to \mathcal{X}(E)$ defined on the natural basis by:

$$(3.1)\qquad \begin{aligned} &J\left(\frac{\partial}{\partial x^i}\right) = \frac{\partial}{\partial y^{(1)i}}, J\left(\frac{\partial}{\partial y^{(1)i}}\right) = \frac{\partial}{\partial y^{(2)i}}, \dots, \\ &J\left(\frac{\partial}{\partial y^{(k-1)i}}\right) = \frac{\partial}{\partial y^{(k)i}}, J\left(\frac{\partial}{\partial y^{(k)i}}\right) = 0. \end{aligned}$$

Theorem 6.3.1 *The following properties hold:*

1° *J is globally defined on E.*

2° *J is a tensor field of type $(1,1)$ on E.*

3° *J is an integrable structure.*

4° $Im\, J = V_1$, $\ker J = V_k$.

5° $rank\|J\| = kn$.

6° $J \overset{k}{\Gamma} = \overset{k-1}{\Gamma}, \dots, J(\overset{2}{\Gamma}) = \overset{1}{\Gamma},\ J \overset{1}{\Gamma} = 0.$

7° $J \circ \cdots \circ J = 0$, $(k+1$ *factors*$)$.

Having J we may introduce

Definition 6.3.1 A k–spray on E is a vector field $S \in \mathcal{X}(E)$ with the property

$$(3.2)\qquad JS = \overset{k}{\Gamma}$$

Of course, there not always exists a vector field S, globally defined on E with the property (3.2). Therefore, the notion of local spray must be introduced. In this case, a vector field $S \in \mathcal{X}(\widetilde{U})$, where $\widetilde{U}$ is an open set in E, with the property $JS = \overset{k}{\Gamma}_{|\widetilde{U}}$ gives us a k–spray on $\widetilde{U}$.

We also have:

Theorem 6.3.2

1° *A k–spray S can be uniquely written in the form*

$$S = \Gamma - (k+1)G^i(x, y^{(1)}, ..., y^{(k)})\frac{\partial}{\partial y^{(k)i}} \tag{3.3}$$

or in the form

$$\begin{aligned} S &= y^{(1)i}\frac{\partial}{\partial x^i} + \cdots + ky^{(k)i}\frac{\partial}{\partial y^{(k-1)i}} - \\ &- (k+1)G^i(x, y^{(1)}, ..., y^{(k)})\frac{\partial}{\partial y^{(k)i}}. \end{aligned} \tag{3.3$'$}$$

2° *With respect to (1.5), the coefficients G^i are transformed as follows:*

$$\begin{aligned} (k+1)\widetilde{G}^i &= (k+1)G^j\frac{\partial \tilde{x}^i}{\partial x^j} - \\ &- \left(y^{(1)j}\frac{\partial \tilde{y}^{(k)i}}{\partial x^j} + \cdots + ky^{(k)j}\frac{\partial \tilde{y}^{(k)i}}{\partial y^{(k-1)j}}\right). \end{aligned} \tag{3.4}$$

3° *If the set of functions $G^i(x, y^{(1)}, ..., y^{(k)})$ are given on every domain of local chart of the manifold E so that (3.4) holds, then the vector field S from (3.3)$'$ is a k–spray.*

A curve $c : I \longrightarrow M$ is called a *k–path* on M if its k–extension $\tilde{c}$ from (1.7) is the integral curve of a k–spray in (3.3)$'$.

Theorem 6.3.3

1° *The spray S from (3.3)$'$ determines a k–path given by the differential equations:*

$$\frac{d^{k+1}x^i}{dt^{k+1}} + (k+1)!G^i\left(x, \frac{dx}{dt}, ..., \frac{1}{k!}\frac{d^kx}{dt^k}\right) = 0 \tag{3.5}$$

2° *If the system of differential equations (3.5) on the base manifold M is given and it has a geometrical meaning (t being a fixed parameter), then the functions $G^i(x, y^{(1)}, ..., y^{(k)})$ from (3.5) are the coefficients of a local k–spray.*

Theorem 6.3.4 *If the base manifold M is paracompact, then on $E = \mathrm{Osc}^k M$ there exist local sprays.*

Proof. The manifold M being paracompact, by Theorem 6.1.1, it follows that the manifold E is paracompact. Let g be a Riemannian metric on M. In a next chapter we shall prove that there exists a prolongation of order k of the Riemannian space $R^n = (M, g)$. Let $z^{(1)i}, ..., z^{(k)i}$ the Liouville d–vector fields on E, of the mentioned prolongation and the function

$$L(x, y^{(1)}, ..., y^{(k)}) = g_{ij}(x) z^{(k)i} z^{(k)j}. \tag{3.6}$$

Considering the functions

$$(k+1)G^i = \frac{1}{2} g^{ij}(x) \left\{ \Gamma\left(\frac{\partial L}{\partial y^{(k)j}}\right) - \frac{\partial L}{\partial y^{(k-1)j}} \right\}$$

we can prove that the equality (3.4) holds. So, taking into account the Theorem 6.3.2, the statement is verified. The explicit form of the d–vectors $z^{(k)i}$ will be given later.

To end this section, let us consider a k–spray S with the coefficients G^i. We set

$$\underset{(1)}{N}{}^i{}_j = \frac{\partial G^i}{\partial y^{(k)j}} \tag{3.7}$$

and we have

Theorem 6.3.5 *The functions $\underset{(1)}{N}{}^i{}_j$ from (3.7) determined by a k–spray S, have the rule of transformation:*

$$\underset{(1)}{\widetilde{N}}{}^i{}_m \frac{\partial \tilde{x}^m}{\partial x^j} = \underset{(1)}{N}{}^m{}_j \frac{\partial \tilde{x}^i}{\partial x^m} - \frac{\partial \tilde{y}^{(1)i}}{\partial x^j}. \tag{3.8}$$

The proof is the same as that of Theorem 2.3.5.

The system of functions $\underset{(1)}{N}{}^i{}_j$ will be used in the concept of nonlinear connection.

6.4 Nonlinear Connections

The notion of nonlinear connection on the total space $E = \mathrm{Osc}^k M$ is perfectly similar with that described in the §4, Ch.2, for the case $k = 2$.

Thus, we have the exact sequence of vector bundles (4.1), Ch.2, and a nonlinear connection on E is a left splitting of this exact sequence. It follows that Theorem 2.4.1 can be repeated here.

Theorem 6.4.1 *There exists a nonlinear connection on the k–osculator bundle E if, and only if, there exists a vector subbundle NE of the tangent bundle TE so that the Whitney sum*

$$TE = NE \oplus VE \tag{4.1}$$

holds.

In (4.1) VE is the vertical subbundle. NE is called a horizontal subbundle.

The fibres of NE and VE on E will be denoted by N_u or $N(u)$ and V_u or $V(u)$, $u \in E$, respectively.

So, (4.1) implies:

$$T_uE = N_uE \oplus V_uE, \quad \forall u \in E. \tag{4.2}$$

We can assert:

Theorem 6.4.2 *A nonlinear connection on the manifold E is a distribution on E, $N : u \in E \longrightarrow N_u \subset T_uE$ which is supplementary to the vertical distribution $V : u \in E \longrightarrow V_u \subset T_uE$.*

N is called a *horizontal distribution.*

According to Theorem 6.4.2, the local dimension of a horizontal distribution N, which determines a nonlinear connection is the same with the dimension of the base manifold M, that is, n.

Exactly as in §4, Ch.2, we can establish:

Theorem 6.4.3 *If the base manifold M is paracompact, then on $E = \mathrm{Osc}^k M$ there exist nonlinear connections.*

Let N be a nonlinear connection and h and v the horizontal and vertical projectors, given by (4.1). We have

$$h + v = I,\ h^2 = h,\ v^2 = v,\ hv = vh = 0. \tag{4.3}$$

As usual, we denote

$$X^H = hX,\ X^V = vX,\ \forall X \in \mathcal{X}(E). \tag{4.4}$$

If follows

$$X = X^H + X^V,\ \forall X \in \mathcal{X}(E), \tag{4.5}$$

the components X^H and X^V being uniquely determined.

A *horizontal lift* is a $\mathcal{F}(M)$–linear map $\ell_h : \mathcal{X}(M) \longrightarrow \mathcal{X}(E)$ with the properties

$$v \circ \ell_h = 0,\ d\pi \circ \ell_h = Id. \tag{4.6}$$

Locally, for any vector field $X \in \mathcal{X}(M)$, it follows that $\ell_h X$ is uniquely determined.

We can prove (cf. Ch.2, §4):

Theorem 6.4.4

1° *There exists a unique local basis adapted to the horizontal distribution N, which is projected by $d\pi$ onto the natural basis $\left(\dfrac{\partial}{\partial x^i}\right)$, $(i = 1, .., n)$ of $\mathcal{X}(M)$. It is given by*

$$\frac{\delta}{\delta x^i} = \ell_h\left(\frac{\partial}{\partial x^i}\right),\quad (i = 1, .., n). \tag{4.7}$$

2° *The linearly independent vector fields $\dfrac{\delta}{\delta x^1}, ..., \dfrac{\delta}{\delta x^n}$ can be uniquely written in the form:*

$$\frac{\delta}{\delta x^i} = \frac{\partial}{\partial x^i} - \underset{(1)}{N}{}^j{}_i \frac{\partial}{\partial y^{(1)j}} - \cdots - \underset{(k)}{N}{}^j{}_i \frac{\partial}{\partial y^{(k)j}}. \tag{4.8}$$

3° *With respect to* (1.5), *the basis* $\left(\frac{\delta}{\delta x^i}\right)$ *is transformed as follows:*

$$\frac{\delta}{\delta x^i} = \frac{\partial \tilde{x}^j}{\delta x^i} \frac{\delta}{\delta \tilde{x}^j}. \tag{4.9}$$

The functions $\underset{(\alpha)}{N}{}^i{}_j(x, y^{(1)}, ..., y^{(k)})$, $(\alpha = 1, ..., k)$ are called the *coefficients of the nonlinear connection* N and $\left(\frac{\delta}{\delta x^i}\right)$, $(i = 1, ..., n)$ is named the *adapted basis* to the horizontal distribution N.

Theorem 6.4.5

1° *With respect to* (1.5), *the coefficients of the nonlinear connection* N *are transformed by the rule*

$$\begin{aligned}
\underset{(1)}{\tilde{N}}{}^i{}_m \frac{\partial \tilde{x}^m}{\partial x^j} &= \underset{(1)}{N}{}^m{}_j \frac{\partial \tilde{x}^i}{\partial x^m} - \frac{\partial \tilde{y}^{(1)i}}{\partial x^j} \\
\underset{(2)}{\tilde{N}}{}^i{}_m \frac{\partial \tilde{x}^m}{\partial x^j} &= \underset{(2)}{N}{}^m{}_j \frac{\partial \tilde{x}^i}{\partial x^m} + \underset{(1)}{N}{}^m{}_j \frac{\partial \tilde{y}^{(1)i}}{\partial x^m} - \frac{\partial \tilde{y}^{(2)i}}{\partial x^j} \\
&\vdots \\
\underset{(k)}{\tilde{N}}{}^i{}_m \frac{\partial \tilde{x}^m}{\partial x^j} &= \underset{(k)}{N}{}^m{}_j \frac{\partial \tilde{x}^i}{\partial x^m} + \underset{(k-1)}{N}{}^m{}_j \frac{\partial \tilde{y}^{(1)i}}{\partial x^m} + \cdots + \\
&\quad + \underset{(1)}{N}{}^m{}_j \frac{\partial \tilde{y}^{(k-1)i}}{\partial x^m} - \frac{\partial \tilde{y}^{(k)i}}{\partial x^j}.
\end{aligned} \tag{4.10}$$

2° *Conversely, if a set of functions* $\left(\underset{(1)}{N}{}^i{}_j, ..., \underset{(k)}{N}{}^i{}_j\right)$ *is given, on every domain of local chart on* E, *so that* (4.10) *holds, then there exists on* E *a unique nonlinear connection* N, *which has as coefficients just the given set of functions.*

6.5 J–Vertical Distributions

Let us consider again a nonlinear connection N. The k–tangent structure J, defined in (3.1), applies to the horizontal distribution N in a vertical distribution $N_1 \subset V_1$ of dimension n, supplementary to the subdistribution V_2. Then it applies to the distribution N_1 in a distribution $N_2 \subset V_2$, supplementary to the distribution V_3 and so on. Setting $N_0 = N,\ V_1 = V$, we can write

$$N_1 = J(N_0), N_2 = J^2(N_0), ..., N_{k-1} = J^{k-1}(N_0) \tag{5.1}$$

and obtain:

Theorem 6.5.1

1° *The following direct decomposition of linear spaces holds:*

$$T_u(E) = N_0(u) \oplus N_1(u) \oplus \cdots \oplus N_{k-1}(u) \oplus V_k(u),\ \ \forall u \in E. \tag{5.2}$$

2° *The adapted basis to the distributions $N_0, ..., N_{k-1}, V_k$ is given by:*

$$\frac{\delta}{\delta x^i}, \frac{\delta}{\delta y^{(1)i}}, ..., \frac{\delta}{\delta y^{(k-1)i}}, \frac{\delta}{\delta y^{(k)i}},\ (i = 1, ..., n), \tag{5.3}$$

where

$$\begin{aligned} &\frac{\delta}{\delta y^{(1)i}} = J\left(\frac{\delta}{\delta x^i}\right), \frac{\delta}{\delta y^{(2)i}} = J^2\left(\frac{\delta}{\delta x^i}\right), ..., \\ &\frac{\delta}{\delta y^{(k-1)i}} = J^{k-1}\left(\frac{\delta}{\delta x^i}\right), \frac{\delta}{\delta y^{(k)i}} = \frac{\partial}{\partial y^{(k)i}}. \end{aligned} \tag{5.3$'$}$$

3° *With respect to* (1.5), *we have:*

$$\frac{\delta}{\delta y^{(\alpha)i}} = \frac{\partial \tilde{x}^j}{\partial x^i} \frac{\delta}{\delta \tilde{y}^{(\alpha)j}},\ (\alpha = 0, 1, ...k; y^{(0)} = x). \tag{5.4}$$

The explicit form of basis from (5.3) is given by:

Proposition 6.5.1 *The adapted basis* (5.3) *is expressed by:*

$$\frac{\delta}{\delta x^i} = \frac{\partial}{\partial x^i} - \underset{(1)}{N}{}^j{}_i \frac{\partial}{\partial y^{(1)j}} - \underset{(2)}{N}{}^j{}_i \frac{\partial}{\partial y^{(2)j}} - \cdots - \underset{(k)}{N}{}^j{}_i \frac{\partial}{\partial y^{(k)j}}$$

$$\frac{\delta}{\delta y^{(1)i}} = \frac{\partial}{\partial y^{(1)i}} - \underset{(1)}{N}{}^j{}_i \frac{\partial}{\partial y^{(2)j}} - \underset{(2)}{N}{}^j{}_i \frac{\partial}{\partial y^{(3)j}} - \cdots - \underset{(k-1)}{N}{}^j{}_i \frac{\partial}{\partial y^{(k)j}}$$

(5.5)

....

$$\frac{\delta}{\delta y^{(k-1)i}} = \frac{\partial}{\partial y^{(k-1)i}} - \underset{(1)}{N}{}^j{}_i \frac{\partial}{\partial y^{(k)j}},$$

$$\frac{\delta}{\delta y^{(k)i}} = \frac{\partial}{\partial y^{(k)i}}.$$

Taking into account the equalities (4.2) and (5.2), we can write for the vertical distribution V :

$$V(u) = N_1(u) \oplus \cdots \oplus N_{k-1}(u) \oplus V_k(u),\ \forall u \in E. \tag{5.6}$$

The vertical distributions $N_0, ..., N_{k-1}$ will be called *J–vertical distributions*. It is obvious that the distributions $N_0, N_1, ..., N_{k-1}, V_k$ have the same local dimension n.

The direct decomposition (5.2) is important in the study of the geometry of total space of the k–osculator bundle. The main geometrical object fields on E will be reported to this direct decomposition.

Let $h, v_1, ..., v_k$ be the projectors determined by (5.2):

$$h + v_1 + \cdots + v_k = I,\ h^2 = h,\ v_\alpha v_\alpha = v_\alpha,\ hv_\alpha = v_\alpha h = 0,\quad (\alpha = 1, ..., k)$$

(5.7)

$$v_\alpha v_\beta = v_\beta v_\alpha = 0,\ (\alpha \neq \beta,\ \alpha, \beta = 1, ..., k).$$

Denoting $X^H = hX,\ X^{V_\alpha} = v_\alpha X\ (\alpha = 1, ..., k),\ \forall X \in \mathcal{X}(E)$, we get, uniquely, the decomposition:

$$X = X^H + X^{V1} + \cdots + V^{V_k},\ \forall X \in \mathcal{X}(E). \tag{5.8}$$

In the adapted basis (5.3) we can write:

$$X^H = X^{(0)i} \frac{\delta}{\delta x^i},\ X^{V_\alpha} = X^{(\alpha)i} \frac{\delta}{\delta y^{(\alpha)i}},\ (\alpha = 1, ..., k). \tag{5.8'}$$

With respect to (1.5), all the coordinates in (5.8)′ are transformed by the same rule:

$$\widetilde{X}^{(\alpha)i} = \frac{\partial \tilde{x}^i}{\partial x^j} X^{(\alpha)j} \quad (\alpha = 0, 1, ..., k). \tag{5.8}''$$

As an application of the foregoing theory, we have:

Theorem 6.5.2 *The following properties hold:*
1° *The distribution $N = N_0$ is integrable if, and only if, :*

$$[X^H, Y^H]^{V_\alpha} = 0, \ (\alpha = 1, ..., k), \ \forall X, Y \in \mathcal{X}(E).$$

2° *The distribution N_α, $(\alpha = 1, ..., k)$ is integrable if, and only if, :*

$$[X^{V_\alpha}, Y^{V_\alpha}]^H = [X^{V_\alpha}, Y^{V_\alpha}]^{V_\beta} = 0, \ (\beta \neq \alpha), \ \forall X, Y \in \mathcal{X}(E).$$

If the distribution N is integrable, we say that the nonlinear connection N is integrable.

6.6 The Dual Coeficients of a Nonlinear Connection N on $\mathrm{Osc}^k M$

The dual basis

$$\delta x^i, \delta y^{(1)i}, ..., \delta y^{(k)i} \tag{6.1}$$

of the adapted basis (5.3) is given by the system of equations:

$$\begin{array}{l} \dfrac{\delta}{\delta x^i} \rfloor \delta x^j = \delta^j{}_i, \dfrac{\delta}{\delta x^i} \rfloor \delta y^{(1)j} = \cdots = \dfrac{\delta}{\delta x^i} \rfloor \delta y^{(k)j} = 0, \\ \cdots\cdots\cdots\cdots\cdots\cdots\cdots\cdots\cdots\cdots\cdots\cdots \\ \dfrac{\delta}{\delta y^{(k)i}} \rfloor \delta x^j = \cdots = \dfrac{\delta}{\delta y^{(k)i}} \rfloor \delta y^{(k-1)j} = 0, \dfrac{\delta}{\delta y^{(k)i}} \rfloor \delta y^{(k)j} = \delta^j{}_i. \end{array} \tag{6.2..}$$

So, (6.1) is called *local cobasis adapted to decomposition* (5.2). A straightforward calculation leads to:

Theorem 6.6.1 *The dual basis* (6.1) *of the adapted basis* (5.3), (5.5) *is given by:*

$$
\begin{aligned}
&\delta x^i = dx^i,\\
&\delta y^{(1)i} = dy^{(1)i} + \underset{(1)}{M}{}^i{}_j dx^j,\\
(6.3)\qquad &\dots\dots\dots\dots\dots\dots\dots\dots\dots\dots\dots\\
&\delta y^{(k)i} = dy^{(k)i} + \underset{(1)}{M}{}^i{}_j dy^{(k-1)j} + \cdots +\\
&+ \underset{(k-1)}{M}{}^i{}_j dy^{(1)j} + \underset{(k)}{M}{}^i{}_j dx^j,
\end{aligned}
$$

where

$$
(6.4)\qquad
\begin{aligned}
&\underset{(1)}{M}{}^i{}_j = \underset{(1)}{N}{}^i{}_j,\ \underset{(2)}{M}{}^i{}_j = \underset{(2)}{N}{}^i{}_j + \underset{(1)}{N}{}^i{}_m \underset{(1)}{M}{}^m{}_j, \dots,\\
&\underset{(k)}{M}{}^i{}_j = \underset{(k)}{N}{}^i{}_j + \underset{(k-1)}{N}{}^i{}_m \underset{(1)}{M}{}^m{}_j + \cdots + \underset{(2)}{N}{}^i{}_m \underset{(k-2)}{M}{}^m{}_j + \underset{(1)}{N}{}^i{}_m \underset{(k-1)}{M}{}^m{}_j.
\end{aligned}
$$

It is clear that, given a nonlinear connection N, with the coefficients $\left(\underset{(1)}{N}{}^i{}_j, \dots, \underset{(k)}{N}{}^i{}_j\right)$, then the system of functions $\left(\underset{(1)}{M}{}^i{}_j, \dots, \underset{(k)}{M}{}^i{}_j\right)$, from (6.3), is uniquely determined by (6.4).

Conversely, we have:

$$
(6.5)\qquad
\begin{cases}
\underset{(1)}{N}{}^i{}_j = \underset{(1)}{M}{}^i{}_j,\ \underset{(2)}{N}{}^i{}_j = \underset{(2)}{M}{}^i{}_j - \underset{(1)}{M}{}^m{}_j \underset{(1)}{N}{}^i{}_m\\
\dots\dots\dots\dots\dots\dots\dots\dots\dots\dots\dots\dots\dots\\
\underset{(k)}{N}{}^i{}_j = \underset{(k)}{M}{}^i{}_j - \underset{(1)}{M}{}^m{}_j \underset{(k-1)}{N}{}^i{}_m - \cdots - \underset{(k-2)}{M}{}^m{}_j \underset{(2)}{N}{}^i{}_m -\\
\qquad\qquad - \underset{(k-1)}{M}{}^m{}_j \underset{(1)}{N}{}^i{}_m.
\end{cases}
$$

The system of functions $(\underset{(1)}{M}{}^i{}_j, \dots, \underset{(k)}{M}{}^i{}_j)$ is called the system of *dual coefficients* of the nonlinear connection N.

From (5.4) and (6.2) we deduce:

Proposition 6.6.1 *With respect to* (1.5), *we have:*

$$
\delta \tilde{x}^i = \frac{\partial \tilde{x}^i}{\partial x^j}\delta x^j,\ \delta \tilde{y}^{(\alpha)i} = \frac{\partial \tilde{x}^i}{\partial x^j}\delta y^{(\alpha)j},\ (\alpha = 1, \dots, k).
$$

Also, by a direct calculation, we obtain:

Theorem 6.6.2

1° *A transformation of coordinates* (1.5) *implies the transformation of dual coefficients:*

$$(6.6)\qquad \begin{cases} \underset{(1)}{M}{}^m{}_j \dfrac{\partial \tilde{x}^i}{\partial x^m} = \underset{(1)}{\tilde{M}}{}^i{}_m \dfrac{\partial \tilde{x}^m}{\partial x^j} + \dfrac{\partial \tilde{y}^{(1)i}}{\partial x^j} \\ \underset{(2)}{M}{}^m{}_j \dfrac{\partial \tilde{x}^i}{\partial x^m} = \underset{(2)}{\tilde{M}}{}^i{}_m \dfrac{\partial \tilde{x}^m}{\partial x^j} + \underset{(1)}{\tilde{M}}{}^i{}_m \dfrac{\partial \tilde{y}^{(1)m}}{\partial x^j} + \dfrac{\partial \tilde{y}^{(2)i}}{\partial x^j} \\ \cdots\cdots\cdots\cdots\cdots\cdots\cdots\cdots\cdots\cdots \\ \underset{(k)}{M}{}^m{}_j \dfrac{\partial \tilde{x}^i}{\partial x^m} = \underset{(k)}{\tilde{M}}{}^i{}_m \dfrac{\partial \tilde{x}^m}{\partial x^j} + \underset{(k-1)}{\tilde{M}}{}^i{}_m \dfrac{\partial \tilde{y}^{(1)m}}{\partial x^j} + \cdots + \\ \qquad\qquad + \underset{(1)}{\tilde{M}}{}^i{}_m \dfrac{\partial \tilde{y}^{(k-1)m}}{\partial x^j} + \dfrac{\partial \tilde{y}^{(k)i}}{\partial x^j}. \end{cases}$$

2° *If, on every domain of local chart on E, a set of functions* $(\underset{(1)}{M}{}^i{}_j, \dots, \underset{(k)}{M}{}^i{}_j)$ *is given, so that* (6.6) *holds, then there exists on E a unique nonlinear connection N, which has as dual coefficients just the given set of functions.*

A first application of the previous theory is the following one. For any function $f \in \mathcal{F}(E)$, we have:

$$(6.7)\qquad df = \frac{\delta f}{\delta x^i}\,\delta x^i + \frac{\delta f}{\delta y^{(1)i}}\,\delta y^{(1)i} + \cdots + \frac{\delta f}{\delta y^{(k)i}}\,\delta y^{(k)i}.$$

Indeed, from (5.5), we can express the natural basis by means of adapted basis in the form:

$$(6.8)\qquad \begin{aligned} \frac{\partial}{\partial x^i} &= \frac{\delta}{\delta x^i} + \underset{(1)}{M}{}^j{}_i \frac{\delta}{\delta y^{(1)j}} + \underset{(2)}{M}{}^j{}_i \frac{\delta}{\delta y^{(2)j}} + \cdots + \underset{(k)}{M}{}^j{}_i \frac{\delta}{\delta y^{(k)j}} \\ \frac{\partial}{\partial y^{(1)i}} &= \frac{\delta}{\delta y^{(1)i}} + \underset{(1)}{M}{}^j{}_i \frac{\delta}{\delta y^{(2)j}} + \cdots + \underset{(k-1)}{M}{}^j{}_i \frac{\delta}{\delta y^{(k)j}} \\ &\;\vdots \\ \frac{\partial}{\partial y^{(k)i}} &= \frac{\delta}{\delta y^{(k)i}}. \end{aligned}$$

Also, from (6.3), we can express the natural cobasis by means of adapted cobasis in the form

$$
(6.9)\qquad
\begin{aligned}
dx^i &= \delta x^i\\
dy^{(1)i} &= \delta y^{(1)i} - \underset{(1)}{N}{}^i{}_j\,\delta x^j\\
dy^{(2)i} &= \delta y^{(2)i} - \underset{(1)}{N}{}^i{}_j\,\delta y^{(1)j} - \underset{(2)}{N}{}^i{}_j\,\delta x^j,\\
&\vdots\\
dy^{(k)i} &= \delta y^{(k)i} - \underset{(1)}{N}{}^i{}_j\,\delta y^{(k-1)j} - \dots -\\
&- \underset{(k-1)}{N}{}^i{}_j\,\delta y^{(1)j} - \underset{(k)}{N}{}^i{}_j\,\delta x^j.
\end{aligned}
$$

Now, the proof of the equality (6.7) is immediate.

Another application of the dual coefficients of a nonlinear connection is given by

Theorem 6.6.3

1° *The Liouville vector fields* $\overset{1}{\Gamma},\dots,\overset{k}{\Gamma}$ *can be expressed in the adapted basis* (5.3) *in the form:*

$$
(6.10)\qquad
\begin{aligned}
\overset{1}{\Gamma} &= z^{(1)i}\frac{\delta}{\delta y^{(k)i}},\\
\overset{2}{\Gamma} &= z^{(1)i}\frac{\delta}{\delta y^{(k-1)i}} + 2z^{(2)i}\frac{\delta}{\delta y^{(k)i}},\\
&\dots\dots\dots\dots\dots\dots\dots\dots\dots\dots\\
\overset{k}{\Gamma} &= z^{(1)i}\frac{\delta}{\delta y^{(1)i}} + 2z^{(2)i}\frac{\delta}{\delta y^{(2)i}} + \dots + kz^{(k)i}\frac{\delta}{\delta y^{(k)i}},
\end{aligned}
$$

where

$$
(6.10)'\qquad
\begin{aligned}
z^{(1)i} &= y^{(1)i}\\
2z^{(2)i} &= 2y^{(2)i} + \underset{(1)}{M}{}^i{}_m\,y^{(1)i}\\
&\dots\dots\dots\dots\dots\dots\dots\dots\dots\dots\\
kz^{(k)i} &= ky^{(k)i} + (k-1)\underset{(1)}{M}{}^i{}_m\,y^{(k-1)i} + \dots + \underset{(k-1)}{M}{}^i{}_m\,y^{(1)i}.
\end{aligned}
$$

2° *With respect to* (1.5), *we have:*

$$
(6.10)''\qquad \tilde{z}^{(\alpha)i} = \frac{\partial \tilde{x}^i}{\partial x^j}\,z^{(\alpha)j}\quad (\alpha = 1,\dots,k).
$$

By virtue of $(6.8)''$ we see that $z^{(1)i}, ..., z^{(k)i}$ have a geometrical character. They will be called the *Liouville distinguished vector fields* (shortly, *Liouville d–vector fields*). These vector fields are important in the theory of higher order Lagrangians.

A field of 1–form $\omega \in \mathcal{X}^*(E)$ can be uniquely written as

$$\omega = \omega^H + \omega^{V_1} + \cdots + \omega^{V_k} \tag{6.11}$$

where

$$\omega^H = \omega \circ h, \omega^{V_1} = \omega \circ v_1, ..., \omega^{V_k} = \omega \circ v_k. \tag{6.11$'$}$$

In adapted cobasis (6.1), we get:

$$\omega^H = \omega^{(0)}{}_i \delta x^i, \omega^{V_\alpha} = \omega^{(\alpha)}{}_i \delta y^{(\alpha)i} \quad (\alpha = 1, ..., k). \tag{6.11$''$}$$

With respect to (1.5), we have:

$$\omega^{(\alpha)}{}_i = \frac{\partial \tilde{x}^m}{\partial x^i} \widetilde{\omega}^{(\alpha)}{}_m, \quad (\alpha = 0, ..., k). \tag{6.11$'''$}$$

In particular, if $f \in \mathcal{F}(E)$, then the 1–form df can be written in the form (6.11). Using (6.7), we determine its components:

$$(df)^H = \frac{\delta f}{\delta x^i} \delta x^i, \ (\delta f)^{V_\alpha} = \frac{\delta f}{\delta y^{(\alpha)i}} \delta y^{(\alpha)i}, (\alpha = 1, ..., k). \tag{6.12}$$

We apply these to determine some special curves related to a non-linear connection N.

Let (Γ, γ) be a parametrized curve in E, $\gamma : I \longrightarrow E$, $\Gamma = Im\ \gamma$. In a local chart $\pi^{-1}(U)$, γ can be represented in the form

$$x^i = x^i(t), \ y^{(\alpha)i} = y^{(\alpha)i}(t), \ t \in I, \quad (\alpha = 1, ..., k). \tag{6.13}$$

The tangent vector field $\frac{d\gamma}{dt}$ can be written according to (5.7):

$$\begin{aligned} \frac{d\gamma}{dt} &= \left(\frac{d\gamma}{dt}\right)^H + \left(\frac{d\gamma}{dt}\right)^{V_1} + \cdots + \left(\frac{d\gamma}{dt}\right)^{V_k} = \\ &= \frac{dx^i}{dt}\frac{\delta}{\delta x^i} + \frac{\delta y^{(1)i}}{dt}\frac{\delta}{\delta y^{(1)i}} + \cdots + \frac{\delta y^{(k)i}}{dt}\frac{\delta}{\delta y^{(k)i}}. \end{aligned} \tag{6.14}$$

Clearly:

$$(6.15)\qquad \begin{aligned} \frac{\delta y^{(1)i}}{dt} &= \frac{dy^{(1)i}}{dt} + \underset{(1)}{M}{}^{i}{}_{j}\frac{dx^{j}}{dt}, \dots, \\ \frac{\delta y^{(k)i}}{dt} &= \frac{dy^{(k)i}}{dt} + \underset{(1)}{M}{}^{i}{}_{j}\frac{dy^{(k-1)j}}{dt} + \dots + \underset{(k)}{M}{}^{i}{}_{j}\frac{dx^{j}}{dt}. \end{aligned}$$

Definition 6.6.1 A parametrized curve (Γ, γ) is horizontal if the vertical components of its tangent field vanish.

Theorem 6.6.4 *A parametrized curve* (6.13) *is horizontal if, and only if, the functions* $\{x^i(t), y^{(1)i}(t), ..., y^{(k)i}(t),\ t \in I\}$ *are solutions of the system of differential equations:*

$$(6.16)\qquad \frac{\delta y^{(1)i}}{dt} = \dots = \frac{\delta y^{(k)i}}{dt} = 0.$$

If the curve $x^i = x^i(t), t \in I$, in the manifold M is given, then (6.16) assures the local existence of the horizontal curves.

In the case when (Γ, γ) is the extension to $\mathrm{Osc}^k M$ of a curve $x^i = x^i(t),\ t \in I$, of the base manifold we introduce

Definition 6.6.2 A parametrized curve (C, c), $c : I \to M$ with the property that its extension to $\mathrm{Osc}^k M$ is horizontal is called an autoparallel curve of the nonlinear connectionN.

Theorem 6.6.5 *An autoparallel curve of the nonlinear connection* N *is characterized by the system of differential equations:*

$$(6.17)\qquad \begin{aligned} & y^{(1)i} = \frac{1}{1!}\frac{dx^i}{dt}, \dots, y^{(k)i} = \frac{1}{k!}\frac{d^k x^i}{dt^k} \\ & \frac{\delta y^{(1)i}}{dt} = 0, \dots, \frac{\delta y^{(k)i}}{dt} = 0. \end{aligned}$$

A theorem of existence and uniqueness for the autoparallel curves of a nonlinear connection N can be easily stated taking into account the previous theorem.

6.7 The Determination of a Nonlinear Connection from a k–Spray. The Structures $\mathbb{P}$ and $\mathbb{F}$

Applying the method used in Chapter 2, we can establish a fundamental result: Any k–spray determines a nonlinear connection. The proofs being adaptations of those given in Ch.2, we present here only the results.

We shall do likewise in the case of the almost product structure $\mathbb{P}$ and almost contact structure $\mathbb{F}$.

Let S be a k–spray

$$(7.1)\qquad \begin{aligned} S = y^{(1)i}\frac{\partial}{\partial x^i} + 2y^{(2)i}\frac{\partial}{\partial y^{(1)i}} + \cdots + ky^{(k)i}\frac{\partial}{\partial y^{(k-1)i}} - \\ -(k+1)G^i(x, y^{(1)}, ..., y^{(k)})\frac{\partial}{\partial y^{(k)i}}. \end{aligned}$$

A first fundamental result, proved by R.Miron and Gh. Atanasiu [197]–[201], is the following:

Theorem 6.7.1 *If the k-spray S, with the coefficients $G^i(x, y^{(1)}, ..., y^{(k)})$ is given, then the set of functions*

$$(7.2)\qquad \begin{aligned} &\underset{(1)}{M}{}^i{}_j = \frac{\partial G^i}{\partial y^{(k)j}} \\ &\underset{(2)}{M}{}^i{}_j = \frac{1}{2}\left(S\,\underset{(1)}{M}{}^i{}_j + \underset{(1)}{M}{}^i{}_m\,\underset{(1)}{M}{}^m{}_j\right) \\ &\ldots\ldots\ldots\ldots\ldots\ldots\ldots\ldots\ldots\ldots \\ &\underset{(k)}{M}{}^i{}_j = \frac{1}{k}\left(S\,\underset{(k-1)}{M}{}^i{}_j + \underset{(1)}{M}{}^i{}_m\,\underset{(k-1)}{M}{}^m{}_j\right) \end{aligned}$$

is the set of dual coefficients of a nonlinear connection N determined only by the k–spray S.

This theorem will be used in the geometry of the higher order Lagrange space.

Now, considering the $\mathcal{F}(E)$–mapping $I\!P:\mathcal{X}(E)\to\mathcal{X}(E)$ defined by

(7.3) $I\!P(X^H)=X^H,\ I\!P(X^{V_\alpha})=-X^{V_\alpha},\ (\alpha=1,...,k),\ \forall X\in\mathcal{X}(E)$

we have

$$I\!P\circ I\!P=I. \tag{7.4}$$

So $I\!P$ is an almost product structure on E, determined by the nonlinear connection N.

Proposition 6.7.1

a. *The almost product structure $I\!P$ can be expressed in the form:*

$$I\!P=I-2(v_1+\cdots+v_k),\quad I\!P=2h-I.$$

b. *rank* $\|I\!P\|=(k+1)n$.

The following theorems can be proved exactly as in §8, Ch.2.

Theorem 6.7.2 *A nonlinear connection N on the manifold E is characterized by the existence of an almost product structure $I\!P$ on E, whose eigensubspaces corresponding to the eigenvalue -1 coincide with the linear spaces of the vertical distribution V on E.*

Theorem 6.7.3 *The almost product structure $I\!P$ determined by the nonlinear connection N is integrable if, and only if, the nonlinear connection N is integrable.*

In §8, Ch.2, we showed that a nonlinear connection N determines a contact structure, too. In the following lines we extend those results to k–osculator bundle.

Let us consider the $\mathcal{F}(E)$–linear mapping $I\!F:\mathcal{X}(\widetilde{E})\longrightarrow\mathcal{X}(\widetilde{E})$ defined on the adapted basis (5.3) by:

$$\begin{aligned} &I\!F\left(\frac{\delta}{\delta x^i}\right)=-\frac{\partial}{\partial y^{(k)i}},\\ &I\!F\left(\frac{\delta}{\delta y^{(1)i}}\right)=\cdots=I\!F\left(\frac{\delta}{\delta y^{(k-1)i}}\right)=0,\\ &I\!F\left(\frac{\partial}{\partial y^{(k)i}}\right)=\frac{\delta}{\delta x^i}. \end{aligned}\tag{7.5}$$

Taking into account (5.4), we can see that $I\!F$ is globally defined on $\widetilde{E}$ if the nonlinear connection N has this property.

Also, we can prove:

Theorem 6.7.4 *The mapping $I\!F$ has the properties:*

1° *$I\!F$ is globally defined on $\widetilde{E}$.*

2° *$I\!F$ is a tensor field of type $(1,1)$ on $\widetilde{E}$.*

3° $\ker I\!F = N_1 \oplus \cdots \oplus N_{k-1}$, $Im\, I\!F = N_0 \oplus V_k$.

4° *$rank\, \|I\!F\| = 2n$.*

5° $I\!F^3 + I\!F = 0$.

Thus, $I\!F$ is an almost $(k-1)n$–contact structure on E determined by the nonlinear connection N.

Let $(\underset{1a}{\xi}, \underset{2a}{\xi}, ..., \underset{(k-1)a}{\xi})$, $(a = 1, ..., n)$, a local basis adapted to the direct decomposition $N_1 \oplus N_2 \oplus \cdots \oplus N_{k-1}$ and $\left(\overset{1a}{\eta}, ..., \overset{(k-1)a}{\eta}\right)$ its dual.

We show that the set

$$(7.5)' \qquad \left(I\!F, \underset{1a}{\xi}, ..., \underset{(k-1)a}{\xi}, \overset{1a}{\eta}, ..., \overset{(k-1)a}{\eta}\right), \ (a = 1, ..., n)$$

is a $(k-1)n$–contact structure.

Indeed, (7.5) imply:

$$(7.6) \qquad \begin{aligned} & I\!F(\underset{\alpha a}{\xi}) = 0, \overset{\alpha a}{\eta}(\underset{\beta b}{\xi}) = \begin{cases} \delta^a{}_b & \text{for } \alpha = \beta \\ 0 & \text{for } \alpha \neq \beta \end{cases} \quad (\alpha, \beta = 1, ..., k-1) \\ & I\!F^2(X) = -X + \sum_{a=1}^{n} \sum_{\alpha=1}^{k-1} \left[\overset{\alpha a}{\eta}(X) \underset{\alpha a}{\xi}\right], \forall X \in \mathcal{X}(E). \end{aligned}$$

Also, one deduces:

$$\overset{\alpha a}{\eta} \circ I\!\!F = 0. \tag{7.6$'$}$$

Let $N_{I\!\!F}$ be the Nijenhuis tensor of the structure $I\!\!F$,

$$N_{I\!\!F}(X,Y) = [I\!\!F X, I\!\!F Y] + I\!\!F^2[X,Y] - I\!\!F[I\!\!F X, Y] - I\!\!F[X, I\!\!F Y], \quad \forall X,Y \in \mathcal{X}(E).$$

The structure (7.5)$'$ is said to be normal if

$$N_{I\!\!F}(X,Y) + \sum_{a=1}^{n}\sum_{\alpha=1}^{k-1} d\overset{\alpha a}{\eta}(X,Y) = 0, \ \forall X,Y \in \mathcal{X}(E). \tag{7.7}$$

We have an extension of Theorem 2.8.4:

Theorem 6.7.5 *The almost* $(k-1)n$*–contact structure* (7.5)$'$ *is normal if, and only if,*

$$N_{I\!\!F}(X,Y) + \sum_{a=1}^{n}\sum_{\alpha=1}^{k-1} d\left(\delta y^{(\alpha)a}\right)(X,Y) = 0.$$

In a next section we shall use the $(k-1)n$–contact structure in the case when a Riemannian metric G on $\widetilde{E}$ is given. We shall obtain a geometrical model for a higher order Lagrange space.

6.8 Problems

1. For the 3–tangent structure J in (3.1), prove the properties expressed in Theorem 6.3.1.

2. Let $z^{(3)i}$ be the Liouville d–vector field of the space $\text{Prol}\,^3\mathcal{R}^n$, where $\mathcal{R}^n = (M, g_{ij}(x))$ is a Riemann space. Namely, $3z^{(3)i} =$ $= 3y^{(3)i} + 2\underset{(1)}{M^i}{}_j y^{(2)j} + \underset{(2)}{M^i}{}_j y^{(1)j}$, where

$$\underset{(1)}{M^i}{}_j = \gamma_j{}^i{}_m(x)y^{(1)m},\ \underset{(2)}{M^i}{}_j = \frac{1}{2}\{\Gamma\underset{(1)}{M^i}{}_j + \underset{(1)}{M^i}{}_m \underset{(1)}{M^m}{}_j\}.$$

(a) Show that $L = g_{ij}(x)z^{(3)i}z^{(3)j}$ is a scalar field on $E = \mathrm{Osc}^3 M$.

(b) Prove that

$$4G^i = \frac{1}{2}g^{ij}\left\{\Gamma\frac{\partial L}{\partial y^{(k)j}} - \frac{\partial L}{\partial y^{(k-1)j}}\right\}$$

are the coefficients of a 3–spray, L being from (a).

(c) For the previous 3–spray, determine the corresponding non-linear connection.

Chapter 7

Linear Connections on $\mathrm{Osc}^k M$

We continue the theory from the previous chapter, defining the notion of N-linear connection. In other words, we consider the linear connections which preserve by parallelism a nonlinear connection N and the k–tangent structure J. These kind of linear connections are compatible with the direct decomposition (1.1) and are very convenient for studying the geometry of total space of the k–osculator bundle $\mathrm{Osc}^k M$.

Following the line of the theory presented in Chapter 3, we give here the results concerning the main geometrical object fields on $\mathrm{Osc}^k M$ without proofs [197]–[202].

7.1 d–Tensors Algebra

Let N be a nonlinear connection on $E = \mathrm{Osc}^k M$ and the J–vertical distributions $N_1 = J(N_0), ..., N_{k-1} = J(N_{k-2})$, $N_0 = N$. Cf. §5, Ch.6, we have the direct decomposition

$$(1.1) \qquad T_u E = N_0(u) \oplus N_1(u) \oplus \cdots \oplus N_{k-1}(u) \oplus V_k(u), \ \forall u \in E.$$

A vector field $X \in \mathcal{X}(E)$ and a 1–form field $\omega \in \mathcal{X}^*(E)$ can be uniquely written, with respect to (1.1):

$$(1.2) \qquad \begin{cases} X = X^H + X^{V_1} + \cdots X^{V_k} \\ \omega = \omega^H + \omega^{V_1} + \cdots + \omega^{V_k} \end{cases}$$

Definition 7.1.1 A d–tensor field (d means distinguished) T on E of type (r,s) is a tensor field with the property

$$(1.3)\quad \begin{gathered} T(\overset{(1)}{\omega},...,\overset{(r)}{\omega},\underset{(1)}{X},...,\underset{(s)}{X}) = T(\overset{(1)}{\omega}{}^{H},...,\overset{(r)}{\omega}{}^{V_k},\underset{(1)}{X}{}^{H},...,\underset{(s)}{X}{}^{V_k}), \\ \forall \underset{(1)}{X},...,\underset{(s)}{X}\in \mathcal{X}(E),\ \forall \overset{(1)}{\omega},...,\overset{(r)}{\omega}\in \mathcal{X}^*(E). \end{gathered}$$

In the adapted basis (5.3) and adapted cobasis (6.3), a d–tensor field T has the components

$$(1.4)\quad T^{i_1...i_r}_{j_1...j_s}(x,y^{(1)},...,y^{(k)}) = T\left(dx^{i_1},...,\delta y^{(k)i_r},\frac{\delta}{\delta x^{j_1}},...,\frac{\delta}{\delta y^{(k)j_s}}\right).$$

Therefore, T can be expressed in the form:

$$(1.5)\quad T = T^{i_1...i_r}_{j_1...j_s}(x,y^{(1)},...,y^{(k)})\frac{\delta}{\delta x^{i_1}}\otimes ...\otimes \frac{\delta}{\delta y^{(k)i_r}}\otimes dx^{j_1}\otimes ...\otimes \delta y^{(k)j_s}.$$

A transformation of coordinate (1.5), Ch.6, on E, by virtue of (5.4), Ch.6, and Proposition 6.6.1, implies:

$$(1.6)\quad \widetilde{T}^{i_1...i_r}_{j_1...j_s} = \frac{\partial \tilde{x}^{i_1}}{\partial x^{h_1}}\cdots\frac{\partial \tilde{x}^{i_r}}{\partial x^{h_r}}\frac{\partial x^{k_1}}{\partial \tilde{x}^{j_1}}\cdots\frac{\partial x^{k_s}}{\partial \tilde{x}^{j_s}}T^{h_1...h_r}_{k_1...k_s}.$$

But this is possible only for the coordinates (1.4) of a d–tensor in adapted basis and in adapted cobasis. The coordinates of a tensor field on E in the natural basis and natural cobasis have complicated the rules of transformation. This is the reason we systematically use the d–tensor fields represented in adapted basis.

Taking into account (1.3) and (1.6), we can prove that the sum and tensor product of d–tensor fields lead to d–tensor fields.

So, $\left(1,\frac{\delta}{\delta x^i},\frac{\delta}{\delta y^{(1)i}},...,\frac{\partial}{\partial y^{(k)i}}\right)$ generate the tensor algebra of d–tensor fields over the ring $\mathcal{F}(E)$. The elements of this algebra give us a system of generators of the tensorial algebra of tensor fields on E.

For instance, the vector fields $X^H, X^{V_1},...,X^{V_k}$ are d–vector fields. The 1–form fields $\omega^H,\omega^{V_1},...,\omega^{V_k}$ are d–covector fields. Every component $\overset{k}{\Gamma}{}^{V_1},...,\overset{k}{\Gamma}{}^{V_k}$ of the Liouville vector field $\overset{k}{\Gamma}$ is a d–vector field.

So, $z^{(1)i}, ..., z^{(k)i}$ are Liouville d–vector fields (Th. 6.6.3).

Another example is given by a Riemannian metric G on E (on $\widetilde{E}$). We assume that we have:

$$(1.7)\qquad G(X^H, Y^{V_\alpha}) = G(X^{V_\alpha}, y^{V_\beta}) = 0,\ (\alpha, \beta = 1, ..., k;\ \alpha \neq \beta).$$

In this case, G can be uniquely written in the form

$$(1.8)\qquad G = G^H + G^{V_1} + \cdots + G^{V_k},$$

where

$$(1.8)'\qquad G^H(X, Y) = G(X^H, Y^H),\ G^{V_\alpha}(X, Y) = G(X^{V_\alpha}, Y^{V_\alpha}),\quad (\alpha = 1, ..., k).$$

Consequently, G^H, G^{V_α}, $(\alpha = 1, ..., k)$ are d–tensor fields. In adapted basis (5.3), (6.3), we have

$$(1.8)''\qquad g^{(\alpha)}{}_{ij} = G\left(\frac{\delta}{\delta y^{(\alpha)i}}, \frac{\delta}{\delta y^{(\alpha)j}}\right),\ (\alpha = 0, 1, ..., k;\ y^{(0)} = x).$$

Therefore, G takes the form

$$(1.9)\ G = g^{(0)}{}_{ij} dx^i \otimes dx^j + g^{(1)}{}_{ij} \delta y^{(1)i} \otimes \delta y^{(1)j} + \cdots + g^{(k)}{}_{ij} \delta y^{(k)i} \otimes \delta y^{(k)j}.$$

The following properties hold:

$$(1.10)\qquad g^{(\alpha)}{}_{ij} = g^{(\alpha)}{}_{ji},\ \mathrm{rank}\, \|g^{(\alpha)}{}_{ij}\| = n, (\alpha = 0, 1, ..., k).$$

With respect to (1.5), Ch.6, we get

$$(1.10)'\qquad \widetilde{g}^{(\alpha)}{}_{ij} = \frac{\partial x^h}{\partial \widetilde{x}^i} \frac{\partial x^k}{\partial \widetilde{x}^j}\, g^{(\alpha)}{}_{hk}.$$

Similar considerations can be made for the tensors of the almost product structure $I\!P$ or for $(k-1)n$–contact structure $I\!F$.

Definition 7.1.2 A covariant d–tensor field g_{ij} on $\widetilde{E}$, symmetric, rank $\|g_{ij}\| = n$, and with constant signature is called a *d–metric structure* on $\widetilde{E}$.

Definition 7.1.3 We call *Sasaki N–lift* of a d–metric structure g_{ij} on $\widetilde{E}$ the pseudo–Riemannian metric on E:

$$G = g_{ij}dx^i \otimes dx^j + g_{ij}\delta y^{(1)i} \otimes \delta y^{(1)j} + \cdots + g_{ij}\delta y^{(k)i} \otimes \delta y^{(k)j}. \tag{1.11}$$

We have

Theorem 7.1.1 *N being a nonlinear connection apriori given and g_{ij} a d–metric structure on $\widetilde{E}$, the set $(I\!F, \underset{1a}{\xi}, ..., \underset{(k-1)a}{\xi}, \overset{1a}{\eta}, ..., \overset{(k-1)a}{\eta}, G)$, where $I\!F$ is $(k-1)n$–contact structure determined by N and G is the Sasaki N-lift of g_{ij}, is a Riemannian $(k-1)n$–contact structure on $\widetilde{E}$.*

Proof. Using (7.5), (7.6) Ch.6, and (1.11) we can easily prove the next condition

$$G(I\!FX, Y) = -G(I\!FY, X), \ \forall X, Y \in \mathcal{X}(E).$$ **q.e.d.**

Definition 7.1.4 The Riemannian $(k-1)n$–contact structure given in the previous theorem determines on $\widetilde{E}$ the $(k-1)n$–contact model of d–metric structure g_{ij}.

Remark. We shall use this model in the geometry of higher order Lagrange spaces.

7.2 N–Linear Connection

In conformity with the theory from the section 2, Ch.3, we can repeat the definition of an N–linear connection in the case of the manifold $E = \mathrm{Osc}^k M$. It is a linear connection D on E with the properties

(1) D preserves by parallelism the horizontal distribution N.

(2) The k–tangent structure J is absolutely parallel with respect to D.

It follows that the Theorem 3.2.1 can be extended to E and then we obtain:

Theorem 7.2.1 *A linear connection D on the manifold E is an N-linear connection if, and only if, the following properties hold:*

$$(D_X Y^H)^{V_\alpha} = 0,\ (D_X Y^{V_\alpha})^H = 0,\ (D_X Y^{V_\alpha})^{V_\beta} = 0, \quad (\alpha \neq \beta,\ \alpha, \beta = 1, ..., k) \tag{2.1}$$

$$(2.1)' \quad D_X(JY^H) = J(D_X Y^H),\ D_X(JY^{V_\alpha}) = JD_X Y^{V_\alpha},\ (\alpha = 1, ..., k)$$

for any $X, Y \in \mathcal{X}(E)$.

Theorem 7.2.2 *For any N-linear connection D we have*

$$D_X h = 0,\ D_X v_\alpha = 0,\ (\alpha = 1, ...k) \tag{2.2}$$

$$D_X I\!\!P = 0. \tag{2.2$'$}$$

Since, $D_X Y$ is $\mathcal{F}(E)$-linear with respect to the argument X, we get

$$D_X Y = D_{X^H} Y + D_{X^{V_1}} Y + \cdots + D_{X^{V_\alpha}} Y.$$

Let us denote:

$$D_X^H = D_{X^H},\ D_X^{V_\alpha} = D_{X^{V_\alpha}},\ (\alpha = 1, ...k). \tag{2.3}$$

Consequently, we can write:

$$D_X Y = D_X^H Y + D_X^{V_1} Y + \cdots + D_X^{V_k} Y. \tag{2.4}$$

The operators $D_X^H, D_X^{V_\alpha}$ are not covariant derivations in the algebra of d-tensors fields, since $D_X^H f = X^H f \neq Xf$ etc. But they have similar properties with the covariant derivatives.

Theorem 7.2.3 *The operators $D_X^H, D_X^{V_\alpha}$ have the following properties:*

1° *The equalities (2.1) hold for d-vector fields $X = X^H$, $X = X^{V_\alpha}$ to which one can add: $D_X^H f = X^H f$, $D_X^{V_\alpha} f = X^{V_\alpha} f$.*

2° $D^H_X(fY) = X^H fY + fD^H_X Y,$

$D^{V_\alpha}_X(fY) = X^{V_\alpha} fY + fD^{V_\alpha}_X Y, \quad (\alpha = 1, ..., k)$

3° $(D^H_X Y)_{|U} = D^H_{X|U} Y_{|U}, (D^{V_\alpha}_X Y)_{|U} = D^{V_\alpha}_{X|U} Y_{|U} \ (\alpha = 1, ..., k)$ *for any open set* $U \subset E$.

4° $D^H_{X+Y} = D^H_X + D^H_Y, \ D^{V_\alpha}_{X+Y} = D^{V_\alpha}_X + D^{V_\alpha}_Y$

$D^H_{fX} = fD^H_X, D^{V_\alpha}_{fX} = fD^{V_\alpha}_X, \ (\alpha = 1, ..., k)$

5° $D^H_X(JY) = JD^H_X Y, \ D^{V_\alpha}_X(JY) = JD^{V_\alpha}_X Y, \ (\alpha = 1, ..., k)$

We can prove that the properties $1° - 5°$ are sufficient to determine an N–linear connection (by means of (2.4)).

Therefore, $D^H_X, D^{V_\alpha}_X$, $(\alpha = 1, ..., k)$ are called *h– and* v_α*–covariant derivatives*, respectively.

According to the Theorem 7.2.3, we extend the action of h– and v_α–covariant derivatives to any tensor field on E, or to any d–tensor field on E. So, (2.7) and (2.8), from Ch.3, hold for $\alpha = 1, ..., k$.

If T is a tensor field on E of type (r, s) we have

$$(D^H_X T)(\overset{(1)}{\omega}, ..., \overset{(r)}{\omega}, \underset{(1)}{X}, ..., \underset{(s)}{X}) = X^H T(\overset{(1)}{\omega}, ..., \overset{(r)}{\omega}, \underset{(1)}{X}, ..., \underset{(s)}{X}) -$$
$$- T(D^H_X \overset{(1)}{\omega}, ..., \overset{(r)}{\omega}, \underset{(1)}{X}, ..., \underset{(s)}{X}) - \cdots - T(\overset{(1)}{\omega}, ..., \overset{(r)}{\omega}, \underset{(1)}{X}, ..., D^H_X \underset{(s)}{X}),$$

(2.5)
$$(D^{V_\alpha} T)(\overset{(1)}{\omega}, ..., \overset{(r)}{\omega}, \underset{(1)}{X}, ..., \underset{(s)}{X}) = X^{V_\alpha} T(\overset{(1)}{\omega}, ..., \overset{(r)}{\omega}, \underset{(1)}{X}, ..., \underset{(s)}{X}) -$$
$$- T(D^{V_\alpha}_X \overset{(1)}{\omega}, ..., \overset{(r)}{\omega}, \underset{(1)}{X}, ..., \underset{(s)}{X}) - \cdots - T(\overset{(1)}{\omega}, ..., \overset{(r)}{\omega}, \underset{(1)}{X}, ..., D^{V_\alpha}_X \underset{(s)}{X}),$$
$$(\alpha = 1, ..., k)$$

In particular, if $\omega \in \mathcal{X}^*(E)$ we have

(2.6)
$$(D^H_X \omega)(Y) = X^H \omega(Y) - \omega(D^H_X Y);$$
$$(D^{V_\alpha} \omega)(Y) = X^{V_\alpha} \omega(Y) - \omega(D^{V_\alpha}_X Y), \ (\alpha = 1, ..., k).$$

The whole theory can be developed in the adapted basis (5.3) and in adapted cobasis (6.3), Ch.6.

7.3 N–Linear Connections in Adapted Basis

An N–linear connection D on E can be represented in the adapted basis (5.3), Ch.6, in the form

$$
(3.1)\quad
\begin{aligned}
&D_{\frac{\delta}{\delta x^j}}\frac{\delta}{\delta x^i}=L^m{}_{ij}\frac{\delta}{\delta x^m},\ D_{\frac{\delta}{\delta x^j}}\frac{\delta}{\delta y^{(\alpha)i}}=L^m{}_{ij}\frac{\delta}{\delta y^{(\alpha)m}},\ (\alpha=1,...,k)\\
&D_{\frac{\delta}{\delta y^{(\beta)j}}}\frac{\delta}{\delta x^i}=\underset{(\beta)}{C}{}^m{}_{ij}\frac{\delta}{\delta x^i},\ D_{\frac{\delta}{\delta y^{(\beta)j}}}\frac{\delta}{\delta y^{(\alpha)i}}=\underset{(\beta)}{C}{}^m{}_{ij}\frac{\delta}{\delta y^{(\alpha)m}},\\
&\qquad\qquad\qquad\qquad\qquad\qquad(\alpha,\beta=1,...,k).
\end{aligned}
$$

Indeed, the vector fields $D_{\frac{\delta}{\delta y^{(\beta)j}}}\frac{\delta}{\delta x^i}$ belong to the distribution N. So, they can be uniquely written, in the adapted basis as in the first equality (3.1). Similarly, by setting $D_{\frac{\delta}{\delta x^j}}\frac{\delta}{\delta y^{(\alpha)i}}=\overset{(\alpha)}{L}{}^m{}_{ij}\frac{\delta}{\delta y^{(\alpha)m}}$ (no sum by α) and using Theorem 7.2.3 we get $\overset{(\alpha)}{L}{}^m{}_{ij}=L^m{}_{ij}$.

Analogously for the other equalities (3.1).

For any N–linear connection D the set of functions

$$
(3.2)\quad L^i{}_{jh}(x,y^{(1)},...,y^{(k)}),\ \underset{(1)}{C}{}^i{}_{jh}(x,y^{(1)},...,y^{(k)}),...,\ \underset{(k)}{C}{}^i{}_{jh}(x,y^{(1)},...,y^{(k)})
$$

is well determined. These functions are called the coefficients of the N–linear connection D and are denoted by $D\Gamma(N)=(L^i_{jh},\underset{(\alpha)}{C}{}^i{}_{jh})$, $(\alpha=1,...,k)$.

It follows, without difficulties (see Theorem 3.3.2):

Theorem 7.3.1 *We have:*

1° *With respect to* (1.5), *Ch.6, the coefficients* (3.2) *of an N–linear are transformed as follows*

$$
(3.3)\quad \widetilde{L}^i{}_{rs}\frac{\partial\widetilde{x}^r}{\partial x^j}\frac{\partial\widetilde{x}^s}{\partial x^h}=L^r{}_{jh}\frac{\partial\widetilde{x}^i}{\partial x^r}-\frac{\partial^2\widetilde{x}^i}{\partial x^j\partial x^h},\ \underset{(\alpha)}{\widetilde{C}}{}^i{}_{rs}\frac{\partial\widetilde{x}^r}{\partial x^j}\frac{\partial\widetilde{x}^s}{\partial x^h}=\underset{(\alpha)}{C}{}^r{}_{jh}\frac{\partial\widetilde{x}^i}{\partial x^r},
$$

$(\alpha=1,...,k)$.

2° *If, on every domain on E, a system of functions* (3.2) *is given, so that* (3.3) *holds, then there exists a unique N–linear connection D on E which satisfies* (3.1).

Example 7.3.1. Let N be a nonlinear connection with the coefficients $\left(\underset{(1)}{N}{}^i{}_j, ..., \underset{(k)}{N}{}^i{}_j\right)$. Setting:

$$(3.4)\qquad B\Gamma(N) = B^i{}_{jh} = \left(\frac{\delta N^i{}_j}{\delta y^{(1)h}}, \underset{(\alpha)}{C}{}^i{}_{jh} = 0\right),\ (\alpha = 1, ..., k),$$

it follows:

Proposition 7.3.1 *$B\Gamma(N)$ from (3.4) determines an N–linear connection.*

This connection is called the *Berwald N-linear connection.* So, by means of the previous proposition, and of Theorem 3.3.3, it follows:

Theorem 7.3.2 *If the base manifold M is paracompact, then on $E = \mathrm{Osc}^k M$ there exist N–linear connections.*

Now, using the coefficients (3.2) of an N–linear connection D we can express in adapted basis the operators of h– and v^α–covariant derivatives. With this end in view, we consider a d–tensor field T locally represented in adapted basis by (1.5).

If $X = X^H = X^i(x, y^{(1)}, ..., y^{(k)})\dfrac{\delta}{\delta x^i}$, then the h–covariant derivation $D^H_X T$ can be calculated by means of (3.1), as follows:

$$(3.5)\quad D^H_X T = X^m T^{i_1...i_r}_{j_1...j_s|m} \frac{\delta}{\delta x^{i_1}} \otimes \cdots \otimes \frac{\delta}{\delta y^{(k)i}} \otimes dx^{j_1} \otimes \cdots \otimes \delta y^{(k)j_s},$$

where

$$(3.6)\qquad T^{i_1...i_r}_{j_1...j_s|m} = \frac{\delta T^{i_1...i_r}_{j_1...j_s}}{\delta x^m} + L^{i_1}{}_{hm} T^{hi_2...i_r}_{j_1...j_s} + \cdots - L^h{}_{j_r m} T^{i_1...i_r}_{j_1...h}.$$

The operator $|$ will be called the *h–covariant derivative*, identical to D^H_X. It has the same properties as in Proposition 1.3.4.

Let us consider the operators $D^{V_\alpha}_X$, for $X = X^{V_\alpha} = \overset{(\alpha)}{X}{}^i \dfrac{\delta}{\delta y^{(\alpha)i}}$, $(\alpha = 1, ..., k)$. Thus (2.5) give us:

$$(3.7)\quad D_X^{V_\alpha} T = \overset{(\alpha)}{X}{}^m T^{i_1\dots i_r}_{j_1\dots j_s} \overset{(\alpha)}{|}_m \frac{\delta}{\delta x^{i_1}} \otimes \dots \otimes \frac{\delta}{\delta y^{(k)i_r}} \otimes dx^{j_1} \otimes \dots \otimes \delta y^{(k)j_s},$$

where

$$(3.8)\quad T^{i_1\dots i_r}_{j_1\dots j_s} \overset{(\alpha)}{|}_m = \frac{\delta T^{i_1\dots i_r}_{j_1\dots j_s}}{\delta y^{(\alpha)m}} + \underset{(\alpha)}{C}{}^{i_1}{}_{hm} T^{hi_2\dots i_r}_{j_1\dots j_s} + \dots - \underset{(\alpha)}{C}{}^{h}{}_{j_s m} T^{i_1\dots i_r}_{j_1\dots h},$$
$$(\alpha = 1, \dots, k).$$

The operators $\overset{(\alpha)}{|}$, in number of k, are called *v^α-covariant derivatives*, with the same denomination as $D_X^{V_\alpha}$. Each of them has similar properties to those of h-covariant derivative. For instance:

$$(3.9)\quad \begin{aligned} &(fT^{\cdots}_{\cdots}) \overset{(\alpha)}{|}_m = \frac{\delta f}{\delta y^{(\alpha)m}} T^{\cdots}_{\cdots} + f T^{\cdots}_{\cdots} \overset{(\alpha)}{|}_m \\ &(\overset{1}{T}{}^{\cdots}_{\cdots} \otimes \overset{2}{T}{}^{\cdots}_{\cdots}) \overset{(\alpha)}{|}_m = \overset{1}{T}{}^{\cdots}_{\cdots} \overset{(\alpha)}{|}_m \otimes \overset{2}{T}{}^{\cdots}_{\cdots} + \overset{1}{T}{}^{\cdots}_{\cdots} \otimes \overset{2}{T}{}^{\cdots}_{\cdots} \overset{(\alpha)}{|}_m . \end{aligned}$$

The operators $|$ and $\overset{(\alpha)}{|}$ commute with the operation of contraction etc.

Example 7.3.2. Let $T^{i_1\dots i_r}_{j_1\dots j_s}$ be a d-tensor field and $B\Gamma(N)$ the Berwald connection (3.4). The h- and v^α-covariant derivatives with respect to $B\Gamma(N)$ are denoted by $\|$ and $\overset{(\alpha)}{\|}$, respectively. We get

$$(3.10)\quad \begin{aligned} &T^{i_1\dots i_r}_{j_1\dots j_s \| m} = \frac{\delta T^{i_1\dots i_r}_{j_1\dots j_s}}{\delta x^m} + B^{i_1}{}_{hm} T^{hi_2\dots i_r}_{j_1\dots j_s} + \dots - B^{h}{}_{j_s m} T^{i_1\dots i_r}_{j_1\dots h} \\ &T^{i_1\dots i_r}_{j_1\dots j_s} \overset{(\alpha)}{\|}_m = \frac{\delta T^{i_1\dots i_r}_{j_1\dots j_s}}{\delta y^{(\alpha)m}}, \ (\alpha = 1, \dots, k) \end{aligned}$$

The deflection tensors of the N-linear connection D with the coefficients (3.2) are given by the h- and v^α-covariant derivatives of the Liouville d-vector fields $z^{(1)i}, \dots, z^{(k)i}$, (§6, Ch.6). They are given by

$$(3.11)\quad \overset{(\alpha)}{D}{}^i{}_j = z^{(\alpha)i}{}_{|j}, \quad \overset{(\beta\alpha)}{d}{}^i{}_j = z^{(\beta)i} \overset{(\alpha)}{|}_j \ (\alpha, \beta = 1, \dots, k).$$

Proposition 7.3.2 *The deflection tensors of the N–linear connection D with the coefficients* (3.2) *can be given in the form:*

$$\overset{(\alpha)}{D}{}^{i}{}_{j} = \frac{\delta z^{(\alpha)i}}{\delta x^{j}} + L^{i}{}_{hj} z^{(\alpha)h}, \quad \overset{(\beta\alpha)}{d}{}^{i}{}_{j} = \frac{\delta z^{(\beta)i}}{\delta y^{(\alpha)j}} + \underset{(\alpha)}{C}{}^{i}{}_{hj} z^{(\beta)h}, \tag{3.12}$$
$$(\alpha, \beta = 1, ..., k).$$

As an application, another important property of the N–linear connection D is obtained.

Theorem 7.3.3 *Any N–linear connection D is compatible with the* $(k-1)n$*–contact structure* $I\!\!F$, *i.e.*

$$D_X I\!\!F = 0, \ \forall X \in \mathcal{X}(E). \tag{3.13}$$

The proof is the same as that of Theorem 3.3.4.

7.4 d–Tensors of Torsion

The theory of the N–linear connections expounded in the previous section can be continued by studying the torsion and curvature of such a connection reported to the direct decomposition (1.1).

Let $I\!\!T$ be the tensor of torsion of an N–linear connection D:

$$I\!\!T(X, Y) = D_X Y - D_Y X - [X, Y], \ \forall X, Y \in \mathcal{X}(E). \tag{4.1}$$

If the vector fields X and Y are written in the form (1.2), we obtain from (4.1) the following vector fields:

$$I\!\!T(X^H, Y^H), I\!\!T(X^H, Y^{V_\alpha}), \ I\!\!T(X^\alpha, Y^\beta), \ (\alpha, \beta = 1, ..., k).$$

Proposition 7.4.1 *The tensor of torsion* $I\!\!T$ *of the N–linear connection D is well determined by the following components, which are d–*

tensor fields of type $(1, 2)$:

$$
\begin{aligned}
&\mathbb{T}(X^H, Y^H) = h\mathbb{T}(X^H, Y^H) + \sum_{\alpha=1}^{k} v_\alpha \mathbb{T}(X^H, Y^H),\\
(4.2)\qquad &\mathbb{T}(X^H, Y^{v_\beta}) = h\mathbb{T}(X^H, Y^{v_\beta}) + \sum_{\alpha=1}^{k} v_\alpha \mathbb{T}(X^H, Y^{v_\beta}),\\
&\mathbb{T}(X^{v_\alpha}, Y^{v_\beta}) = \sum_{\gamma=1}^{k} v_\gamma \mathbb{T}(X^{v_\alpha}, Y^{v_\beta}),\ (\alpha, \beta = 1, ...k).
\end{aligned}
$$

Indeed, the first and second equalities are evident. For the third it is sufficient to prove that $h\mathbb{T}(X^{v_\alpha}, Y^{v_\beta}) = 0$. Taking into account (2.1), (4.1) and Theorem 7.2.3, this property holds. The d–tensors from the right side of (4.2) are called the *d-tensors of torsion* of the N–linear connection D. In order to express these d–tensors it is necessary to study the Lie brackets of the vector fields of the adapted basis (5.3), Ch.6.

Theorem 7.4.1 *The Lie brackets of the vector fields* $\left\{\frac{\delta}{\delta x^i}, \frac{\delta}{\delta y^{\alpha)i}}\right\}$ $(\alpha = 1, ..., k)$ *are given by*

$$
\begin{aligned}
&\left[\frac{\delta}{\delta x^j}, \frac{\delta}{\delta x^h}\right] = \underset{(01)}{R}{}^{i}{}_{jh}\frac{\delta}{\delta y^{(1)i}} + \cdots + \underset{(0k)}{R}{}^{i}{}_{jh}\frac{\delta}{\delta y^{(k)i}}\\
(4.3)\qquad &\left[\frac{\delta}{\delta x^j}, \frac{\delta}{\delta y^{(\alpha)h}}\right] = \underset{(\alpha 1)}{B}{}^{i}{}_{jh}\frac{\delta}{\delta y^{(1)i}} + \cdots + \underset{(\alpha k)}{B}{}^{i}{}_{jh}\frac{\delta}{\delta y^{(k)i}}\\
&\left[\frac{\delta}{\delta y^{(\alpha)j}}, \frac{\delta}{\delta y^{(\beta)h}}\right] = \underset{(\alpha\beta)}{\overset{(1)}{C}}{}^{i}{}_{jh}\frac{\delta}{\delta y^{(1)i}} + \cdots + \underset{(\alpha\beta)}{\overset{(k)}{C}}{}^{i}{}_{jh}\frac{\delta}{\delta y^{(k)i}}
\end{aligned}
$$

where the coefficients from the right hands have similar expressions with those from Theorem 3.4.3. *For instance, we have*

$$
(4.3)'\qquad \underset{(\alpha\alpha)}{\overset{(\alpha)}{C}} = 0,\ (\alpha = 1, ..., k)
$$

It is easy to prove that $\underset{(01)}{R}{}^{i}{}_{jh}, ..., \underset{(0k)}{R}{}^{i}{}_{jh}$ are d–tensor fields of type $(1, 2)$. So, from (4.3), we can establish

Theorem 7.4.2 *The nonlinear connection N is integrable if, and only if, the following equations hold:*

$$\underset{(01)}{R}{}^{i}{}_{jh} = \cdots = \underset{(0k)}{R}{}^{i}{}_{jh} = 0.$$

Now, we can calculate the components of d–tensors of torsion, written in the adapted basis. We get, from (4.2)

$$\begin{aligned}
&h\mathbb{T}\left(\frac{\delta}{\delta x^h},\frac{\delta}{\delta x^j}\right)=\underset{(00)}{T}{}^{i}{}_{jh}\frac{\delta}{\delta x^i};\; v_\alpha\mathbb{T}\left(\frac{\delta}{\delta x^h},\frac{\delta}{\delta x^j}\right)=\underset{(0\alpha)}{T}{}^{i}{}_{jh}\frac{\delta}{\delta y^{(\alpha)i}};\\
&h\mathbb{T}\left(\frac{\delta}{\delta x^h},\frac{\delta}{\delta y^{(\beta)j}}\right)=\underset{(\beta 0)}{T}{}^{i}{}_{jh}\frac{\delta}{\delta x^i};\; v_\alpha\mathbb{T}\left(\frac{\delta}{\delta x^h},\frac{\delta}{\delta y^{(\beta)j}}\right)=\underset{(\beta\alpha)}{T}{}^{i}{}_{jh}\frac{\delta}{\delta y^{(\alpha)i}};\\
&h\mathbb{T}\left(\frac{\delta}{\delta y^{(\alpha)h}},\frac{\delta}{\delta y^{(\beta)j}}\right)=0;\; v_\gamma\mathbb{T}\left(\frac{\delta}{\delta y^{(\alpha)h}},\frac{\delta}{\delta y^{(\beta)j}}\right)=\underset{(\beta\alpha)}{\overset{\gamma}{T}}{}^{i}{}_{jh}\frac{\delta}{\delta y^{(\gamma)i}};\\
&\qquad\qquad\qquad\qquad\qquad\qquad\qquad\qquad\qquad\qquad(\alpha,\beta,\gamma=1,...,k).
\end{aligned}\tag{4.4}$$

The d–tensors of torsion, from (4.4), can be easily calculated using (4.1), (4.2) and (4.3). In the case $k = 2$, the d–tensors of torsion are given in Theorem 3.4.5.

For instance, in (4.4), we have:

$$\underset{(00)}{T}{}^{i}{}_{jh} = L^i{}_{jh} - L^i{}_{hj},\quad \underset{(\alpha\alpha)}{\overset{\alpha}{T}}{}^{i}{}_{jh} = \underset{(\alpha)}{C}{}^{i}{}_{jh} - \underset{(\alpha)}{C}{}^{i}{}_{hj},\ (\alpha = 1,...,k).\tag{4.5}$$

Indeed, from (4.4), (4.3) and (4.3)$'$

$$\underset{(00)}{T}{}^{i}{}_{jh}\frac{\delta}{\delta x^i} = h\left\{D_{\frac{\delta}{\delta x^h}}\frac{\delta}{\delta x^j} - D_{\frac{\delta}{\delta x^j}}\frac{\delta}{\delta x^h} - \sum_{\alpha=1}^{k}\underset{(\alpha\alpha)}{R}{}^{i}{}_{jh}\frac{\delta}{\delta y^{(\alpha)i}}\right\} = L^i{}_{jh} - L^i{}_{hj}.$$

Similarly, we prove the last part of (4.5),

$$\begin{aligned}
\underset{(\alpha\alpha)}{\overset{\alpha}{T}}{}^{i}{}_{jh}\frac{\delta}{\delta y^{(\alpha)i}} &= v_\alpha\left\{D_{\frac{\delta}{\delta y^{(\alpha)h}}}\frac{\delta}{\delta y^{(\alpha)j}} - D_{\frac{\delta}{\delta y^{(\alpha)j}}}\frac{\delta}{\delta y^{(\alpha)h}} - \underset{(\alpha\alpha)}{\overset{\alpha}{C}}{}^{i}{}_{jh}\frac{\delta}{\delta y^{(\alpha)i}}\right\} =\\
&= \underset{(\alpha)}{C}{}^{i}{}_{jh} - \underset{(\alpha)}{C}{}^{i}{}_{hj}.
\end{aligned}$$

We denote:

$$\underset{(0)}{T}{}^{i}{}_{jh} = \underset{(00)}{T}{}^{i}{}_{jh},\quad \underset{(\alpha)}{S}{}^{i}{}_{jh} = \underset{(\alpha\alpha)}{\overset{\alpha}{T}}{}^{i}{}_{jh}\tag{4.6}$$

and we have

$$\underset{(0)}{T}{}^i{}_{jh} = L^i{}_{jh} - L^i{}_{hj},\ \underset{(\alpha)}{S}{}^i{}_{jh} = \underset{(\alpha)}{C}{}^i{}_{jh} - \underset{(\alpha)}{C}{}^i{}_{hj}\ (\alpha = 1, ..., k).$$

They will be called h–torsion tensor and v_α–torsion tensor, respectively.

7.5 d–Tensors of Curvature

The curvature tensor $I\!R$ of the N–linear connection D is given by

(5.1) $I\!R(X,Y)Z = (D_X D_Y - D_Y D_X)Z - D_{[X,Y]}Z,\ \forall X, Y, Z \in \mathcal{X}(E).$

Theorem 7.2.3 and (2.1)$'$ imply

Proposition 7.5.1 *For any N–linear connection D, the curvature tensor $I\!R$ satisfies the following identities*

(5.2) $J[R(X,Y)Z] = R(X,Y)(JZ), ..., J^k[R(X,Y)Z] = I\!R(X,Y)(J^k Z),$

where $J^\alpha = J \circ \cdots \circ J$ (α factors).

Proposition 7.5.2

1° *For any vector fields $X, Y, Z \in \mathcal{X}(E)$ with $Z^{V_\alpha} = J^\alpha Z^H$, we have*

(5.3) $$I\!R(X,Y)Z^{V_\alpha} = J^\alpha(I\!R(X,Y)Z^H),\ (\alpha = 1, ..., k).$$

2° *The essential components of the curvature tensor field $I\!R$ are $I\!R(X,Y)Z^H$.*

3° *The vector field $I\!R(X,Y)Z^H$ is a horizontal one.*

4° *The following properties hold:*

(5.3) $$\begin{aligned} &v_\beta[I\!R(X,Y)Z^H] = 0,\ h[I\!R(X,Y)Z^\beta] = 0 \\ &v_\alpha[I\!R(X,Y)Z^{v_\beta}] = 0,\ (\alpha \neq \beta),\ (\alpha, \beta = 1, ..., k). \end{aligned}$$

Thus, the curvature tensor $I\!R$ of the N–linear connection gives rise to the d–vector fields

(5.4) $$\begin{gathered} R(X^H, Y^H)Z^H, I\!R(X^{V_\alpha}, Y^H)Z^H, I\!R(X^{V_\beta}, Y^{V_\alpha})Z^H, \\ (\alpha, \beta = 1, ..., k;\ \beta \leq \alpha) \end{gathered}$$

and to those obtained from these applying the operator J of the k–tangent structure:

$$(5.4)' \quad \begin{aligned} &J^{\gamma}\{[I\!R(X^H, Y^H]Z^H\} = I\!R(X^H, Y^H)Z^{V_\gamma};\\ &J^{\gamma}\{[I\!R(X^{V_\alpha}, Y^H]Z^H\} = I\!R(X^{V_\alpha}, Y^H)Z^{V_\gamma};\\ &J^{\gamma}\{I\!R(X^{V_\beta}, Y^{V_\alpha}\}Z^H = I\!R(X^{V_\beta}, Y^{V_\alpha}\}Z^{V_\gamma},\\ &(\alpha, \beta, \gamma = 1, ..., k;\ \beta \leq \alpha). \end{aligned}$$

The d–tensors fields (5.4), (5.4)$'$ are called the *d–tensors of curvature* of the N–linear connection D.

We have also the following extension of Theorem 3.5.1:

Theorem 7.5.1 *The following properties* 1° *and* 2° *hold.*

1° *The d–tensors of curvature* (5.4) *have the expressions*

$$(5.5) \quad \begin{aligned} I\!R(X^H, Y^H)Z^H &= [D^H_X, D^H_Y]Z^H - D^H_{[X^H, Y^H]}Z^H -\\ &- \sum_{\gamma=1}^{k} D^{V_\gamma}_{[X^H, Y^H]}Z^H,\\ I\!R(X^{V_\alpha}, Y^H)Z^H &= [D^{V_\alpha}_X, D^H_Y]Z^H - D^H_{[X^{V_\alpha}, Y^H]}Z^H -\\ &- \sum_{\gamma=1}^{k} D^{V_\gamma}_{[X^{V_\alpha}, Y^H]}Z^H,\\ I\!R(X^{V_\beta}, Y^{V_\alpha})Z^H &= [D^{V_\beta}_X, D^{V_\alpha}_Y]Z^H - \sum_{\gamma=1}^{k} D^{V_\gamma}_{[X^{V_\beta}, Y^{V_\alpha}]}Z^H,\\ &\qquad (\alpha, \beta = 1, ..., k;\ \beta \leq \alpha). \end{aligned}$$

2° *The d–tensor field* (5.4)$'$ *are obtained from the previous d–tensors applying the operators* $J, J^2, ..., J^k$ *and setting* $J^\gamma Z^H = Z^{V_\gamma}$, $(\gamma = 1, ..., k)$.

Using the previous results we can express the Bianchi identities (5.7), Ch.3, for an N–linear connection D by means of the d–tensors of curvature and torsion.

In the applications, it is suitable to consider the equalities (5.5) as Ricci identities.

Theorem 7.5.2 *Any N-linear connection D satisfies the following Ricci identities:*

$$
(5.6)\quad
\begin{aligned}
&D^H_X D^H_Y Z^H - D^H_Y D^H_X Z^H = I\!R(X^H, Y^H)Z^H + D^H_{[X^H,Y^H]} Z^H + \\
&\qquad\qquad + \sum_{\gamma=1}^{k} D^{V_\gamma}_{[X^H,Y^H]} Z^H, \\
&D^{V_\alpha}_X D^H_Y Z^H - D^H_Y D^{V_\alpha}_X Z^H = I\!R(X^{V_\alpha}, Y^H)Z^H + D^H_{[X^{V_\alpha},Y^H]} Z^H + \\
&\qquad\qquad + \sum_{\gamma=1}^{k} D^{V_\gamma}_{[X^{V_\alpha},Y^H]} Z^H, \\
&D^{V_\beta}_X D^{V_\alpha}_Y Z^H - D^{V_\alpha}_Y D^{V_\beta}_X Z^H = I\!R(X^{V_\beta}, Y^{V_\alpha})Z^H + \sum_{\gamma=1}^{k} D^{V_\gamma}_{[X^{V_\beta},Y^{V_\alpha}]} Z^H,
\end{aligned}
$$

as well as the identities obtained from previously applying the mapping $J^\gamma = J \circ \cdots \circ J$ *(*γ*–terms).*

7.6 The d–Tensors of Curvature in Adapted Basis

The local expressions of the d–tensors of curvature of an N–linear connection D on $E = \mathrm{Osc}^k M$ can be obtained starting from Theorem 7.5.1.

We consider D in the local adapted basis $\left(\dfrac{\delta}{\delta x^i}, \dfrac{\delta}{\delta y^{(\alpha)i}}, \ \alpha = 1, ..., k\right)$ and the d–tensors (5.4) in this basis in the known form (see, Ch.3):

$$
(6.1)\quad
\begin{aligned}
&I\!R\left(\frac{\delta}{\delta x^m}, \frac{\delta}{\delta x^j}\right)\frac{\delta}{\delta x^h} = R_h{}^i{}_{jm}\frac{\delta}{\delta x^i}, \\
&I\!R\left(\frac{\delta}{\delta y^{(\alpha)m}}, \frac{\delta}{\delta x^j}\right)\frac{\delta}{\delta x^h} = \underset{(\alpha)}{P}{}_h{}^i{}_{jm}\frac{\delta}{\delta x^i} \\
&I\!R\left(\frac{\delta}{\delta y^{(\beta)m}}, \frac{\delta}{\delta y^{(\alpha)m}}\right)\frac{\delta}{\delta x^h} = \underset{(\beta\alpha)}{S}{}_h{}^i{}_{jm}\frac{\delta}{\delta x^i} \\
&\qquad (\alpha, \beta = 1, ..., k;\ \beta \le \alpha).
\end{aligned}
$$

Note that the others d–tensors fields $(5.4)'$ have the same coefficients

$$(6.2)\qquad R_{h}{}^{i}{}_{jm},\ \underset{(\alpha)}{P}{}_{h}{}^{i}{}_{jm},\ \underset{(\beta\alpha)}{S}{}_{h}{}^{i}{}_{jm}\ (\alpha,\beta=1,...,k;\ \beta\leq\alpha).$$

So, we can conclude that the d–tensor fields (6.2) characterize the d–tensors of curvature of an N–linear connection D. Therefore, (6.2) will be called the d–tensors of curvature.

Now, using Theorem 7.5.1 and formulae (3.1), (4.3), by a straightforward calculation we get:

Theorem 7.6.1 *For any N-linear connection D with the coefficients $D\Gamma(N)=(L^{i}{}_{jk},\underset{(\alpha)}{C}{}^{i}{}_{jk},(\alpha=1,...,k))$ the d-tensors of curvature* (6.2) *have the expressions:*

$$(6.3)\qquad \begin{aligned}
R_{h}{}^{i}{}_{jm} &= \frac{\delta L^{i}{}_{hj}}{\delta x^{m}}-\frac{\delta L^{i}{}_{jm}}{\delta x^{j}}+L^{p}{}_{hj}L^{i}{}_{pm}-L^{p}{}_{hm}L^{i}{}_{pj}+\sum_{\gamma=1}^{k}\underset{(\gamma)}{C}{}^{i}{}_{hp}\underset{(0\gamma)}{R}{}^{p}{}_{jm}\\
\underset{(\alpha)}{P}{}_{h}{}^{i}{}_{jm} &= \frac{\delta L^{i}{}_{hj}}{\delta y^{(\alpha)m}}-\frac{\delta \underset{(\alpha)}{C}{}^{i}{}_{hm}}{\delta x^{j}}+L^{p}{}_{hj}\underset{(\alpha)}{C}{}^{i}{}_{pm}-\underset{(\alpha)}{C}{}^{p}{}_{hm}L^{i}{}_{pj}+\\
&\qquad +\sum_{\gamma=1}^{k}\underset{(\gamma)}{C}{}^{i}{}_{hp}\underset{(\alpha\gamma)}{B}{}^{p}{}_{jm}\\
\underset{(\beta\alpha)}{S}{}_{h}{}^{i}{}_{jm} &= \frac{\delta \underset{(\alpha)}{C}{}^{i}{}_{hj}}{\delta y^{(\beta)m}}-\frac{\delta \underset{(\beta)}{C}{}^{i}{}_{hm}}{\delta y^{(\alpha)j}}+\underset{(\alpha)}{C}{}^{p}{}_{hj}\underset{(\beta)}{C}{}^{i}{}_{pm}-\underset{(\beta)}{C}{}^{p}{}_{hm}\underset{(\alpha)}{C}{}^{i}{}_{pj}+\\
&\qquad +\sum_{\gamma=1}^{k}\underset{(\gamma)}{C}{}^{i}{}_{hp}\underset{(\alpha\beta)}{\overset{\gamma}{C}}{}^{p}{}_{jm}.
\end{aligned}$$

For $k=2$, these formulae have been written in Theorem 3.5.3. In view of applications, it is useful to determine the Ricci identities in the adapted basis.

So, using Theorem 7.5.2 we can formulate:

Theorem 7.6.2 *For any N-linear connection D, with the coefficients $D\Gamma(N)=(L^{i}{}_{jk},\underset{(\alpha)}{C}{}^{i}{}_{jk},(\alpha=1,...,k))$ the following Ricci identities:*

$$
\begin{aligned}
&X^i{}_{|j|h} - X^i{}_{|h|j} = X^m R_m{}^i{}_{jh} - \underset{(0)}{T}{}^m{}_{jh} X^i{}_{|m} - \sum_{\gamma=1}^{k} \underset{(0\gamma)}{R}{}^m{}_{jh} X^i \overset{(\gamma)}{|}_m,\\
&X^i{}_{|j} \overset{(\alpha)}{|}_h - X^i \overset{(\alpha)}{|}_{h|j} = X^m \underset{(\alpha)}{P}{}_m{}^i{}_{jh} - \underset{(\alpha)}{C}{}^m{}_{jh} X^i{}_{|m} +\\
&\qquad\qquad + L^m{}_{hj} X^i \overset{(\alpha)}{|}_m - \sum_{\gamma=1}^{k} \underset{(\alpha\gamma)}{B}{}^m{}_{jh} X^i \overset{(\gamma)}{|}_m,\\
&X^i \overset{(\alpha)}{|}_j \overset{(\beta)}{|}_h - X^i \overset{(\beta)}{|}_h \overset{(\alpha)}{|}_j = X^m \underset{(\beta\alpha)}{S}{}_m{}^i{}_{jh} - \underset{(\beta)}{C}{}^m{}_{jh} X^i \overset{(\alpha)}{|}_m +\\
&\qquad\qquad + \underset{(\alpha)}{C}{}^m{}_{hj} X^i \overset{(\beta)}{|}_m - \sum_{\gamma=1}^{k} \underset{(\alpha\beta)}{\overset{(\gamma)}{C}}{}^m{}_{jh} X^i \overset{(\gamma)}{|}_m,\\
&\qquad\qquad\qquad (\alpha, \beta = 1, ..., k;\ \beta \leq \alpha)
\end{aligned}
\tag{6.4}
$$

hold for any d–vector fields X^i.

Proof. We can establish the previous identities using (5.6) or directly using the h– and v_α–covariant derivatives

$$
X^i{}_{|j} = \frac{\delta X^i}{\delta x^j} + X^m L^i{}_{mj},\quad X^i \overset{(\alpha)}{|}_j = \frac{\delta X^i}{\delta y^{(\alpha)j}} + X^m \underset{(\alpha)}{C}{}^i{}_{mj}
$$

and

$$
\begin{aligned}
X^i{}_{|j|h} &= \frac{\delta X^i{}_{|j}}{\delta x^h} + X^m{}_{|j} L^i{}_{mh} - X^i{}_{|m} L^m{}_{jh}\\
X^i{}_{|j} \overset{(\alpha)}{|}_h &= \frac{\delta X^i{}_{|j}}{\delta y^{(\alpha)h}} + X^m{}_{|j} \underset{(\alpha)}{C}{}^i{}_{mh} - X^i{}_{|m} \underset{(\alpha)}{C}{}^m{}_{jh}\\
X^i \overset{(\alpha)}{|}_j \overset{(\beta)}{|}_h &= \frac{\delta X^i \overset{(\alpha)}{|}_j}{\delta y^{(\beta)h}} + X^m \overset{(\alpha)}{|}_j \underset{(\beta)}{C}{}^i{}_{mh} - X^i \overset{(\alpha)}{|}_m \underset{(\beta)}{C}{}^m{}_{jh}.
\end{aligned}
$$

By a straightforward calculation we go just to (6.4). **q.e.d.**

Remark. In the right hands of the identities(6.4) the coefficients of X^i, $X^i{}_{|j}$ and $X^i \overset{(\alpha)}{|}_j$ are the d–tensor fields. To see this it is sufficient to note that

$$\underset{(\alpha\alpha)}{T}{}^{i}{}_{jh} = L^{i}{}_{jh} - \underset{(\alpha\alpha)}{B}{}^{i}{}_{hj}, \quad \overset{\beta}{\underset{(\beta\alpha)}{T}}{}^{i}{}_{jh} = \underset{(\alpha)}{C}{}^{i}{}_{jh} - \overset{\beta}{\underset{(\beta\alpha)}{C}}{}^{i}{}_{hj} \tag{6.5}$$

are d–tensors of torsion.

Of course, the Ricci identities can be extended to every d–tensor field $T^{i_1 \dots i_r}_{j_1 \dots j_s}$. We write down them for a d–tensor field $K^{i}{}_{j}$, only. We have:

$$K^{i}{}_{j|h|m} - K^{i}{}_{j|m|h} = K^{s}{}_{j}R_{s}{}^{i}{}_{hm} - K^{i}{}_{s}R_{j}{}^{s}{}_{hm} - \\ - \underset{(0)}{T}{}^{s}{}_{jm}K^{i}{}_{j|s} - \sum_{\gamma=1}^{k} \underset{(0\gamma)}{R}{}^{s}{}_{hm}K^{i}{}_{j}\overset{(\gamma)}{|}_{s}$$

$$K^{i}{}_{j|h}\overset{(\alpha)}{|}_{m} - K^{i}{}_{j}\overset{(\alpha)}{|}_{m|h} = K^{s}{}_{j}\underset{(\alpha)}{P}{}_{s}{}^{i}{}_{hm} - K^{i}{}_{s}\underset{(\alpha)}{P}{}_{j}{}^{s}{}_{hm} - \\ - \underset{(\alpha)}{C}{}^{s}{}_{hm}K^{i}{}_{j|s} + L^{s}{}_{mh}K^{i}{}_{j}\overset{(\alpha)}{|}_{s} - \sum_{\gamma=1}^{k} \underset{(\alpha\gamma)}{B}{}^{s}{}_{hm}K^{i}{}_{j}\overset{(\gamma)}{|}_{s} \tag{6.6}$$

$$K^{i}{}_{j}\overset{(\alpha)}{|}_{h}\overset{(\beta)}{|}_{m} - K^{i}{}_{j}\overset{(\beta)}{|}_{m}\overset{(\alpha)}{|}_{h} = K^{s}{}_{j}\underset{(\alpha\beta)}{S}{}_{s}{}^{i}{}_{hm} - K^{i}{}_{s}\underset{(\alpha\beta)}{S}{}_{j}{}^{s}{}_{hm} - \\ - \underset{(\beta)}{C}{}^{s}{}_{hm}K^{i}{}_{j}\overset{(\alpha)}{|}_{s} + \underset{(\alpha)}{C}{}^{s}{}_{hm}K^{i}{}_{j}\overset{(\beta)}{|}_{s} - \sum_{\gamma=1}^{k} \overset{\gamma}{\underset{(\alpha\beta)}{T}}{}^{s}{}_{hm}K^{i}{}_{j}\overset{(\gamma)}{|}_{s}, \\ (\alpha, \beta = 1, ..., k;\ \beta \leq \alpha).$$

Let us consider a d–tensor field g_{ij} which satisfies the properties

$$g_{ij|k} = 0, \quad g_{ij}\overset{(\alpha)}{|}_{k} = 0, \quad (\alpha = 1, ..., k) \tag{6.7}$$

Proposition 7.6.1 *If the equations* (6.7) *hold, then the following equations are verified*

$$g_{sj}R_{i}{}^{s}{}_{hm} + g_{is}R_{j}{}^{s}{}_{hm} = 0, \quad g_{sj}\underset{(\alpha)}{P}{}_{i}{}^{s}{}_{hm} + g_{is}\underset{(\alpha)}{P}{}_{j}{}^{s}{}_{hm} = 0 \\ g_{sj}\underset{(\alpha\beta)}{S}{}_{i}{}^{s}{}_{hm} + g_{is}\underset{(\alpha\beta)}{S}{}_{j}{}^{s}{}_{hm} = 0, \quad (\alpha, \beta = 1, ..., k;\ \beta \leq \alpha). \tag{6.8}$$

We apply the Ricci identities (6.6) to a scalar field $L(x, y^{(1)}, ..., y^{(k)})$.

Proposition 7.6.2 *For any scalar field* L *and any* N*–linear connection* D*, the following identities are verified*

$$L_{|h|m} - L_{|m|h} = \underset{(0)}{T}{}^s{}_{hm} L_{|s} - \sum_{\gamma=1}^{k} \underset{(0\gamma)}{R}{}^s{}_{hm} L \overset{(\gamma)}{|}_s$$

$$L_{|h} \overset{(\alpha)}{|}_m - L \overset{(\alpha)}{|}_{m|h} = - \underset{(\alpha)}{C}{}^s{}_{hm} L_{|s} + L^s{}_{hm} L \overset{(\alpha)}{|}_s - \sum_{\gamma=1}^{k} \underset{(\alpha\gamma)}{B}{}^s{}_{hm} L \overset{(\gamma)}{|}_s$$

$$L \overset{(\alpha)}{|}_h \overset{(\beta)}{|}_m - L \overset{(\beta)}{|}_m \overset{(\alpha)}{|}_h = - \underset{(\beta)}{C}{}^s{}_{hm} L \overset{(\alpha)}{|}_s + \underset{(\alpha)}{C}{}^s{}_{mh} L \overset{(\beta)}{|}_s -$$

$$- \sum_{\gamma=1}^{k} \overset{(\gamma)}{\underset{(\beta\alpha)}{C}}{}^s{}_{hm} L \overset{(\gamma)}{|}_s, \ (\alpha, \beta = 1, ..., k;\ \beta \leq \alpha).$$

Remarks.

1° If $L_{|h} = 0$, then we have

$$\sum_{\gamma=1}^{k} \underset{(0\gamma)}{R}{}^s{}_{hm} L \overset{(\gamma)}{|}_s = 0. \tag{6.9}$$

2° If $L \overset{(\alpha)}{|}_h = 0$ for some indices α, then

$$\overset{(1)}{\underset{(\alpha\beta)}{C}}{}^s{}_{hm} L \overset{(1)}{|}_s + \cdots + \overset{(\alpha-1)}{\underset{(\alpha\beta)}{C}}{}^s{}_{hm} L \overset{(\alpha-1)}{|}_s + \overset{(\alpha+1)}{\underset{(\alpha\beta)}{C}}{}^s{}_{hm} L \overset{(\alpha+1)}{|}_s + \cdots$$
$$+ \overset{(k)}{\underset{(\alpha\beta)}{C}}{}^s{}_{hm} L \overset{(k)}{|}_s = 0. \tag{6.9)'$$

Finally, applying the Ricci identities (6.6) to the Liouville d–vectors $z^{(1)i}, ..., z^{(k)i}$, we obtain some fundamental identities.

Theorem 7.6.3 *For any N–linear connection D, the deflection tensor fields* $\overset{(\alpha)}{D}{}^i{}_j$, $\overset{(\beta\alpha)}{d}{}^i{}_j$, *from* (3.11) *satisfy the following identities:*

$$
\begin{aligned}
&\overset{(\alpha)}{D}{}^i{}_{h|m} - \overset{(\alpha)}{D}{}^i{}_{m|h} = z^{(\alpha)s} R_s{}^i{}_{hm} - \underset{(0)}{T}{}^s{}_{hm} \overset{(\alpha)}{D}{}^i{}_s - \\
&\qquad - \sum_{\gamma=1}^{k} \underset{(0\gamma)}{R}{}^s{}_{hm} \overset{(\alpha\gamma)}{d}{}^i{}_s \\
&\overset{(\alpha)}{D}{}^i{}_h \overset{(\beta)}{\Big|}_m - \overset{(\alpha\beta)}{d}{}^i{}_{m|h} = z^{(\alpha)s} \underset{(\beta)}{P}{}_s{}^i{}_{hm} - \underset{(\beta)}{C}{}^s{}_{hm} \overset{(\alpha)}{D}{}^i{}_s + \\
(6.10)\quad &\qquad + L^s{}_{mh} \overset{(\alpha\beta)}{d}{}^i{}_s - \sum_{\gamma=1}^{k} \underset{(\beta\gamma)}{M}{}^s{}_{hm} \overset{(\alpha\gamma)}{d}{}^i{}_s \\
&\overset{(\alpha\beta)}{d}{}^i{}_h \overset{(\delta)}{\Big|}_m - \overset{(\alpha\delta)}{d}{}^i{}_m \overset{(\beta)}{\Big|}_h = z^{(\alpha)s} \underset{(\beta\delta)}{S}{}_s{}^i{}_{hm} - \underset{(\delta)}{C}{}^s{}_{hm} \overset{(\alpha\beta)}{d}{}^i{}_s + \\
&\qquad + \underset{(\beta)}{C}{}^s{}_{mh} \overset{(\alpha\delta)}{d}{}^i{}_s - \sum_{\gamma=1}^{k} \underset{(\beta\delta)}{\overset{\gamma}{T}}{}^s{}_{hm} \overset{(\alpha\gamma)}{d}{}^i{}_s, \\
&\qquad (\delta \leq \beta,\ \beta, \delta = 1, ..., k).
\end{aligned}
$$

An important particular case is when the following equations are verified

$$
(6.11)\qquad z^{(\alpha)i}{}_{|j} = 0,\ z^{(\alpha)i} \overset{(\gamma)}{\Big|}_j = \delta^i{}_j\ (\alpha, \gamma = 1, ..., k)
$$

For $k = 1$, these are the Cartan conditions for an N–connection D.

Proposition 7.6.3 *If the Liouville d–vector fields* $z^{(1)i}, ..., z^{(k)i}$ *satisfy the conditions* (6.11), *then we have*

$$
\begin{aligned}
&z^{(\alpha)s} R_s{}^i{}_{hm} = \sum_{\gamma=1}^{k} \underset{(0\gamma)}{R}{}^i{}_{hm} \\
(6.11)'\quad &z^{(\alpha)s} \underset{(\beta)}{P}{}_s{}^i{}_{hm} = \sum_{\gamma=1}^{k} \underset{(\beta\gamma)}{B}{}^i{}_{hm} - L^i{}_{mr} \\
&z^{(\alpha)s} \underset{(\beta\delta)}{S}{}_s{}^i{}_{hm} = \underset{(\delta)}{C}{}^i{}_{hm} - \underset{(\beta)}{C}{}^i{}_{mh} + \sum_{\gamma=1}^{k} \underset{(\beta\delta)}{\overset{(\gamma)}{C}}{}^i{}_{hm} \\
&\qquad (\delta \leq \beta,\ \beta, \delta, \alpha = 1, ..., k)
\end{aligned}
$$

Of course, the N–linear connections D which satisfy (6.11) are very particular. We say that D is of Cartan type.

7.7 The Structure Equations

Following the same method as in Chapter 3, we shall study the structure equations of an N–linear connection D on $E = \text{Osc}^k M$, whose coefficients are $D\Gamma(N) = (L^i{}_{jh}, \underset{(1)}{C}{}^i{}_{jh}, ..., \underset{(k)}{C}{}^i{}_{jh})$.

Let us consider a smooth curve $\gamma : I \longrightarrow E$ and its tangent vector field $\dot{\gamma}$. The covariant differential is expressed by

$$DX = D_{\dot{\gamma}}X \cdot dt, \ \forall X \in \mathcal{X}(E). \tag{7.1}$$

The vector field X is decomposed in the known form

$$\begin{aligned} X = X^H + X^{V_1} + \cdots + X^{V_k} = X^{(0)i}\frac{\delta}{\delta x^i} + \\ + X^{(1)i}\frac{\delta}{\delta y^{(1)i}} + \cdots + X^{(k)i}\frac{\delta}{\delta y^{(k)i}}. \end{aligned} \tag{7.2}$$

We get for DX

$$\begin{aligned} DX = \{dX^{(0)i} + X^{(0)m}\omega^i{}_m\}\frac{\delta}{\delta x^i} + \{dX^{(1)i} + X^{(1)m}\omega^i{}_m\}\frac{\delta}{\delta y^{(1)i}} + \\ + \cdots + \{dX^{(k)i} + X^{(k)m}\omega^i{}_m\}\frac{\delta}{\delta y^{(k)i}} \end{aligned} \tag{7.1}$$

where 1–forms $\omega^i{}_j$ are the same in each parantheses. It gives us the *connection 1–forms* of the N–linear connection D. Exactly as in Chapter 3, we obtain:

$$\omega^i{}_j = L^i{}_{jh}dx^h + \underset{(1)}{C}{}^i{}_{jh}\delta y^{(1)h} + \cdots + \underset{(k)}{C}{}^i{}_{jh}\delta y^{(k)h}. \tag{7.3}$$

For the determination of the equations of structure of D, we first prove:

Lemma 7.7.1 *The exterior differentials of the 1–forms* $dx^i, \delta y^{(1)i}, ...,$ $\delta y^{(k)i}$ *are expressed as follows:*

$$
(7.4)\quad
\begin{aligned}
d(dx^i) &= 0\\
d(\delta y^{(\alpha)i}) &= \frac{1}{2}\underset{(0\alpha)}{R}{}^{i}{}_{jm} dx^m \wedge dx^j + \sum_{\gamma=1}^{k} \underset{(\gamma\alpha)}{B}{}^{i}{}_{jm}\delta y^{(\gamma)m}\wedge dx^j +\\
&+ \sum_{\beta,\gamma=1}^{k} \underset{(\alpha\gamma)}{\overset{(\beta)}{C}}{}^{i}{}_{jm}\delta y^{(\gamma)m}\wedge \delta y^{(\beta)j},
\end{aligned}
$$

where the coefficients are the same as in the Lie brackets $\left[\frac{\delta}{\delta y^{(\alpha)i}}, \frac{\delta}{\delta y^{(\beta)j}}\right]$, $(\alpha, \beta = 0, ..., k,\ y^{(0)} = x)$ *and where* (4.3)′ *holds and the coefficients of* $\delta y^{(\beta)j}\wedge\delta y^{(\beta)m}$ *are skewsymmetric.*

The proof is based on the formula

$$
(7.5)\quad d(\delta y^{(\alpha)i})\left(\frac{\delta}{\delta y^{(\beta)j}}, \frac{\delta}{\delta y^{(\gamma)m}}\right) = -\delta y^{(\alpha)i}\left(\left[\frac{\delta}{\delta y^{(\beta)j}}, \frac{\delta}{\delta y^{(\gamma)m}}\right]\right),
$$
$$
(\beta, \gamma = 0, 1, ..., k; \alpha = 1, ..., k;\ y^{(0)i} = x^i).
$$

Applying (7.5), we get

$$
d(\delta y^{(\alpha)i})\left(\frac{\delta}{\delta x^j}, \frac{\delta}{\delta x^m}\right) =
$$
$$
= -(\delta y^{(\alpha)i})\left(\underset{(01)}{R}{}^{s}{}_{jm}\frac{\delta}{\delta y^{(1)s}} + \cdots + \underset{(0k)}{R}{}^{s}{}_{jm}\frac{\delta}{\delta y^{(k)s}}\right) = -\underset{(0\alpha)}{R}{}^{s}{}_{jm}.
$$

We also deduce

$$
d(\delta y^{(\alpha)i})\left(\frac{\delta}{\delta x^j}, \frac{\delta}{\delta y^{(\gamma)m}}\right) =
$$
$$
= -\delta y^{(\alpha)i}\left(\underset{(\gamma 1)}{B}{}^{s}{}_{jm}\frac{\delta}{\delta y^{(1)s}} + \cdots + \underset{(\gamma k)}{B}{}^{s}{}_{jm}\frac{\delta}{\delta y^{(k)s}}\right) - \underset{(\gamma\alpha)}{B}{}^{i}{}_{jm}
$$

and so on. Now, we can assert

Theorem 7.7.1 *The structure equations of an N–linear connection D on the total space of a k–osculator bundle $E = \text{Osc}^k M$ are given by*

$$
(7.6)\qquad
\begin{aligned}
&d(dx^i) - dx^m \wedge \omega^i{}_m = -\overset{(0)}{\Omega}{}^i\\
&d(\delta y^{(\alpha)i}) - \delta y^{(\alpha)m} \wedge \omega^i{}_m = -\overset{(\alpha)}{\Omega}{}^i\\
&d\omega^i{}_j - \omega^m{}_j \wedge \omega^i{}_m = -\Omega^i{}_j,
\end{aligned}
$$

where the 2–forms of torsion are:

$$
\overset{(0)}{\Omega}{}^i = \frac{1}{2}\underset{(0)}{T}{}^i{}_{jh}dx^j \wedge dx^h + \underset{(1)}{C}{}^i{}_{jh}dx^j \wedge \delta y^{(1)h} + \cdots + \underset{(k)}{C}{}^i{}_{jh}dx^j \wedge \delta y^{(k)h}
$$

$$
(7.7)\qquad
\begin{aligned}
\overset{(\alpha)}{\Omega}{}^i = {}&\frac{1}{2}\underset{(0\alpha)}{R}{}^i{}_{jh}dx^j \wedge dx^h + \sum_{\gamma=1}^{k}\underset{(\gamma\alpha)}{B}{}^i{}_{jm}dx^j \wedge \delta y^{(\gamma)m} +\\
&+\sum_{\gamma=1}^{k}\overset{\beta}{\underset{(\alpha\gamma)}{C}}{}^i{}_{jm}\delta y^{(\beta)j} \wedge \delta y^{(\gamma)m} - (L^i{}_{jm}dx^j \wedge \delta y^{(\alpha)m} +\\
&+\underset{(1)}{C}{}^i{}_{jm}\delta y^{(1)j} \wedge \delta y^{(\alpha)m} + \cdots + \underset{(k)}{C}{}^i{}_{jm}\delta y^{(k)j} \wedge \delta y^{(\alpha)m}).
\end{aligned}
$$

and where the 2–forms of curvature are

$$
(7.8)\qquad
\begin{aligned}
\Omega^i{}_j = {}&\frac{1}{2}R_j{}^i{}_{pq}dx^p \wedge dx^q + \sum_{\alpha=1}^{k}\underset{(\alpha)}{P}{}_j{}^i{}_{pq}dx^p \wedge \delta y^{(\alpha)q} +\\
&+\sum_{\alpha,\beta=1}^{k}\underset{(\alpha\beta)}{S}{}_j{}^i{}_{pq}\delta y^{(\alpha)p} \wedge \delta y^{(\beta)q}.
\end{aligned}
$$

Proof. The form (7.6) is required by the properties of the N–linear connection D. Namely, D preserves by parallelism the horizontal distribution $N_0 = N$ and each J–vertical distribution $N_1, ..., N_{k-1}$, as well.

A long but not difficult calculus of the left hands of (7.6) gives us, using Lemma 7.7.1, the expressions (7.7) of the 2–forms of torsions and the 2–forms of curvature (7.8). **q.e.d.**

Now, Bianchi identities of an N–linear connection D given by its coefficients $D\Gamma(N) = (L^i{}_{jh}, \underset{(1)}{C}{}^i{}_{jh}, ..., \underset{(k)}{C}{}^i{}_{jh})$, in adapted basis can be derived from (7.6) applying the operator of exterior differentiation and calculating $d\overset{0}{\Omega}{}^i, d\overset{(\alpha)}{\Omega}{}^i, d\Omega^i{}_j$ modulo the system (7.6).

7.8 Problems

1. Write the expressions of the d–tensors of torsion and curvature of an N–linear connection D on $E = \mathrm{Osc}^3 M$.

2. Write the d–tensors of curvature and torsion for the Berwald connection $B\Gamma(N) = (B^i{}_{jk}, 0, ..., 0)$.

3. Derive the Bianchi identities of an N–linear connection D on E by means of the d–tensors of torsion and curvature of D.

4. Write down the Ricci identity for the Berwald connection $B\Gamma(N)$.

5. Derive the structure equations of the Berwald connection $B\Gamma(N)$.

Chapter 8

Lagrangians of Order k. Applications to Higher–Order Analytical Mechanics

As we have remarked in Chapter 4, the foundations of the higher order Lagrangian Mechanics were studied by M. Crampin et al., M. de Léon et al., D. Krupka, D. Grigore, K. Kondo, R. Miron and many others [see References].

We shall present here, using the geometrical theory of the k-osculator bundle, a study of the Lagrangians $L(x, y^{(1)}, ..., y^{(k)})$. It concerns the variational problem, the Craig–Synge covectors, the higher order energies, the laws of conservations and the relations between Euler–Lagrange equations and Jacobi–Hamilton equations. We shall proceed as in our paper [187].

8.1 Lagrangians of Order k. Zermello Conditions

A Lagrangian of order k, $(k \in N^*)$, is a mapping $L : E = \mathrm{Osc}^k M \to R$. L is called *differentiable* if it is of class C^∞ on $\widetilde{E}$ and continuous in the points $(x, 0, y^{(2)}, ..., y^{(k)})$ of the manifold E.

The Hessian of a differentiable Lagrangian L, with respect to $y^{(k)i}$, on $\widetilde{E}$ has the elements $2g_{ij}$:

$$g_{ij} = \frac{1}{2}\frac{\partial^2 L}{\partial y^{(k)i}\partial y^{(k)j}}. \tag{1.1}$$

We can see, exactly as in Ch.1 or 4, that $g_{ij}(x, y^{(1)}, ..., y^{(k)})$ is a d–tensor field, covariant of order 2, symmetric. As in the Ch.4, we can show that the following theorem holds.

Theorem 8.1.1 *If the manifold M is paracompact, then on* $\mathrm{Osc}^k M$ *there exist the differentiable Lagrangians.*

If

$$\operatorname{rank} \|g_{ij}(x, y^{(1)}, ..., y^{(k)})\| = n \quad on \quad \widetilde{E} \tag{1.2}$$

we say that $L(x, y^{(1)}, ..., y^{(k)})$ is a *regular Lagrangian.*

Example 8.1.1. Let γ_{ij} be a Riemannian structure on M and $z^{(k)i}$ the Liouville d–vector field constructed with $\gamma_{ij}(x)$ only. The existence of $z^{(k)i}$ will be proved in the last part of Chapter 9. Let us consider

$$\begin{aligned} L(x, y^{(1)}, ..., y^{(k)}) &= \gamma_{ij}(x) z^{(k)i}(x) z^{(k)j} + \\ &+ a_i(x, y^{(1)}, ..., y^{(k-1)}) z^{(k)i} + b(x, y^{(1)}, ..., y^{(k-1)}), \end{aligned} \tag{1.3}$$

where $a_i(x, y^{(1)}, ..., y^{(k-1)})$ is a d–covector field (with the property $\dfrac{\partial a_i}{\partial y^{(k)j}} = 0$) and $b_i(x, y^{(1)}, ..., y^{(k-1)})$ is a scalar field.

Taking into account that $\dfrac{\partial z^{(k)i}}{\partial y^{(k)j}} = \delta^i{}_j$ it follows that L from (1.3) is a regular Lagrangian of order k on E. This is a generalization of a very well known Lagrangian from electrodynamics (6.4), Ch.1.

For the begining we study the higher order differentiable Lagrangians without the regularity condition.

The Lie derivatives of a differentiable Lagrangian $L(x, y^{(1)}, ..., y^{(k)})$ with respect to the Liouville vector fields $\overset{1}{\Gamma}, ..., \overset{k}{\Gamma}$ determine the scalars

$$I^1(L) = \pounds_{\overset{1}{\Gamma}} L, ..., I^k(L) = \pounds_{\overset{k}{\Gamma}} L. \tag{1.4}$$

These are differentiable functions on $\widetilde{E}$, called the *main invariants* of the Lagrangian L, because of their importance in this theory. They have the expressions:

$$(1.5)\quad \begin{cases} I^1(L) = y^{(1)i}\dfrac{\partial L}{\partial y^{(k)i}}, I^2(L) = y^{(1)i}\dfrac{\partial L}{\partial y^{(k-1)i}} + 2y^{(2)i}\dfrac{\partial L}{\partial y^{(k)i}}, ..., \\ I^k(L) = y^{(1)i}\dfrac{\partial L}{\partial y^{(1)i}} + 2y^{(2)i}\dfrac{\partial L}{\partial y^{(2)i}} + \cdots + ky^{(k)i}\dfrac{\partial L}{\partial y^{(k)i}} \end{cases}$$

Let us consider a smooth parametrized curve $c : [0,1] \longrightarrow M$ represented in a domain of local chart by $x^i = x^i(t), t \in [0,1]$. Its extension $\widetilde{c} : [0,1] \longrightarrow \widetilde{E}$ is given by (1.7), Ch.6.

The integral of action for $L(x, y^{(1)}, ..., y^{(k)})$ is

$$(1.6)\qquad I(c) = \int_0^1 L(x(t), \frac{dx(t)}{dt}, ..., \frac{1}{k!}\frac{d^k x(t)}{dt^k})dt.$$

Exactly as in Theorem 1.2, Ch.4, we can prove:

Theorem 8.1.2 *The necessary conditions for the integral of action $I(c)$ be independent on the parametrization of the curve c are*

$$(1.7)\qquad I^1(L) = \cdots = I^{k-1}(L) = 0,\ I^k(L) = L.$$

We call (1.7) the Zermello conditions (see §1, Ch.4).

We also have:

Theorem 8.1.3 *If the differentiable Lagrangian L of order k, $k > 1$, satisfies the Zermello conditions, then*

$$\text{rank}\, \|g_{ij}(x, y^{(1)}, ..., y^{(k)})\| < n \quad \text{on } \widetilde{E}.$$

8.2 Variational Problem

The variational problem involving the functional $I(c)$, from (1.6), will be studied as a natural extension of the theory expounded in §2, Ch.4. So, we shall present only a few proofs.

Let $c : [0,1] \longrightarrow M$ be the curve considered in the previous section, and $\widetilde{c} : [0,1] \longrightarrow E$ its extension to $\mathrm{Osc}^k M$. On the open set U we consider the curves

$$c_\varepsilon : t \in [0,1] \longrightarrow (x^i(t) + \varepsilon V^i(t)) \in M, \tag{2.1}$$

where ε is a real number, sufficiently small in absolute value so that $\mathrm{Im}\, c_\varepsilon \subset U$, $V^i(t) = V^i(x(t))$ being a regular vector field on U, restricted to the curve c. We assume that all curves c_ε have the same endpoints $c(0)$ and $c(1)$ with the curve c and their osculator spaces of order $1, ..., k-1$ coincident at the points $c(0)$ and $c(1)$. This means:

$$V^i(0) = V^i(1),\ \frac{d^\alpha V^i}{dt^\alpha}(0) = \frac{d^\alpha V^i}{dt^\alpha}(1),\ (\alpha = 1, ..., k-1). \tag{2.2}$$

The integral of action $I(c_\varepsilon)$ for the Lagrangians $L(x, y^{(1)}, ..., y^{(k)})$ is as follows:

$$I(c_\varepsilon) = \int_0^1 L(x + \varepsilon V, \frac{dx}{dt} + \varepsilon \frac{dV}{dt}, ..., \frac{1}{k!}(\frac{d^k x}{dt^k} + \varepsilon \frac{d^k V}{dt^k}))dt.$$

A necessary condition for $I(c)$ to be an extremal value for $I(c_\varepsilon)$ is

$$\frac{dI(c_\varepsilon)}{d\varepsilon}\Big|_{\varepsilon=0} = 0. \tag{$*$}$$

We have

$$\frac{dI(c_\varepsilon)}{d\varepsilon} = \int_0^1 \frac{d}{d\varepsilon}[L(x + \varepsilon V, \frac{dx}{dt} + \varepsilon \frac{dV}{dt}, ..., \frac{1}{k!}(\frac{d^k x}{dt^k} + \varepsilon \frac{d^k V}{dt^k}))dt]$$

and the Taylor expansion of L at the point $\varepsilon = 0$ gives

$$\frac{dI(c_\varepsilon)}{d\varepsilon}\Big|_{\varepsilon=0} = \int_0^1 (\frac{\partial L}{\partial x^i} V^i + \frac{\partial L}{\partial y^{(1)i}} \frac{dV^i}{dt} + \cdots + \frac{1}{k!} \frac{\partial L}{\partial y^{(k)i}} \frac{d^k V^i}{dt^k})dt.$$

Now, putting

$$\begin{aligned} &I_V^1(L) = V^i \frac{\partial L}{\partial y^{(k)i}},\ I_V^2(L) = V^i \frac{\partial L}{\partial y^{(k-1)i}} + \frac{dV^i}{dt} \frac{\partial L}{\partial y^{(k)i}}, ..., \\ &I_V^k(L) = V^i \frac{\partial L}{\partial y^{(1)i}} + \frac{dV^i}{dt} \frac{\partial L}{\partial y^{(2)i}} + \cdots + \frac{1}{(k-1)!} \frac{d^{k-1} V^i}{dt^{k-1}} \frac{\partial L}{\partial y^{(k)i}} \end{aligned} \tag{2.3}$$

and

$$(2.4)\qquad \overset{\circ}{E}_i (L) = \frac{\partial L}{\partial x^i} - \frac{d}{dt}\frac{\partial L}{\partial y^{(1)i}} + \cdots + (-1)^k \frac{1}{k!}\frac{d^k}{dt^k}\frac{\partial L}{\partial y^{(k)i}}$$

one deduces a very important identity:

$$(2.5)\qquad \begin{aligned} &\frac{\partial L}{\partial x^i}V^i + \frac{\partial L}{\partial y^{(1)i}}\frac{dV^i}{dt} + \cdots + \frac{1}{k!}\frac{\partial L}{\partial y^{(k)i}}\frac{d^kV^i}{dt^k} = \\ &= \overset{\circ}{E}_i (L)V^i + \frac{d}{dt}I_V^k(L) - \frac{1}{2!}\frac{d^2}{dt^2}I_V^{k-1}(L) + \cdots + \\ &+ (-1)^{k-1}\frac{1}{k!}\frac{d^k}{dt^k}I_V^1(L). \end{aligned}$$

Also, using (3.2), we have

$$(**)\qquad I_V^\alpha(L)(c(0)) = I_V^\alpha(L)(c(1)) = 0,\ (\alpha = 1, ..., k).$$

Consequently, we can write

$$(2.6)\qquad \begin{aligned} \frac{dI(c_\varepsilon)}{d\varepsilon}\Big|_{\varepsilon=0} &= \int_0^1 \overset{\circ}{E}_i (L)V^i dt + \int_0^1 \frac{d}{dt}\{I_V^k(L) - \\ &- \frac{1}{2!}\frac{d}{dt}I_V^{k-1}(L) + \cdots + (-1)^{k-1}\frac{1}{k!}\frac{d^{k-1}}{dt^{k-1}}I_V^1(L)\}dt. \end{aligned}$$

By means of $(**)$, it follows that

$$(2.6)\qquad \frac{dI(c_\varepsilon)}{d\varepsilon}\Big|_{\varepsilon=0} = \int_0^1 \overset{\circ}{E}_i (L)V^i dt.$$

Now, taking into account the fact that V^i is an arbitrary vector field, then $(2.6)'$ and $(*)$ lead to the following result:

Theorem 8.2.1 *In order that the integral of action $I(c)$ be an extremal value for the functionals $I(c_\varepsilon)$ it is necessary that the following Euler–Lagrange equations hold:*

$$(2.7)\qquad \begin{aligned} &\overset{\circ}{E}_i (L) := \frac{\partial L}{\partial x^i} - \frac{d}{dt}\frac{\partial L}{\partial y^{(1)i}} + \cdots + (-1)\frac{1}{k!}\frac{d^k}{dt^k}\frac{\partial L}{\partial y^{(k)i}} = 0 \\ &y^{(1)i} = \frac{dx^i}{dt}, ..., y^{(k)i} = \frac{1}{k!}\frac{d^kx^i}{dt^k}. \end{aligned}$$

The curves $c : [0,1] \longrightarrow M$, solutions of equations (2.7) are called extremal curves of the integral of action $I(c)$.

The equality (2.6)′ implies the following result:

Theorem 8.2.2 $\overset{\circ}{E}_i\,(L)$ *is a d–covector field.*

Remarks.

1° Starting from (2.4) and using the rules of transformation of natural basis, (6.6), Ch.6, we can directly prove the last theorem.

2° The equation $\overset{\circ}{E}_i\,(L) = 0$, has a geometrical meaning.

8.3 Operators $\dfrac{d_V}{dt}, I^1_V, ..., I^k_V$

For further examination of the identity (2.5) we can introduce some new operators frequently used in the study of higher–order Lagrangians.

Let $c : t \in [0,1] \to M$ be a smooth curve, $\widetilde{c}$ as in (1.7), Ch.4, its extension to $\mathrm{Osc}^k M$ and $V^i(x^i(t))$ a differentiable vector field along c.

In Section 10, Chapter 6, we proved:

Lemma 8.3.1 *The mapping* $S_V : c \longrightarrow \mathrm{Osc}^k M$*, defined by*

$$(3.1)\qquad \begin{aligned} &x^i = x^i(t),\ t \in [0,1] \\ &y^{(1)i} = V^i(x(t)), 2y^{(2)i} = \frac{dV^i}{dt}, ..., ky^{(k)i} = \frac{1}{(k-1)!}\frac{d^{k-1}V^i}{dt^{k-1}} \end{aligned}$$

is a section of the projection $\pi : \mathrm{Osc}^k M \longrightarrow M$ *along the curve c.*

The identity (2.5) suggests the introduction of the following operator along the curve c:

$$(3.2)\qquad \frac{d_V}{dt} = V^i\frac{\partial}{\partial x^i} + \frac{dV^i}{dt}\frac{\partial}{\partial y^{(1)i}} + \cdots + \frac{1}{k!}\frac{d^kV^i}{dt^k}\frac{\partial}{\partial y^{(k)i}}.$$

The importance of this operator results from:

Theorem 8.3.1 *The operator* d_V/dt *has the following properties:*

1° *It is invariant with respect to the coordinate transformations* (1.5), *Ch.*6.

2° *For any differentiable Lagrangian* $L(x, y^{(1)}, ..., y^{(k)})$, $\dfrac{d_V L}{dt}$ *is a scalar field.*

3° d_V/dt *behaves as a derivative operator, i.e.*

$$\text{(3.3)}\quad \begin{aligned} &\frac{d_V(L+L')}{dt} = \frac{d_V L}{dt} + \frac{d_V L'}{dt}, \quad \frac{d_V}{dt}(aL) = a\frac{d_V L}{dt}, \quad a \in R, \\ &\frac{d_V}{dt}(L \cdot L') = \frac{d_V L}{dt} \cdot L' + L \cdot \frac{d_V L'}{dt}. \end{aligned}$$

4° *If* $V^i = \dfrac{dx^i}{dt}$, *then*

$$\text{(3.4)}\quad \begin{aligned} \frac{d_V L}{dt} = \frac{dL}{dt} &= \frac{\partial L}{\partial x^i} y^{(1)i} + 2\frac{\partial L}{\partial y^{(1)i}} y^{(2)i} + \cdots + \\ &+ k\frac{\partial L}{\partial y^{(k-1)i}} y^{(k)i} + \frac{\partial L}{\partial y^{(k)i}} \frac{dy^{(k)i}}{dt} \\ y^{(1)i} &= \frac{dx^i}{dt}, ..., y^{(k)i} = \frac{1}{k!}\frac{d^k x^i}{dt^k}. \end{aligned}$$

Therefore, $\dfrac{d_V}{dt}$ is called the *total derivative* in the direction of vector field V^i.

Now, let us consider the operators:

$$\text{(3.5)}\quad \begin{aligned} &I_V^1 = V^i \frac{\partial}{\partial y^{(k)i}}, I_V^2 = V^i \frac{\partial}{\partial y^{(k-1)i}} + \frac{dV^i}{dt}\frac{\partial}{\partial y^{(k)i}}, ..., \\ &I_V^k = V^i \frac{\partial}{\partial y^{(1)i}} + \frac{dV^i}{dt}\frac{\partial}{\partial y^{(2)i}} + \cdots + \frac{1}{(k-1)!}\frac{d^{k-1}V^i}{dt^{k-1}}\frac{\partial}{\partial y^{(k)i}}. \end{aligned}$$

Theorem 8.3.2 *The following properties hold:*

1° $I_V^1, ..., I_V^k$ *are vector fields along the curve* c.

2° $I_V^k = J(\frac{d_V}{dt}), I_V^{k-1} = J(I_V^k), ..., I_V^1 = J(I_V^2), 0 = J(I_V^1)$.

3° $I_V^1(L), ..., I_V^k(L)$ *are the scalars* (2.3).

4° *If* $V^i = \frac{dx^i}{dt}$, *then* $I_V^1, ..., I_V^k$ *are the Liouville vector fields* $\overset{1}{\Gamma}, ..., \overset{k}{\Gamma}$ *along the curve c.*

Finally, the identity (2.5) takes a new form:

Theorem 8.3.3 *Along a smooth curve c on the manifold M we have:*
1°

$$\frac{d_V L}{dt} = V^i \overset{\circ}{E}_i(L) + \frac{d}{dt} I_V^k(L) - \frac{1}{2!}\frac{d^2}{dt^2} I_V^{k-1}(L) + \cdots + (-1)^{k-1}\frac{1}{k!}\frac{d^k}{dt^k} I_V^1(L). \tag{3.6}$$

2° *If* $V^i = \frac{dx^i}{dt}$:

$$\frac{dL}{dt} = \frac{dx^i}{dt} \overset{\circ}{E}_i(L) + \frac{d}{dt} I^k(L) - \frac{1}{2!}\frac{d^2}{dt^2} I_V^{k-1}(L) + \cdots + (-1)^{k-1}\frac{1}{k!}\frac{d^k}{dt^k} I^1(L). \tag{3.7}$$

Corollary 8.3.1 *The Lagrangian* $L(x, y^{(1)}, ..., y^{(k)})$ *is constant along an extremal curve c of integral of action* $I(c)$ *if, and only if, along this curve we have*

$$I^k(L) - \frac{1}{2!}\frac{dI^{k-1}(L)}{dt} + \cdots + (-1)^{k-1}\frac{1}{k!}\frac{d^{k-1}I^1(L)}{dt^{k-1}} = const.$$

8.4 Craig–Synge Covectors

To the covectors field $\overset{\circ}{E}_i(L)$ along a smooth curve c we shall associate other k–covector fields $\overset{1}{E}_i(L), ..., \overset{k}{E}_i(L)$ introduced by Craig and Synge [63], [285]. These fields are useful in the geometry of higher order Lagrange spaces, to be studied in Chapter 10.

Let us consider a smooth curve $c : [0,1] \longrightarrow M$ and along c the operators

$$\begin{aligned}
&\overset{\circ}{E}_i = \frac{\partial}{\partial x^i} - \frac{d}{dt}\frac{\partial}{\partial y^{(1)i}} + \cdots + (-1)^k \frac{1}{k!}\frac{d^k}{dt^k}\frac{\partial}{\partial y^{(k)i}},\\
&\overset{1}{E}_i = \sum_{\alpha=1}^{k} (-1)^{\alpha} \frac{1}{\alpha!} \binom{\alpha}{\alpha-1} \frac{d^{\alpha-1}}{dt^{\alpha-1}} \frac{\partial}{\partial y^{(\alpha)i}},\\
(4.1)\qquad &\overset{2}{E}_i = \sum_{\alpha=2}^{k} (-1)^{\alpha} \frac{1}{\alpha!} \binom{\alpha}{\alpha-2} \frac{d^{\alpha-2}}{dt^{\alpha-2}} \frac{\partial}{\partial y^{(\alpha)i}},\\
&\dots\dots\dots\dots\dots\dots\dots\dots\dots\dots\dots\dots\\
&\overset{k}{E}_i = (-1)^{\alpha} \frac{1}{k!} \frac{\partial}{\partial y^{(k)i}},
\end{aligned}$$

These operators act R–linearly on the R–linear space of Lagrangians $L(x, y^{(1)}, ..., y^{(k)})$. We shall prove that $\overset{1}{E}_i (L), ..., \overset{k}{E}_i (L)$ are d–covector fields.

To this aim, we first prove

Lemma 8.4.1 *For any differentiable Lagrangian $L(x, y^{(1)}, ..., y^{(k)})$ and any differentiable function $\phi(t)$ along the curve c we have*

$$(4.2)\qquad \overset{\circ}{E}_i (\phi L) = \phi \overset{\circ}{E}_i (L) + \frac{d\phi}{dt} \overset{1}{E}_i (L) + \cdots + \frac{d^k \phi}{dt^k} \overset{k}{E}_i (L).$$

From which we can prove without difficulties

Theorem 8.4.1 *For any differentiable Lagrangian $L(x, y^{(1)}, ..., y^{(k)})$ along a smooth curve c, $\overset{1}{E}_i (L). ..., \overset{k}{E}_i (L)$ are d–covector fields.*

Remarks.

1° We shall see in a next chapter the importance of the d–covector fields $\overset{k-1}{E}_i (L)$.

2° Evidently $\dfrac{\partial L}{\partial y^{(k)i}}$ is a d–covector field.

3° Theorem 8.4.1 can be directly proved, after a long calculation.

Exactly as in Chapter 4, we can establish the following results.

Theorem 8.4.2 *For any differentiable Lagrangians* $L(x, y^{(1)}, ..., y^{(k)})$ *and* $F(x, y^{(1)}, ..., y^{(k-1)})$*, along a smooth curve c, we have*

$$(4.3)\qquad \begin{array}{l} \overset{\circ}{E}_i (L + \frac{dF}{dt}) = \overset{\circ}{E}_i (L),\ \overset{\circ}{E}_i (\frac{dF}{dt}) = 0, \\ \overset{1}{E}_i (\frac{dF}{dt}) = -\overset{\circ}{E}_i (F), ..., \overset{k}{E}_i (\frac{dF}{dt}) = -\overset{k-1}{E}_i (F). \end{array}$$

The previous property extends a known result of Carathéodory [52].

Theorem 8.4.3 *The integrals of action*

$$(4.4)\qquad \begin{aligned} I(c) &= \int_0^1 L(x, \frac{dx}{dt}, ..., \frac{1}{k!}\frac{d^k x}{dt^k})dt \\ I'(c) &= \int_0^1 [L(x, \frac{dx}{dt}, ..., \frac{1}{k!}\frac{d^k x}{dt^k}) + \\ &+ \frac{dF}{dt}(x, \frac{dx}{dt}, ..., \frac{1}{(k-1)!}\frac{d^{k-1} x}{dt^{k-1}})]dt \end{aligned}$$

have the same extremal curves, for any differentiable Lagrangian F with the property $\frac{\partial F}{\partial y^{(k)i}} = 0$.

8.5 Energies of Higher Order

We have introduced the energies of higher order in the Section 5 of Chapter 4 for the Lagrangians of second order. Here we extend the mentioned theory to the Lagrangians of order $k > 1$. So we shall give few proofs.

Definition 8.5.1 We call *energies or order* $k, k-1, ..., 1$ of the Lagrangian $L(x, y^{(1)}, ..., y^{(k)})$, with respect to the curve c, the following invariants

$$
\begin{aligned}
\mathcal{E}_c^k(L) &= I^k(L) - \frac{1}{2!}\frac{dI^{k-1}(L)}{dt} + \cdots + \\
&\qquad\qquad + (-1)^{k-1}\frac{1}{k!}\frac{d^{k-1}I^1(L)}{dt^{k-1}} - L, \\
\mathcal{E}_c^{k-1}(L) &= -\frac{1}{2!}I^{k-1}(L) + \frac{1}{3!}\frac{dI^{k-2}(L)}{dt} + \cdots + \\
&\qquad\qquad + (-1)^{k-1}\frac{1}{k!}\frac{d^{k-2}I^1(L)}{dt^{k-2}}, \\
\mathcal{E}_c^{k-2}(L) &= \frac{1}{3!}I^{k-2}(L) - \frac{1}{4!}\frac{d}{dt}I^{k-3}(L) + \cdots + \\
&\qquad\qquad + (-1)^{k-1}\frac{1}{k!}\frac{d^{k-3}I^1(L)}{dt^{k-3}}, \\
&\cdots\cdots\cdots\cdots\cdots\cdots\cdots\cdots\cdots\cdots\cdots\cdots \\
\mathcal{E}_c^1(L) &= (-1)^{k-1}\frac{1}{k!}I^1(L).
\end{aligned}
\tag{5.1}
$$

The dependence of these invariants on the curve c is obvious. A first result:

Proposition 8.5.1 *The following identities hold:*

$$
\begin{aligned}
&\mathcal{E}_c^k(L) - \frac{d}{dt}\mathcal{E}_c^{k-1}(L) = I^k(L) - L \\
&\mathcal{E}_c^{k-1}(L) - \frac{d}{dt}\mathcal{E}_c^{k-2}(L) = -\frac{1}{2}I^{k-1}(L) \\
&\cdots\cdots\cdots\cdots\cdots\cdots\cdots\cdots\cdots\cdots\cdots\cdots \\
&\mathcal{E}_c^2(L) - \frac{d}{dt}\mathcal{E}_c^1(L) = (-1)^{k-2}\frac{1}{(k-1)!}I^2(L).
\end{aligned}
\tag{5.2}
$$

As we shall see, the energies $\mathcal{E}_c^k(L), \ldots, \mathcal{E}_c^1(L)$ are involved in a Noether theory of symmetries of the higher order Lagrangians. With this end in view, we state the following result:

Lemma 8.5.1 *For any differentiable Lagrangian $L(x, y^{(1)}, \ldots, y^{(k)})$ and any differentiable function $\tau: M \to R$ along a smooth curve $c: [0,1] \to M$ we have*

$$\frac{d\tau}{dt}L-[\frac{d\tau}{dt}I^k(L)+\frac{1}{2!}\frac{d^2\tau}{dt^2}I^{k-1}(L)+\cdots+\frac{1}{k!}\frac{d^k\tau}{dt^k}I^1(L)]=$$

$$(5.3)\qquad =\tau\frac{d\mathcal{E}_c^k(L)}{dt}+\frac{d}{dt}\{-\tau\mathcal{E}_c^k(L)+\frac{d\tau}{dt}\mathcal{E}_c^{k-1}(L)-$$

$$-\frac{d^2\tau}{dt^2}\mathcal{E}_c^{k-2}(L)+\cdots+(-1)^k\frac{d^{k-1}\tau}{dt^{k-1}}\mathcal{E}_c^1(L)\}.$$

Proof. The right–hand of this equation, by means of (5.2), successively becomes:

$$-\frac{d\tau}{dt}\{\mathcal{E}^k(L)-\frac{d\mathcal{E}_c^{k-1}(L)}{dt}\}+\frac{d^2\tau}{dt^2}\{\mathcal{E}_c^{k-1}(L)-\frac{d\mathcal{E}_c^{k-1}(L)}{dt}\}+\cdots+$$

$$+(-1)^{k-1}\frac{d^{k-1}\tau}{dt^{k-1}}\{\mathcal{E}_c^2(L)-\frac{d\mathcal{E}_c^1(L)}{dt}\}+(-1)^k\frac{d^k\tau}{dt^k}\mathcal{E}_c^1(L)=$$

$$=-\frac{d\tau}{dt}\{I^k(L)-L\}-\frac{1}{2!}\frac{d^2\tau}{dt^2}I^{k-1}(L)-\frac{1}{3!}\frac{d^3\tau}{dt^3}I^{k-2}(L)-\cdots-$$

$$-\frac{1}{(k-1)!}\frac{d^{k-1}\tau}{dt^{k-1}}I^2(L)-\frac{1}{k!}\frac{d^k\tau}{dt^k}I^1(L)=$$

$$=\frac{d\tau}{dt}L-[\frac{d\tau}{dt}I^k(L)+\frac{d^2\tau}{dt^2}I^{k-1}(L)+\cdots+\frac{1}{k!}\frac{d^k\tau}{dt^k}I^1(L)].$$

An important result obtained by Andreas et al. [13], Léon et al. [164], is given as follows:

Theorem 8.5.1 *For any Lagrangian* $L(x,y^{(1)},...,y^{(k)})$ *along a smooth curve* $c:[0,1]\longrightarrow(x^i(t))\in M$ *we have*

$$(5.4)\qquad \frac{d\mathcal{E}_c^k(L)}{dt}=-\overset{\circ}{E}_i(L)\frac{dx^i}{dt}.$$

Indeed, from (5.1), we get

$$\frac{d\mathcal{E}_c^k(L)}{dt}=\frac{d}{dt}\{I^k(L)-\frac{1}{2!}\frac{dI^{k-1}(L)}{dt}+\cdots+(-1)^{k-1}\frac{1}{k!}\frac{d^{k-1}I^1(L)}{dt^{k-1}}\}-\frac{dL}{dt}.$$

Substituting here $\frac{dL}{dt}$ from (3.4) and performing the obvious reductions, we get (5.4).

An immediate consequence of the last theorem is the following law of conservation:

Theorem 8.5.2 *For any Lagrangian $L(x, y^{(1)}, ..., y^{(k)})$, the energy of order k, $\mathcal{E}_c^k(L)$ is conserved along every curve which is solution of the Euler–Lagrange equations $\overset{\circ}{E}_i\,(L) = 0$.*

8.6 Noether Theorems

Let $L(x, y^{(1)}, ..., y^{(k)})$ be a differentiable Lagrangian, $c\colon [0,1] \to M$ a smooth curve and $I(c), I'(c)$ the integrals of action (4.4), where $F(x, y^{(1)}, ..., y^{(k-1)})$ is an arbitrary differentiable Lagrangian. Theorem 8.4.3 shows that $I(c)$ and $I'(c)$ have the same extremal curves.

Therefore, we can formulate:

Definition 8.6.1 A symmetry of the differentiable Lagrangian $L(x, y^{(1)}, ..., y^{(k)})$ is a C^∞–diffeomorphism $\varphi : M \times R \longrightarrow M \times R$, which preserves the variational principle of the integral of action $I(c)$ from (4.4).

Generally, the variational principle is considered on an open set $U \subset M$. So, we can consider the notion of local symmetry of the Lagrangian L, taking φ as local diffeomorphism.

Therefore, in the following considerations we study the infinitesimal symmetries, given on an open set $U \times (a, b) \subset M \times R$ in the form:

$$(6.1) \qquad \begin{aligned} x'^i &= x^i + \varepsilon V^i(x,t), \ (i = 1, ..., n) \\ t' &= t + \varepsilon\tau(x,t) \end{aligned}$$

where ε is a real number, sufficiently small in absolute value so that the points (x, t) and (x', t') belong to the same set $U \times (a, b)$, where the curve

$$c : t \in [0,1] \longrightarrow (x^i(t), t) \in U \times (a, b)$$

is defined. Throughout the following calculation terms of order $\varepsilon^2, \varepsilon^3, ...$ will be neglected. Of course $V^i(x, t)$ is a vector field on the open set $U \times (a, b)$. In the end points $c(0)$ and $c(1)$, we assume that $V^i(x(t), t) = V^i(t)$ satisfies the conditions (2.2).

The inverse transformation of the local diffeomorphism (6.1) is given by

$$x^i = x'^i - \varepsilon V^i(x,t),\ t = t' - \varepsilon V^i(x,t).$$

Exactly as in Chapter 4, we find that the mapping

$$S_V : c \longrightarrow \mathrm{Osc}^k M \times R$$

defined by

$$(6.2)\qquad \begin{aligned} x^i = x^i(t), y^{(1)i} = V^i(x(t),t), 2y^{(2)i} = \frac{dV^i}{dt}, ..., \\ ky^{(k)i} = \frac{1}{(k-1)!}\frac{d^{k-1}V^i}{dt^{k-1}},\quad t = t \end{aligned}$$

is a section of the mapping $\widetilde{\pi} : (u,t) \in \mathrm{Osc}^k M \times R \longrightarrow (\pi(u), t) \in M \times R$. We also observe that the Lemma 8.5.1 is valid, for the function $\tau(x,t)$ restricted to the curve c.

The infinitesimal transformation (6.1) is a symmetry for the differentiable Lagrangian $L(x, y^{(1)}, ..., y^{(k)})$ if, and only if, for any differentiable function $F(x, y^{(1)}, ..., y^{(k-1)})$ the following equation holds:

$$(6.3)\qquad \begin{aligned} & L(x', \frac{dx'}{dt'}, ..., \frac{1}{k!}\frac{d^k x'}{dt'^k})dt' = \\ & = \{L(x, \frac{dx}{dt}, ..., \frac{1}{k}\frac{d^k x}{dt^k}) + \frac{dF}{dt}(x, \frac{dx}{dt}, ..., \frac{1}{(k-1)!}\frac{d^{k-1}x}{dt^{k-1}})\}dt \end{aligned}$$

From (6.1) we deduce

$$(6.4)\qquad \begin{aligned} & \frac{dt'}{dt} = 1 + \varepsilon\frac{d\tau}{dt} \\ & \frac{dx'^i}{dt'} = \frac{dx^i}{dt} + \varepsilon\varphi^{(1)i}, \\ & \frac{1}{2!}\frac{d^2x'^i}{dt'^2} = \frac{1}{2!}(\frac{d^2x^i}{dt^2} + \varepsilon\varphi^{(2)i}), ..., \frac{1}{k!}\frac{d^k x'^i}{dt'^k} = \\ & \qquad\qquad = \frac{1}{k!}(\frac{d^k x^i}{dt^k} + \varepsilon\varphi^{(k)i}), \end{aligned}$$

where we have put

$$
(6.5)\qquad
\begin{aligned}
\varphi^{(1)i} &= \frac{dV^i}{dt} - \frac{dx^i}{dt}\frac{d\tau}{dt}\\
\varphi^{(2)i} &= \frac{d^2V^i}{dt^2} - \binom{2}{1}\frac{d^2x^i}{dt^2}\frac{d\tau}{dt} - \binom{2}{2}\frac{dx^i}{dt}\frac{d^2\tau}{dt^2},\\
&\cdots\cdots\cdots\cdots\cdots\cdots\cdots\cdots\cdots\cdots\cdots\cdots\\
\varphi^{(k)i} &= \frac{d^kV^i}{dt^k} - \binom{k}{1}\frac{d^kx^i}{dt^k}\frac{d\tau}{dt} - \binom{k}{2}\frac{d^{k-1}x^i}{dt^{k-1}}\frac{d^2\tau}{dt^2} -\\
&\qquad - \cdots - \binom{k}{k}\frac{dx^i}{dt}\frac{d^k\tau}{dt^k}.
\end{aligned}
$$

By virtue of (6.4), (6.5), the equality (6.3), neglecting the terms in $\varepsilon^2, \varepsilon^3, \ldots$ and putting $\phi = \varepsilon F$, leads to

$$
(6.6)\qquad L\frac{d\tau}{dt} + \frac{\partial L}{\partial x^i}V^i + \frac{\partial L}{\partial y^{(1)i}}\varphi^{(1)i} + \cdots + \frac{1}{k!}\frac{\partial L}{\partial y^{(k)i}}\varphi^{(k)i} = \frac{d\phi}{dt}.
$$

Conversely, if (6.6) holds, for L, V^i, τ and c given, then putting $\phi(x, y^{(1)}, \ldots, y^{(k-1)}) = \varepsilon F(x, y^{(1)}, \ldots, y^{(k-1)})$ the equality (6.3) is satisfied for the infinitesimal transformation (6.1) neglecting the terms of order ≥ 2 in ε.

But $\varphi^{(1)i}, \ldots, \varphi^{(k)i}$ are expressed in (6.5). It follows that the equality (6.6) is equivalent to

$$
(6.7)\qquad
\begin{aligned}
&V^i\frac{\partial L}{\partial x^i} + \frac{dV^i}{dt}\frac{\partial L}{\partial y^{(1)i}} + \cdots + \frac{1}{k!}\frac{d^kV^i}{dt^k}\frac{\partial L}{\partial y^{(k)i}} +\\
&+\{L\frac{d\tau}{dt} - [I^k(L)\frac{d\tau}{dt} + \frac{1}{2!}I^{k-1}(L)\frac{d^2\tau}{dt^2} + \cdots +\\
&+\frac{1}{k!}I^1(L)\frac{d^k\tau}{dt^k}]\} = \frac{d\phi}{dt}.
\end{aligned}
$$

Using the operator (3.2), we can state the following result:

Theorem 8.6.1 *A necessary and sufficient condition that an infinitesimal transformation* (6.1) *be a symmetry for the Lagrangian* $L(x, y^{(1)},$

..., $y^{(k)}$) along the smooth curve c is that the left side of the equality

$$\begin{aligned}(6.8)\qquad &\frac{d_V L}{dt} + \{L\frac{d\tau}{dt} - [I^k(L)\frac{d\tau}{dt} + \frac{1}{2!}I^{k-1}(L)\frac{d^2\tau}{dt^2} + \cdots + \\ &+\frac{1}{k!}I^1(L)\frac{d^k\tau}{dt^k}]\} = \frac{d\phi}{dt}.\end{aligned}$$

be of the form $\frac{d}{dt}\phi(x, y^{(1)}, ..., y^{(k-1)})$ *along* c.

Theorem 8.3.3 and Lemma 8.5.1 show that (6.8) is equivalent to:

$$\begin{aligned}(6.9)\qquad &V^i \overset{\circ}{E}_i(L) + \frac{d}{dt}I_V^k(L) - \frac{1}{2!}\frac{d^2}{dt^2}I^{k-1}(L) + \cdots + \\ &+(-1)^{k-1}\frac{1}{k!}\frac{d^k}{dt^k}I_V^1(L) + \tau\frac{d}{dt}\mathcal{E}_c^k(L) + \\ &\frac{d}{dt}[-\tau\mathcal{E}_c^k(L) + \frac{d\tau}{dt}\mathcal{E}_c^{k-1}(L) - \cdots + \\ &+(-1)^{k-1}\frac{d^{k-1}\tau}{dt^{k-1}}\mathcal{E}_c^1(L)] = \frac{d\phi}{dt}.\end{aligned}$$

By theorem 8.5.2, $\overset{\circ}{E}_i(L) = 0$ implies $\frac{d\mathcal{E}_c^k(L)}{dt} = 0$ and (6.9) leads to:

Theorem 8.6.2 (Noether [187]) *For any infinitesimal symmetry (6.1) (which satisfies (6.8)) of a Lagrangian $L(x, y^{(1)}, ..., y^{(k)})$ and for any function $\phi(x, y^{(1)}, ..., y^{(k-1)})$, the function*

$$\begin{aligned}(6.10)\qquad &\mathcal{F}^k(L,\phi) := I_V^k(L) - \frac{1}{2!}\frac{d}{dt}I_V^{k-1}(L) + \cdots + \\ &+(-1)^{k-1}\frac{1}{k!}\frac{d^{k-1}}{dt^{k-1}}I_V^1(L)) - \tau\mathcal{E}_c^k(L) + \frac{d\tau}{dt}\mathcal{E}_c^{k-1}(L) - \cdots + \\ &+(-1)^k\frac{d^{k-1}\tau}{dt^{k-1}}\mathcal{E}_c^1(L) - \phi\end{aligned}$$

is conserved along the solution curves of the Euler–Lagrange equation $\overset{\circ}{E}_i(L) = 0$.

The functions $\mathcal{F}^k(L,\phi)$ in (6.10) contain the relative invariants $I_V^1(L),...,I_V^k(L)$, the energies of order $1,...,k$, $\mathcal{E}_c^1(L),...,\mathcal{E}_c^k(L)$ and the function $\phi(x,y^{(1)},...,y^{(k-1)})$. In the case $k=2$, from (6.10) we obtain the functions (6.9), Ch.4.

In particular, if the Zermello conditions (1.7) are satisfied, then the energies $\mathcal{E}_c^1(L),...,\mathcal{E}_c^k(L)$ vanish and we have a simpler form of the Noether theorem:

Theorem 8.6.3 *For any infinitesimal symmetry* (6.1) *of a Lagrangian* $L(x,y^{(1)},...,y^{(k)})$, *which satisfies the Zermello conditions* (1.7) *and for any* C^∞*-function* $\phi(x,y^{(1)},...,y^{(k-1)})$, *the following function*

$$\mathcal{F}^k(L,\phi) := I_V^k(L) - \frac{1}{2}\frac{d}{dt}I_V^{k-1}(L) + \cdots + \\ +(-1)^{k-1}\frac{1}{k!}\frac{d^{k-1}}{dt^{k-1}}I_V^1(L) - \phi, \tag{6.10$'$}$$

is conserved along the solution curves of the Euler–Lagrange equation $\overset{\circ}{E_i}(L)=0$.

Nice applications can be done for the Lagrangians of the form (5.4), Ch.4, in the higher–order electrodynamics.

Remark. The previous theory can be reformulated without difficulties, in the case when the transformations (6.1) are the symmetries of a Lagrangian $L(x,y^{(1)},...,y^{(k)})$ and, simultaneously, the infinitesimal transformations of a Lie group.

8.7 Jacobi–Ostrogradski Momenta

We extend to $\mathrm{Osc}^k M$ the notion of Jacobi–Ostrogradski momenta introduced in the Section 7 of Chapter 4. The justifications of the statements from this theory will be omitted, because they were given in the case $k=2$.

Let us consider the energy of order k, from (5.1):

$$\mathcal{E}_c^k(L) = I^k(L) - \frac{1}{2!}\frac{dI^{k-1}(L)}{dt} + \cdots + \\ +(-1)^{k-1}\frac{1}{k!}\frac{dk-1}{dt^{k-1}}I^1(L) - L \tag{7.1}$$

along a parametrized curve c. Remarking that $\mathcal{E}_c^k(L)$ is a polynomial function of degree one in $\frac{dx^i}{dt}, \ldots, \frac{d^k x^i}{dt^k}$, we can write:

$$\mathcal{E}_c^k(L) = p_{(1)i}\frac{dx^i}{dt} + p_{(2)i}\frac{d^2x^i}{dt^2} + \cdots + p_{(k)i}\frac{d^kx^i}{dt^k} - L, \tag{7.2}$$

where

$$\begin{aligned} p_{(1)i} &= \frac{\partial L}{\partial y^{(1)i}} - \frac{1}{2!}\frac{d}{dt}\frac{\partial L}{\partial y^{(2)i}} + \cdots + (-1)^{k-1}\frac{1}{k!}\frac{d^{k-1}}{dt^{k-1}}\frac{\partial L}{\partial y^{(k)i}}, \\ p_{(2)i} &= \frac{1}{2!}\frac{\partial L}{\partial y^{(2)i}} - \frac{1}{3!}\frac{d}{dt}\frac{\partial L}{\partial y^{(3)i}} + \cdots + (-1)^{k-2}\frac{1}{k!}\frac{d^{k-2}}{dt^{k-2}}\frac{\partial L}{\partial y^{(k)i}}, \\ &\cdots\cdots\cdots\cdots\cdots\cdots \\ p_{(k)i} &= \frac{1}{k!}\frac{\partial L}{\partial y^{(k)i}}. \end{aligned} \tag{7.3}$$

Using the rule of transformation (2.4), Ch.6, of the natural basis of the module $\mathcal{X}(E)$ and the definition (7.3) of $p_{(1)i}, \ldots, p_{(k)i}$, we can deduce:

Proposition 8.7.1 *With respect to the transformations of coordinate, ((1.5), Ch.6) on E, $p_{(1)i}, \ldots, p_{(k)i}$ is transformed as follows:*

$$\begin{aligned} p_{(1)i} &= \frac{\partial \widetilde{y}^{(1)m}}{\partial y^{(1)i}}\widetilde{p}_{(1)m} + \cdots + \frac{\partial \widetilde{y}^{(k)m}}{\partial y^{(1)i}}\widetilde{p}_{(k)m}, \\ p_{(2)i} &= \frac{\partial \widetilde{y}^{(2)m}}{\partial y^{(2)i}}\widetilde{p}_{(2)m} + \cdots + \frac{\partial \widetilde{y}^{(k)m}}{\partial y^{(2)i}}\widetilde{p}_{(k)m}, \\ &\cdots\cdots\cdots\cdots\cdots\cdots \\ p_{(k)i} &= \frac{\partial \widetilde{y}^{(k)m}}{\partial y^{(k)i}}\widetilde{p}_{(k)m}. \end{aligned} \tag{7.4}$$

Therefore, $p_{(1)i}, \ldots, p_{(k)i}$ are called the *Jacobi–Ostragradski momenta.*

Taking into account the expressions of the momenta (7.3) and the form (2.7) of the covector field $\overset{\circ}{E}_i(L)$ we get

Lemma 8.7.1 *The following identities hold:*

$$(7.5)\qquad \begin{aligned} \frac{dp_{(1)i}}{dt} &= \frac{\partial L}{\partial x^i} - \overset{\circ}{E_i}(L), \\ \frac{dp_{(2)i}}{dt} &= \frac{\partial L}{\partial y^{(1)i}} - p_{(1)i}, \\ &\cdots\cdots\cdots\cdots\cdots\cdots \\ \frac{dp_{(k)i}}{dt} &= \frac{1}{(k-1)!}\frac{\partial L}{\partial y^{(k-1)i}} - p_{(k-1)i}. \end{aligned}$$

Now, we can establish (see M. de Leon et al. [164]):

Theorem 8.7.1 *Along every solution curve of the Euler–Lagrange equations $\overset{\circ}{E_i}(L) = 0$, the following Hamilton–Jacobi equations hold:*

$$(7.6)\qquad \begin{aligned} \frac{\partial \mathcal{E}^k_c(L)}{\partial p_{(\alpha)i}} &= \frac{d^\alpha x^i}{dt^\alpha} \quad (\alpha = 1, ..., k), \\ \frac{\partial \mathcal{E}^k_c}{\partial x^i} &= -\frac{dp_{(1)i}}{dt}, \\ \frac{d\mathcal{E}^k_c}{\partial y^{(\alpha)i}} &= -\alpha!\frac{dp_{(\alpha+1)i}}{dt}, \quad (\alpha = 1, ..., k-1). \end{aligned}$$

Proof. Deriving (7.2) with respect to $p_{(\alpha)i}$ we obtain the first equation (7.6). For $\overset{\circ}{E_i}(L) = 0$, the first identity (7.5) leads to the second equation (7.6). The third equation (7.6) is a consequence of (7.2) and of the Lemma 8.7.1. **q.e.d.**

Remark. A theory of fields, based on the Hamilton–Jacobi equations can be found in the book of M. de Léon and P. Rodrigues [164].

Some considerations on the energy of order $k-1$ can be done. $\mathcal{E}^{k-1}_c(L)$ being given by (5.1), the first formula (5.2) and the expression of (7.2) of $\mathcal{E}^k_c(L)$ lead to:

Proposition 8.7.2 *Along a smooth curve c, the energy of order $k-1$, $\mathcal{E}^{k-1}_c(L)$ of a differentiable Lagrangian L, has the property*

$$(7.7)\qquad \frac{d\mathcal{E}^{k-1}_c}{dt} = p_{(1)i}\frac{dx^i}{dt} + \cdots + p_{(k)i}\frac{d^k x^i}{dt^k} - I^k(L).$$

But along a curve c, the main invariant $I^k(L)$ has the form:

$$I^k(L) = \frac{\partial L}{\partial y^{(1)i}}\frac{dx^i}{dt} + \cdots + \frac{k}{k!}\frac{\partial L}{\partial y^{(k)i}}\frac{d^k x^i}{dt^k}.$$

So, taking into consideration the previous proposition, we obtain:

Theorem 8.7.2 *The energy of order $k-1$, $\mathcal{E}_c^{k-1}(L)$ of a differentiable Lagrangian $L(x, y^{(1)}, ..., y^{(k)})$ is conserved along a smooth curve $c : I \longrightarrow M$ if, and only if, along c we have*

$$(7.8) \qquad (p_{(1)i} - \frac{\partial L}{\partial y^{(1)i}})\frac{dx^i}{dt} + \cdots + (p_{(k)i} - \frac{k}{k!}\frac{\partial L}{\partial y^{(k)i}})\frac{d^k x^i}{dt^k} = 0.$$

The Jacobi–Ostragradski momenta $p_{(1)i}, ..., p_{(k)i}$ allow us to define the 1–forms

$$(7.9) \qquad \begin{aligned} p_{(1)} &= p_{(1)i}dx^i + p_{(2)i}dy^{(1)i} + \cdots + p_{(k)i}dy^{(k-1)i} \\ p_{(2)} &= p_{(2)i}dx^i + p_{(3)i}dy^{(1)i} + \cdots + p_{(k)i}dy^{(k-2)i} \\ &\cdots\cdots\cdots\cdots\cdots\cdots\cdots\cdots\cdots\cdots\cdots\cdots \\ p_{(k)} &= p_{(k)i}dx^i. \end{aligned}$$

Proposition 8.7.3 *With respect to a transformation of coordinates on $E = \mathrm{Osc}^k M$ we have*

$$(7.9)' \qquad \widetilde{p}_{(\alpha)} = p_{(\alpha)} \ (\alpha = 1, ..., k).$$

Indeed, using (7.4), we obtain:

$$\begin{aligned} & p_{(1)i}dx^i + p_{(2)i}dy^{(1)i} + \cdots + p_{(k)i}dy^{(k)i} = \\ & = (\frac{\partial \widetilde{y}^{(1)m}}{\partial y^{(1)i}}\widetilde{p}_{(1)m} + \cdots + \frac{\partial \widetilde{y}^{(k)m}}{\partial y^{(1)i}}\widetilde{p}_{(k)m})dx^i + \\ & +(\frac{\partial \widetilde{y}^{(2)m}}{\partial y^{(2)i}}\widetilde{p}_{(2)m} + \cdots + \frac{\partial \widetilde{y}^{(k)m}}{\partial y^{(2)i}}\widetilde{p}_{(k)m})dy^{(1)i} + \cdots + \\ & +\frac{\partial \widetilde{y}^{(k)m}}{\partial y^{(k)i}}\widetilde{p}_{(k)m}dy^{(k-1)i} = \widetilde{p}_{(1)m}(\frac{\partial \widetilde{y}^{(1)m}}{\partial y^{(1)i}}dx^i) + \\ & +\widetilde{p}_{(2)m}(\frac{\partial \widetilde{y}^{(2)m}}{\partial y^{(1)i}}dx^i + \frac{\partial \widetilde{y}^{(2)m}}{\partial y^{(2)i}}dy^{(1)i}) + \cdots + \\ & +\widetilde{p}_{(k)m}(\frac{\partial \widetilde{y}^{(k)m}}{\partial y^{(1)i}}dx^i + \cdots + \frac{\partial \widetilde{y}^{(k)m}}{\partial y^{(k)i}}dy^{(k-1)i}) = \\ & = \widetilde{p}_{(1)m}d\widetilde{x}^m + \cdots + \widetilde{p}_{(k)m}dy^{(k-1)m} = \widetilde{p}_{(1)} \text{ etc.} \end{aligned}$$

We observe that $p_{(k)}$ is an 1–form field on $\widetilde{E}$, while $p_{(1)},...,p_{(k-1)}$ are 1–forms along the considered curve c.

The exterior differential of the 1–form $p_{(k)}$ is given by

$$
\begin{aligned}
dp_{(k)i} &= \frac{1}{2}\left(\frac{\delta p_{(k)i}}{\delta x^j} - \frac{\delta p_{(k)j}}{\delta x^i}\right)dx^j \wedge dx^i + \\
&+ \sum_{\alpha=1}^{k-1} \frac{\delta p_{(k)i}}{\delta y^{(\alpha)j}}\delta y^{(\alpha)j} \wedge dx^i + \theta,
\end{aligned} \tag{7.10}
$$

where N is a nonlinear connection arbitrarily fixed and

$$\theta = \frac{2}{k!} g_{ji}\delta y^{(k)j} \wedge dx^i. \tag{7.11}$$

Proposition 8.7.4 *θ from (7.11) is a 2–form globally defined on $\widetilde{E}$, if the Lagrangian L and the nonlinear connection N are globally defined on $\widetilde{E}$.*

Indeed, in this case, with respect to a transformation of coordinate on $\widetilde{E}$, we obtain $\theta = \widetilde{\theta}$.

Proposition 8.7.5

1° *The equation $dp_{(k)} = \theta$ holds if, and only if, the tensorial equations*

$$\frac{\delta p_{(k)i}}{\delta x^j} - \frac{\delta p_{(k)j}}{\delta x^i} = 0, \frac{\delta p_{(k)i}}{\delta y^{(\alpha)j}} = 0 \quad (\alpha = 1, ..., k-1).$$

are verified.

2° *In this case, the exterior differential of the 2–form θ vanishes.*

8.8 Regular Lagrangians. Canonical Nonlinear Connection

In Section 1 of the present chapter, we defined the notion of regular Lagrangian as being a differentiable Lagrangian $L(x, y^{(1)}, ..., y^{(k)})$ for which its Hessian, with respect to $y^{(k)i}$ is nonsingular. That is the condition (1.2) is verified.

The Example 8.1.1 shows that: If M is a paracompact manifold then on $\mathrm{Osc}^k M$ there exist regular Lagrangians.

The d–tensor field $g_{ij}(x, y^{(1)}, ..., y^{(k)})$ of a regular Lagrangian L is called *fundamental.*

A regular Lagrangian L determines some geometrical object fields on $\widetilde{E}$ depending only on L. One of them is a k–spray. Indeed, let g^{ij} be the contravariant of the fundamental tensor g_{ij}. We can assert:

Theorem 8.8.1 *For any regular Lagrangian* $L(x, y^{(1)}, ..., y^{(k)})$ *there exist k–sprays determined only by the Lagrangian L. One of them has the coefficients:*

$$(k+1)G^i = \frac{1}{2} g^{ij} \{\Gamma(\frac{\partial L}{\partial y^{(k)j}}) - \frac{\partial L}{\partial y^{(k-1)j}}\} \tag{8.1}$$

where Γ *is the operator* (2.6), *Ch.*6.

Proof. The Craig–Synge covector $\overset{(k-1)}{E}{}_i(L)$, (cf. (4.1)) is as follows

$$\overset{(k-1)}{E}{}_i(L) = (-1)^{k-1} \frac{1}{(k-1)!} \{\frac{\partial L}{\partial y^{(k-1)i}} - \frac{d}{dt} \frac{\partial L}{\partial y^{(k)i}}\}.$$

Along a smooth curve $c : I \longrightarrow M$ it can be expressed in the form

$$\begin{aligned} \overset{(k-1)}{E}{}_i(L) = (-1)^{k-1} \frac{1}{(k-1)!} \{ & \frac{\partial L}{\partial y^{(k-1)i}} - \Gamma(\frac{\partial L}{\partial y^{(k)i}}) - \\ & - \frac{2}{k!} g_{ij} \frac{d^{k+1} x^i}{dt^{k+1}}\}, \end{aligned} \tag{8.2}$$

Γ being the operator (2.6), Ch.6.

But the differential equations $g^{ij} \overset{(k-1)}{E}{}_i(L) = 0$ have a geometrical meaning and, from (8.2), it follows that these equations can be written in the form (3.5), Ch.6:

$$\frac{d^{k+1} x^i}{dt^{k+1}} + (k+1)! G^i(x, y^{(1)}, ..., y^{(k)}) = 0,$$

where $(k+1)G^i$ are as in (8.1). Now, applying Theorem 6.3.3, the enunciated property is proved. **q.e.d.**

The k–spray S with coefficients (8.1) is as follows:

$$(8.3)\quad S = y^{(1)i}\frac{\partial L}{\partial x^i} + 2y^{(2)i}\frac{\partial L}{\partial y^{(1)i}} + \cdots + ky^{(k)i}\frac{\partial L}{\partial y^{(k-1)i}} - (k+1)G^i\frac{\partial}{\partial y^{(k)i}},$$

where $(k+1)G^i$ are from (8.1). S depending only on the Lagrangian L, it will be called *canonical.*

Clearly, if $L(x, y^{(1)}, ..., y^{(k)})$ is globally defined on $\mathrm{Osc}^k M$, because the geometrical character of the canonical k–spray S, it has the same property.

Taking into account Theorem 6.7.1, we can formulate:

Theorem 8.8.2 *If $L(x, y^{(1)}, ..., y^{(k)})$ is a regular Lagrangian, then on $\widetilde{E}$ there exist nonlinear connections N determinated only by the Lagrangian L. One of them has the following dual coefficients:*

$$\underset{(1)}{M}{}^i{}_j = \frac{1}{2(k+1)}\frac{\partial}{\partial y^{(k)j}}\{g^{ir}[\Gamma(\frac{\partial L}{\partial y^{(k)r}}) - \frac{\partial L}{\partial y^{(k-1)r}}]\}$$

$$(8.4)\qquad \underset{(2)}{M}{}^i{}_j = \frac{1}{2}\{S\,\underset{(1)}{M}{}^i{}_j + \underset{(1)}{M}{}^i{}_m\,\underset{(1)}{M}{}^m{}_j\}$$

....

$$\underset{(k)}{M}{}^i{}_j = \frac{1}{k}\{S\,\underset{(k-1)}{M}{}^i{}_j + \underset{(1)}{M}{}^i{}_m\,\underset{(k-1)}{M}{}^m{}_j\}$$

where S is the canonical spray (8.3), *with the coefficients* (8.1).

This theorem is important in the construction of the geometry of Lagrange spaces of order k.

Of course, the nonlinear connection N determined by the dual coefficients (8.4), depending only on the Lagrangian L, will be called canonical, too.

Now, we can construct the Liouville d–vector fields $z^{(1)i}, ..., z^{(k)i}$ with the help of canonical nonlinear connection N.

So, by virtue of (8.4), we have the d–vector field

$$(8.5)\qquad kz^{(k)i} = ky^{(k)i} + (k-1)\underset{(1)}{M}{}^i{}_m y^{(k-1)m} + \cdots + \underset{(k-1)}{M}{}^i{}_m y^{(1)m}.$$

which depends only on the Lagrangian L.

8.9 Problems

1. Prove directly that $\overset{\circ}{E}_i (L), ..., \overset{k}{E}_i (L)$, are d–covector fields.

2. Show that the integrals of action $I(c)$ and $I'(c)$ from (4.4) have the same extremal curves.

3. Show that the equations (4.3) hold.

4. Calculate the higher–order energies for the Lagrangian L from (5.4), Ch.4, where $z^{(k)i}$ is given by (8.5).

5. Write the function $\mathcal{F}^3(L, \phi)$ from the Noether Theorem 6.2 in the case of the differentiable Lagrangian (5.4), Ch.4 of the electrodynamics of order 3.

6. For a regular Lagrangian $L(x, y^{(1)}, y^{(2)}, y^{(3)})$ determine:

   - The canonical 3–spray.
   - The canonical 3–nonlinear connection.
   - The form $\overset{\circ}{E}_j (L) = 0$ of the Euler–Lagrange equation.

Chapter 9

Prolongation of the Riemannian, Finslerian and Lagrangian Structures to the k–Osculator Bundle

In this chapter we shall give a solution of the difficult problem of the prolongation to the manifold $\text{Osc}^k M$ of the Riemannian, Finslerian or Lagrangian structures, defined on the base manifold M.

Several geometers, as E. Bompiani, Ch. Ehresman, A. Morimoto, S. Kobayashi, R. Miron and Gh. Atanasiu have studied this problem for the Riemannian structures. The problem of prolongation of the Finslerian or Lagrangian structures, from the tangent bundle TM to the k-osculator bundle $\text{Osc}^k M$ appeared for the first time in our joint papers with Gh. Atanasiu [see References].

9.1 Prolongation to $\text{Osc}^k M$ of the Riemannian Structures

Let $\mathcal{R}^n = (M, g)$ be a Riemannian space, g being a Riemannian metric defined on M, having the local coordinates $g_{ij}(x)$, $x \in U \subset M$. We extend g_{ij} to $\pi^{-1}(U) \subset E = \text{Osc}^k M$, setting

$$(g_{ij} \circ \pi)(u) = g_{ij}(x), \quad \forall u \in \pi^{-1}(U), \ \pi(u) = x.$$

In this case $g_{ij} \circ \pi$ gives a d-tensor field on E. We also denoted it by g_{ij}.

The problem of prolongation of the Riemannian space $\mathcal{R}^n = (M, g)$ can be stated as follows: the Riemannian structure g on the manifold M being apriori given, determines a Riemannian structure G on the manifold $\mathrm{Osc}^k M$, so that G be provided only by the structure g.

In other words, the problem is to find a lifting $G(x, y^{(1)}, ..., y^{(k)})$ of the structure $g(x)$, so that G be a Riemannian structure on $\mathrm{Osc}^k M$ and G depend only on g. Of course, this lifting will be of Sasaki type.

Therefore, first of all, we determine a canonical nonlinear connection on $\widetilde{\mathrm{Osc}^k M} = \widetilde{E}$ depending only on g.

We denote by $\gamma^m_{ij}(x)$ the Christoffel symbols of g and prove:

Theorem 9.1.1 *There exist nonlinear connections N on $\widetilde{E}$ determined only by the Riemannian structure $g(x)$. One of them has the following coefficients:*

$$\underset{(1)}{M}{}^i{}_j = \gamma^i{}_{jm}(x) y^{(1)m},$$

$$\underset{(2)}{M}{}^i{}_j = \frac{1}{2}\left(\Gamma \underset{(1)}{M}{}^i{}_j + \underset{(1)}{M}{}^i{}_m \underset{(1)}{M}{}^m{}_j\right), \tag{1.1}$$

$$\cdots\cdots\cdots\cdots\cdots\cdots\cdots\cdots\cdots\cdots\cdots$$

$$\underset{(k)}{M}{}^i{}_j = \frac{1}{k}\left(\Gamma \underset{(k-1)}{M}{}^i{}_j + \underset{(1)}{M}{}^i{}_m \underset{(k-1)}{M}{}^m{}_j\right),$$

where Γ is the known operator

$$\Gamma = y^{(1)i}\frac{\partial}{\partial x^i} + \cdots + k y^{(k)i}\frac{\partial}{\partial y^{(k-1)i}}. \tag{1.2}$$

Proof. Remarking that $\underset{(1)}{M}{}^i{}_j$ depends only on the structure g and looking at the expression (1.2) of the operator Γ it follows that: $\underset{(1)}{M}{}^i{}_j$, $\underset{(2)}{M}{}^i{}_j, ..., \underset{(k)}{M}{}^i{}_j$ are provided only by the structure g.

We also notice that:

$$(*)\qquad \begin{aligned} &\underset{(1)}{M}{}^i{}_j = \underset{(1)}{M}{}^i{}_j(x, y^{(1)}),\\ &\underset{(2)}{M}{}^i{}_j = \underset{(2)}{M}{}^i{}_j(x, y^{(1)}, y^{(2)}),\\ &\dots\dots\dots\dots\dots\dots\dots\dots\\ &\underset{(k-1)}{M}{}^i{}_j = \underset{(k-1)}{M}{}^i{}_j(x, y^{(1)}, ..., y^{(k-1)}). \end{aligned}$$

Applying the Theorem 6.6.2, it follows that $\left(\underset{(1)}{M}{}^i{}_j, ..., \underset{(k)}{M}{}^i{}_j\right)$ are the dual coefficients of a nonlinear connection if, and only if, with respect to a transformation of local coordinates on $\widetilde{E}$, we have

$$(**)\qquad \begin{aligned} &\underset{(1)}{M}{}^m{}_j \frac{\partial \widetilde{x}^i}{\partial x^m} = \underset{(1)}{\widetilde{M}}{}^i{}_m \frac{\partial \widetilde{x}^m}{\partial x^j} + \frac{\partial \widetilde{y}^{(1)i}}{\partial x^j}\\ &\underset{(2)}{M}{}^m{}_j \frac{\partial \widetilde{x}^i}{\partial x^m} = \underset{(2)}{\widetilde{M}}{}^i{}_m \frac{\partial \widetilde{x}^m}{\partial x^j} + \underset{(1)}{\widetilde{M}}{}^i{}_m \frac{\partial \widetilde{y}^{(1)m}}{\partial x^j} + \frac{\partial \widetilde{y}^{(2)i}}{\partial x^j}\\ &\dots\dots\dots\dots\dots\dots\dots\dots\dots\dots\dots\dots\\ &\underset{(k)}{M}{}^m{}_j \frac{\partial \widetilde{x}^i}{\partial x^m} = \underset{(k)}{\widetilde{M}}{}^i{}_m \frac{\partial \widetilde{x}^m}{\partial x^j} + \underset{(k-1)}{\widetilde{M}}{}^i{}_m \frac{\partial \widetilde{y}^{(1)m}}{\partial x^j} + \dots +\\ &\qquad\qquad + \underset{(1)}{\widetilde{M}}{}^i{}_m \frac{\partial \widetilde{y}^{(k-1)m}}{\partial x^j} + \frac{\partial \widetilde{y}^{(k)i}}{\partial x^m}. \end{aligned}$$

We shall prove that $\underset{(1)}{M}{}^m{}_j, ..., \underset{(k-1)}{M}{}^m{}_j$ from (1.1) verify the conditions $(**)$.

Indeed, the Christoffel symbols $\gamma^i{}_{jm}(x)$ are transformed, with respect to (1.5), Ch.6, as follows:

$$\gamma^s{}_{jm} \frac{\partial \widetilde{x}^i}{\partial x^s} = \widetilde{\gamma}^i{}_{pq} \frac{\partial \widetilde{x}^p}{\partial x^j} \frac{\partial \widetilde{x}^q}{\partial x^m} + \frac{\partial^2 \widetilde{x}^i}{\partial x^j \partial x^m}.$$

We also have $y^{(1)m} = \dfrac{\partial x^m}{\partial \widetilde{x}^r} \widetilde{y}^r$. These last two equalities show that $\underset{(1)}{M}{}^i{}_j = \gamma^i{}_{jm}(x) y^{(1)m}$ verifies the first equality $(**)$.

Now, by virtue of $(*)$ and of the Lemma 6.2.2, it follows that, with respect to a transformation of local coordinates (1.5), Ch.6, on E, we have

$$(***)\qquad \Gamma \underset{(\beta)}{M}{}^m{}_j = \widetilde{\Gamma}\, \underset{(\beta)}{\widetilde{M}}{}^m{}_j, \quad (\beta = 1, ..., k-1).$$

Now, by induction, we assume that for $\underset{(1)}{M}{}^i{}_j, ..., \underset{(\alpha-1)}{M}{}^i{}_j$ $(\alpha - 1 < k)$, the rules $(**)$ are verified. Then, from the action of the operator Γ on the equality from $(\alpha-1)$–line of $(**)$ we get

$$\Gamma\left(\underset{(\alpha-1)}{M}{}^m{}_j \frac{\partial \widetilde{x}^i}{\partial x^m}\right) =$$
$$= \Gamma\left(\underset{(\alpha-1)}{\widetilde{M}}{}^i{}_m \frac{\partial \widetilde{x}^m}{\partial x^j} + \cdots + \underset{(1)}{\widetilde{M}}{}^i{}_m \frac{\partial \widetilde{y}^{(\alpha-2)m}}{\partial x^j} + \frac{\partial \widetilde{y}^{(\alpha-1)i}}{\partial x^j}\right).$$

This equality leads to

$$\begin{aligned}
&\left(\Gamma \underset{(\alpha-1)}{M}{}^m{}_j\right)\frac{\partial \widetilde{x}^i}{\partial x^m} + \underset{(\alpha-1)}{M}{}^m{}_j \frac{\partial \widetilde{y}^{(1)i}}{\partial x^m} = \left(\widetilde{\Gamma}\, \underset{(\alpha-1)}{\widetilde{M}}{}^i{}_m\right)\frac{\partial \widetilde{x}^m}{\partial x^j} + \\
(a)\qquad & + \cdots + \left(\widetilde{\Gamma}\, \underset{(\alpha-1)}{\widetilde{M}}{}^i{}_m\right)\frac{\partial \widetilde{y}^{(\alpha-2)m}}{\partial x^j} + \\
& + \underset{(\alpha-1)}{\widetilde{M}}{}^i{}_m \frac{\partial \widetilde{y}^{(1)m}}{\partial x^j} + \cdots + (\alpha-1)\, \underset{(1)}{\widetilde{M}}{}^i{}_m \frac{\partial \widetilde{y}^{(\alpha-1)m}}{\partial x^j} + \alpha \frac{\partial \widetilde{y}^{(\alpha)i}}{\partial x^j}.
\end{aligned}$$

Also, from the rules of transformation of $\underset{(1)}{M}{}^i{}_j, ..., \underset{(\alpha-1)}{M}{}^i{}_j$ we obtain:

$$\begin{aligned}
&\left(\underset{(1)}{M}{}^m{}_r \underset{(\alpha-1)}{M}{}^r{}_j\right)\frac{\partial \widetilde{x}^i}{\partial x^m} = \\
(b)\qquad & = \underset{(1)}{M}{}^i{}_s\left\{\underset{(\alpha-1)}{\widetilde{M}}{}^s{}_r \frac{\partial \widetilde{x}^r}{\partial x^j} + \cdots + \underset{(1)}{\widetilde{M}}{}^s{}_r \frac{\partial \widetilde{y}^{(\alpha-2)r}}{\partial x^j} + \frac{\partial \widetilde{y}^{(\alpha-1)s}}{\partial x^j}\right\} + \\
& + \frac{\partial x^m}{\partial \widetilde{x}^s}\frac{\partial \widetilde{y}^{(1)i}}{\partial x^m}\left\{\underset{(\alpha-1)}{\widetilde{M}}{}^s{}_r \frac{\partial \widetilde{x}^r}{\partial x^j} + \cdots + \underset{(1)}{\widetilde{M}}{}^s{}_r \frac{\partial \widetilde{y}^{(\alpha-2)r}}{\partial x^j} + \frac{\partial \widetilde{y}^{(\alpha-1)s}}{\partial x^j}\right\}.
\end{aligned}$$

Therefore, the rule of transformation for

$$\underset{(\alpha)}{M}{}^i{}_j = \frac{1}{\alpha}\left\{\Gamma \underset{(\alpha-1)}{M}{}^i{}_j + \underset{(1)}{M}{}^i{}_m \underset{(\alpha-1)}{M}{}^m{}_j\right\}$$

is obtained from (a) and (b). After some calculations, it follows:

$$\underset{(\alpha)}{M}{}^{m}{}_{j}\frac{\partial \widetilde{x}^{i'}}{\partial x^{m}} = \underset{(\alpha)}{\widetilde{M}}{}^{i}{}_{m}\frac{\partial \widetilde{x}^{m}}{\partial x^{j}} + \cdots + \underset{(1)}{\widetilde{M}}{}^{i}{}_{m}\frac{\partial \widetilde{y}^{(\alpha-1)m}}{\partial x^{j}} + \frac{\partial \widetilde{y}^{(\alpha)i}}{\partial x^{j}}.$$

Concluding, $\underset{(1)}{M}{}^{i}{}_{j}, \dots, \underset{(k)}{M}{}^{i}{}_{j}$ from (1.1) are the dual coefficients of a nonlinear connection N, constructed only with the help of the Riemannian structure g. **q.e.d.**

The previous nonlinear connection is a canonical one for the problem of prolongation of g to E. It has an important property.

Theorem 9.1.2 *The canonical nonlinear connection N, with the dual coefficients* (1.1) *is integrable if, and only if, the Riemannian space $\mathcal{R}^n$ is locally flat.*

Proof. Assuming that the Riemannian space $\mathcal{R}^n = (M, g)$ is locally flat, it follows that in every point $x \in M$ there exists a local chart (U, φ), where the Christoffel symbols $\gamma^{i}{}_{jm}(x)$, $x \in U$, vanish. It follows, from (1.1), that the dual coefficients $\underset{(1)}{M}{}^{i}{}_{j}, \dots, \underset{(k)}{M}{}^{i}{}_{j}$ vanish on $\pi^{-1}(U)$. Applying Theorem 7.4.2, it follows that N is integrable.

Conversely, if the canonical nonlinear connection N is integrable, by means of the same Theorem 7.4.2, on a domain of local chart $\pi^{-1}(U)$ on E, we obtain:

$$(*) \qquad \underset{(01)}{R}{}^{i}{}_{jm} = \frac{\delta \underset{(1)}{M}{}^{i}{}_{j}}{\delta x^{m}} - \frac{\delta \underset{(1)}{M}{}^{i}{}_{m}}{\delta x^{i}} = y^{(1)s} r_{s}{}^{m}{}_{ij}(x) = 0,$$

where $r_{j}{}^{i}{}_{mh}(x)$ is the curvature tensor field of the Riemannian structure $g(x)$.

Deriving $(*)$ with respect to $y^{(1)i}$ on $\pi^{-1}(U)$ we deduce $r_{s}{}^{m}{}_{ij}(x) = 0$, $x \in U \subset M$. Consequently, the Riemannian space $\mathcal{R}^n$ is locally flat. **q.e.d.**

Now, we can use the canonical nonlinear connection N with the dual coefficients $\left(\underset{(1)}{M}{}^{i}{}_{j}, \dots, \underset{(k)}{M}{}^{i}{}_{j}\right)$ from the equations (1.1) and the coefficients $\left(\underset{(1)}{N}{}^{i}{}_{j}, ..., \underset{(k)}{N}{}^{i}{}_{j}\right)$ determined from dual coefficients for constructing the adapted basis $\left(\frac{\delta}{\delta x^{i}}, ..., \frac{\delta}{\delta y^{(k)}}\right)$ and adapted cobasis $(dx^{i}, \delta y^{(1)i}, ..., \delta y^{(k)i})$, which depend only on the dual coefficients of N.

Theorem 9.1.3 *The pair* $\mathrm{Prol}^{k}\mathcal{R}^{n} = \left(\widetilde{\mathrm{Osc}^{k}M}, G\right)$, *where*

$$(1.3)\quad G = g_{ij}(x)dx^{i}\otimes dx^{j} + g_{ij}(x)\delta y^{(1)i}\otimes\delta y^{(1)j} + \cdots + g_{ij}(x)\delta y^{(k)i}\otimes\delta y^{(k)j},$$

is a Riemannian space of dimension $(k+1)n$, *whose metric structure* G *depends only on the structure* $g(x)$ *of the apriori given Riemann space* $\mathcal{R}^{n} = (M, g)$.

Proof. Taking into account the rules of transformation of $g_{ij}, dx^{i}, \delta y^{(1)i}, ..., \delta y^{(k)i}$, with respect to (1.5), Ch.6, it follows that G is a tensor field on $\widetilde{\mathrm{Osc}^{k}M}$, of the type $(0,2)$, symmetric. From (1.3), we deduce that rank $\|G\| = (k+1)n$, and G is positively defined. So G is a Riemannian structure on $\widetilde{\mathrm{Osc}^{k}M}$. The form (1.3) of G and Theorem 9.1.2 imply that G is constructed only by means of the Riemannian structure g. So, the theorem is proved. **q.e.d.**

The existence of the Riemannian space $\mathrm{Prol}^{k}\mathcal{R}^{n} = (\widetilde{\mathrm{Osc}^{k}M}, G)$ solves the enunciated problem. This space is called the *prolongation of order k of the space* $\mathcal{R}^{n} = (M, g)$. Also, we say that G is *Sasaki N–lift* of the Riemannian structure g.

The geometry of the space $\mathrm{Prol}^{k}\mathcal{R}^{n}$ can be studied by means of the method given in Chapters 6 and 7.

We shall prove here only the following theorem:

Theorem 9.1.4 *There exists a unique N-linear connection D, compatible with the Riemannian structure G,* (1.3) *for which the h– and v_α–torsions vanish. The coefficients* $D\Gamma(N) = (L^{i}{}_{jm}, \underset{(1)}{C}{}^{i}{}_{jm}, ..., \underset{(k)}{C}{}^{i}{}_{jm})$ *of D are given by*

$$(1.4)\qquad L^{i}{}_{jm} = \gamma^{i}{}_{jm},\quad \underset{(\alpha)}{C}{}^{i}{}_{jm} = 0,\quad (\alpha = 1, ..., k).$$

Proof. The condition $D_X G = 0$, $\forall X \in \mathcal{X}(\widetilde{E})$ and (1.3) lead to the tensorial equations

$$g_{ij|m} = 0,\ g_{ij} \overset{(\alpha)}{\Big|}_{m} = 0,\quad (\alpha = 1, ..., k), \tag{1.5}$$

and conversely.

In the conditions $L^i{}_{jm} = L^i{}_{mj}$, $\underset{(\alpha)}{C}{}^i{}_{jm} = \underset{(\alpha)}{C}{}^i{}_{mj}$, $(\alpha = 1, ..., k)$, these equations have a unique solution $(L^i{}_{jm}, \underset{(\alpha)}{C}{}^i{}_{jm})$, $(\alpha = 1, ..., k)$, and this is (1.4). **q.e.d.**

Remark. The Lionville vector field $z^{(k)i}$, (8,5), Ch.8, corresponding to the canonical nonlinear connection N, (1.1) is used in Example 8.1.1.

9.2 Prolongation to the k–Osculator Bundle of the Finslerian Structures

The prolongation of a Finsler structure can be introduced in a similar way. Let $F(x, y^{(1)})$ be a fundamental function of a Finsler space $F^n = (M, F)$ and

$$g_{ij}(x, y^{(1)}) = \frac{1}{2}\ \frac{\partial^2 F^2}{\partial y^{(1)i} \partial y^{(1)j}}, \tag{2.1}$$

its fundamental tensor field (see §11, Ch.1).

Assume that $g_{ij}(x, y^{(1)})$ is positively defined.

The problem is to determine a Riemannian structure G on $\widetilde{\mathrm{Osc}^k M}$ which depends only on the fundamental tensor $g_{ij}(x, y^{(1)})$ of the Finsler space F^n.

A solution of this problem can be obtained by the same method as in the case of the prolongation of Riemannian spaces, used in the previous section.

First of all, we extend the d-tensor field g_{ij} to $\widetilde{E} = \widetilde{\mathrm{Osc}^k M}$, setting

$$(g_{ij} \circ \pi_1^k)(u) = g_{ij}(x, y^{(1)}),\ \forall u \in \widetilde{\mathrm{Osc}^k M},\ \pi_1^k(u) = (x, y^{(1)}) \tag{2.2}$$

and we identify $g_{ij} \circ \pi_1^k$ and g_{ij}.

We denote by $\gamma^i{}_{jm}(x, y^{(1)})$ the Christoffel symbols of the d-tensor field $g_{ij}(x, y^{(1)})$.

The canonical 2-spray of F^n is given by

$$\frac{d^2x^i}{dt^2} + 2G^i\left(x, \frac{dx}{dt}\right) = 0$$

where

$$G^i = \frac{1}{2}\gamma^i{}_{jm}y^{(1)j}y^{(1)m}. \tag{2.3}$$

The Cartan nonlinear connection on $\widetilde{TM}$ has the following coefficients:

$$G^i{}_j = \frac{\partial G^i}{\partial y^{(1)j}}. \tag{2.4}$$

Now we can prove

Theorem 9.2.1 *There exist nonlinear connections on $\widetilde{E}$ determined only by fundamental tensor $g_{ij}(x, y^{(1)})$ of the Finsler space $F^n = (M, F)$. One of them has the following dual coefficients:*

$$\begin{aligned} \underset{(1)}{M}{}^i{}_j &= G^i{}_j \\ \underset{(2)}{M}{}^i{}_j &= \frac{1}{2}\left(\Gamma G^i{}_j + G^i{}_m \underset{(1)}{M}{}^m{}_j\right) \\ &\cdots\cdots\cdots\cdots\cdots\cdots , \\ \underset{(k)}{M}{}^i{}_j &= \frac{1}{k}\left(\Gamma \underset{(k-1)}{M}{}^i{}_j + G^i{}_m \underset{(k-1)}{M}{}^m{}_j\right) \end{aligned} \tag{2.5}$$

where Γ is the operator (1.2).

Proof. First, we notice that:

$$\begin{aligned} \underset{(1)}{M}{}^i{}_j &= G^i{}_j(x, y^{(1)}), \\ \underset{(2)}{M}{}^i{}_j &= \underset{(2)}{M}{}^i{}_j(x, y^{(1)}, y^{(2)}), \\ &\cdots\cdots\cdots\cdots\cdots\cdots \\ \underset{(k-1)}{M}{}^i{}_j &= \underset{(k-1)}{M}{}^i{}_j(x, y^{(1)}, \ldots, y^{(k-1)}). \end{aligned} \tag{2.6}$$

Next, with respect to (1.5), Ch.6, $\underset{(1)}{M}{}^i{}_j$ is transformed as follows

$$\underset{(1)}{M}{}^m{}_j \frac{\partial \widetilde{x}^i}{\partial x^m} = \widetilde{M}^i{}_m \frac{\partial \widetilde{x}^m}{\partial x^j} + \frac{\partial \widetilde{y}^{(1)i}}{\partial x^j}. \tag{2.7}$$

Also, according to Lemma 6.2.2, we have

$$\Gamma \underset{(\alpha)}{M}{}^i{}_j = \widetilde{\Gamma}\ \underset{(\alpha)}{\widetilde{M}}{}^i{}_j,\ (\alpha = 1, ..., k-1). \tag{2.8}$$

To prove the theorem, one needs to show that $\underset{(1)}{M}{}^i{}_j, ..., \underset{(k)}{M}{}^i{}_j$ from (2.5) is transformed by the rule (**) from the previous section. The first of these equalities is verified since (2.7) holds. The others can be established just like those in the proof of Theorem 9.1.1, by induction, taking into account (2.6), (2.7) and (2.8).

The conclusion is that $\underset{(1)}{M}{}^i{}_j, ..., \underset{(k)}{M}{}^i{}_j$ from (2.5) are the dual coefficients of a nonlinear connection. Looking to (2.3), (2.4) it results, by means of (2.5), that $\underset{(1)}{M}{}^i{}_j, ..., \underset{(k)}{M}{}^i{}_j$ depend only on the fundamental tensor g_{ij} of the Finsler space F^n. **q.e.d.**

A first consequence of the previous theorem is as follows.

Proposition 9.2.1 *If the Finsler space F^n is locally Minkowski, then the nonlinear connection with the dual coefficients* (2.5) *is integrable.*

Proof. In the case when the Finsler space F^n is locally Minkowski, there exists, around every point $x \in M$, an open set U with the property that $G^i{}_j = 0$ on $\pi^{-1}(U) \subset TM$. It follows, from (2.5), that $\underset{(1)}{M}{}^i{}_j = \cdots = \underset{(k)}{M}{}^i{}_j = 0$ on $(\pi_1^{(k)})^{-1}(U) \subset \mathrm{Osc}^k M$. In this case, all tensors $\underset{(0\alpha)}{R}{}^i{}_{jk}$ vanish, for $(\alpha = 1, ..., k)$. Applying Theorem 7.4.2, the integrability of the mentioned nonlinear connection becomes apparent. **q.e.d.**

Let $\left(\frac{\delta}{\delta x^i}, \frac{\delta}{\delta y^{(1)i}}, ..., \frac{\delta}{\delta y^{(k)i}}\right)$ and $(dx^i, \delta y^{(1)i}, ..., \delta y^{(k)i})$ be the adapted basis and adapted cobasis coefficients (2.5).

Theorem 9.2.2 *The pair* $\mathrm{Prol}^k F^n = (\widetilde{\mathrm{Osc}^k M}, G)$, $k \geq 2$, *where*

$$G = g_{ij}(x, y^{(1)})dx^i \otimes dx^j + g_{ij}(x, y^{(1)})\delta y^{(1)i} \otimes \delta y^{(1)j} + \cdots + \\ + g_{ij}(x, y^{(1)})\delta y^{(k)i} \otimes \delta y^{(k)j} \tag{2.9}$$

is a Riemannian space of dimension $(k+1)n$, *whose metric structure* G *depends only on the fundamental tensor field* $g_{ij}(x, y^{(1)})$ *of the apriori given Finsler space* $F^n = (M, F)$.

Proof. A transformation of coordinate (5.1), Ch.6, has as effect the transformations

$$\widetilde{g}_{ij} = \frac{\partial x^p}{\partial \widetilde{x}^i}\frac{\partial x^q}{\partial \widetilde{x}^j} g_{pq}, \ \delta \widetilde{y}^{(\alpha)i} = \frac{\partial \widetilde{x}^i}{\partial x^p}\delta y^{(\alpha)p}, \ (\alpha = 0, 1, ..., k, y^{(0)} = x).$$

Therefore, G is a symmetric covariant tensor field on $\widetilde{\mathrm{Osc}^k M}$. Because g_{ij} is positively defined, it follows that G is positively defined, too. Consequently, G is a Riemannian structure on $\widetilde{\mathrm{Osc}^k M}$. Evidently, it is constructed by means of $g_{ij}(x, y^{(1)})$ only. **q.e.d.**

This Riemannian space $\mathrm{Prol}^k F^n = (\widetilde{\mathrm{Osc}^k M}, G)$ is called the *prolongation of order k of the Finsler space* F^n. We say that G is Sasaki–N–lift of the metric tensor g_{ij}.

The differential geometry of $\mathrm{Prol}^k F^n$ can be studied using the general theory of the k–osculator bundle $(\mathrm{Osc}^k M, \pi, M)$ endowed with the nonlinear connection N, from (2.5), and with the Riemannian structure G.

This theory needs the following theorem:

Theorem 9.2.3 *There exists a unique N-linear connection D compatible with the Riemannian structure G,* (2.9) *for which h- and* v_α*-torsions vanish. The coefficients* $D\Gamma(N) = (L^i{}_{jk}, \underset{(1)}{C}{}^i{}_{jm}, ..., \underset{(k)}{C}{}^i{}_{jm})$ *of D are given by*

$$L^i{}_{jm} = F^i{}_{jm}, \ \underset{(1)}{C}{}^i{}_{jm} = C^i{}_{jm}, \underset{(2)}{C}{}^i{}_{jm} = \cdots = \underset{(k)}{C}{}^i{}_{jm} = 0, \tag{2.10}$$

where $(F^i{}_{jm}, C^i{}_{jm})$ *are the coefficients of the Cartan connection of the Finsler space* $F^n = (M, F)$.

Proof. The condition $D_X G = 0,\ \forall X \in \mathcal{X}(\widetilde{E})$ is equivalent to the tensorial equations:

$$g_{ij|m} = 0,\ g_{ij} \overset{(\alpha)}{\Big|}_{m} = 0,\ (\alpha = 1, ..., k). \tag{2.11}$$

Therefore, $D\Gamma(N)$ must satisfy the equations

$$(*) \qquad L^i{}_{jm} = L^i{}_{mj},\ \underset{(\alpha)}{C}{}^i{}_{jm} = \underset{(\alpha)}{C}{}^i{}_{mj},\ (\alpha = 1, ..., k),$$

$$(**) \qquad \begin{aligned} &\frac{\delta g_{ij}}{\delta x^m} - L^s{}_{im} g_{sj} - L^s{}_{jm} g_{is} = 0 \\ &\frac{\delta g_{ij}}{\delta y^{(\alpha)m}} - \underset{(\alpha)}{C}{}^s{}_{im} g_{sj} - \underset{(\alpha)}{C}{}^s{}_{jm} g_{is} = 0. \end{aligned}$$

It is not difficult to see that the previous system has a unique solution and this is

$$\begin{aligned} L^i{}_{jm} &= \frac{1}{2} g^{is} \left(\frac{\delta g_{sm}}{\delta x^j} + \frac{\delta g_{sj}}{\delta x^m} - \frac{\delta g_{jm}}{\delta x^s} \right) \\ \underset{(\alpha)}{C}{}^i{}_{jm} &= \frac{1}{2} g^{is} \left(\frac{\delta g_{sm}}{\delta y^{(\alpha)j}} + \frac{\delta g_{sj}}{\delta y^{(\alpha)m}} - \frac{\delta g_{jm}}{\delta y^{(\alpha)s}} \right),\ (\alpha = 1, ..., k). \end{aligned} \tag{2.12}$$

But g_{ij} depends only on x^i and $y^{(1)i}$. This means that

$$\frac{\delta g_{ij}}{\delta x^m} = \frac{\partial g_{ij}}{\partial x^m} - G^s_m \frac{\partial g_{ij}}{\partial y^{(1)s}}.$$

Hence $L^i{}_{jk} = F^i{}_{jk}(x, y^{(1)})$ are just the h-coefficients of the Cartan connection of the space F^n.

Further on, we deduce $\underset{(1)}{C}{}^i{}_{jk} = C^i{}_{jk}$, because $\dfrac{\delta g_{ij}}{\delta y^{(1)m}} = \dfrac{\partial g_{ij}}{\partial y^{(1)m}}$. These are the v-coefficients of Cartan connection of F^n. Finally, from $\dfrac{\delta g_{ij}}{\delta y^{(\alpha)m}} = 0,\ (\alpha = 2, ..., k)$ it follows $\underset{(2)}{C}{}^i{}_{jm} = \cdots = \underset{(k)}{C}{}^i{}_{jm} = 0$. The theorem is proved. **q.e.d.**

9.3 Prolongation to $\text{Osc}^k M$ of a Lagrangian Structure

Let us consider a Lagrange space $L^n = (M, L(x, y^{(1)}))$. We assume that its fundamental tensor

$$g_{ij}(x, y^{(1)}) = \frac{1}{2} \frac{\partial^2 L}{\partial y^{(1)i} \partial y^{(1)j}} \tag{3.1}$$

is positively defined.

The problem of the prolongation of L^n to $\text{Osc}^k M$ can be formulated as follows: Determine a Riemannian structure G on $\widetilde{E} = \widetilde{\text{Osc}^k M}$, which depends only on the Lagrangian $L(x, y^{(1)})$ of the space L^n.

Let $\overset{\circ}{N}$ be the canonical nonlinear connection of L^n (see §.9, Ch.1). Its coefficients are

$$\overset{\circ}{N}{}^i{}_j = \frac{\partial G^i}{\partial y^{(1)j}}, \; G^i = \frac{1}{4} g^{is} \left(\frac{\partial^2 L}{\partial y^s \partial x^m} y^{(1)m} - \frac{\partial L}{\partial x^s} \right). \tag{3.2}$$

The canonical $\overset{\circ}{N}$-connection of the space L^n has the coefficients

$$\begin{aligned} \overset{\circ}{L}{}^i{}_{jm} &= \frac{1}{2} g^{is} \left(\frac{\overset{\circ}{\delta} g_{sj}}{\delta x^m} + \frac{\overset{\circ}{\delta} g_{ms}}{\delta x^j} - \frac{\overset{\circ}{\delta} g_{jm}}{\delta x^s} \right) \\ \overset{\circ}{C}{}^i{}_{jm} &= \frac{1}{2} g^{is} \left(\frac{\partial g_{sj}}{\partial y^{(1)m}} + \frac{\partial g_{ms}}{\partial y^{(1)j}} - \frac{\partial g_{jm}}{\partial y^{(1)s}} \right). \end{aligned} \tag{3.3}$$

We can extend the fundamental tensor $g_{ij}(x, y^{(1)})$ to $\widetilde{\text{Osc}^k M}$, setting

$$(g_{ij} \circ \pi_1^k)(x, y^{(1)}, ..., y^{(k)}) = g_{ij}(x, y^{(1)}). \tag{3.4}$$

As in the previous section we can prove

Theorem 9.3.1 *There exist nonlinear connections N on $\widetilde{E}$ determined only by the Lagrangian $L(x, y^{(1)})$ of Lagrange space $L^n = (M, L)$. One of them has the dual coefficients (2.5), where $\underset{(1)}{M}{}^i{}_j = \overset{\circ}{N}{}^i{}_j$ is from (3.2).*

Using the adapted basis, constructed with the previous nonlinear connection, we can consider the Sasaki N–lift G of the fundamental tensor field (3.1).

Obviously, it is of the form (2.9). We can state:

Theorem 9.3.2 *The pair* $\text{Prol}^k L^n = (\widetilde{\text{Osc}^k M}, G)$, *where* G *is Sasaki* N*-lift of the fundamental tensor* $g_{ij}(x, y^{(1)})$ *from* (3.1) *of the Lagrange space* $L^n = (M, L(x, y^{(1)})$ *is a Riemannian space of dimension* $(k+1)m$, *whose metric structure* G *depends only on the Lagrangian* $L(x, y^{(1)})$.

Theorem 9.3.3 *There exists a unique* N*-linear connection* D *compatible with Riemannian structure* G *of the space* $\text{Prol}^k L^n = (\widetilde{\text{Osc}^k M}, G)$, *for which* h*- and* v_α*-torsions vanish. The coefficients* $D\Gamma(N) = (L^i{}_{jm}, \underset{(1)}{C}{}^i{}_{jm}, ..., \underset{(k)}{C}{}^i{}_{jm})$ *of* D *are given by*

$$L^i{}_{jm} = \overset{\circ}{L}{}^i{}_{jm}, \ \underset{(1)}{C}{}^i{}_{jm} = \overset{\circ}{C}{}^i{}_{jm}, \ \underset{(2)}{C}{}^i{}_{jm} = \cdots = \underset{(k)}{C}{}^i{}_{jm} = 0,$$

where $(\overset{\circ}{L}{}^i{}_{jk}, \overset{\circ}{C}{}^i{}_{jk})$ *are the coefficients* (3.1) *of the canonical connection of the Lagrange space* $L^n = (M, L)$.

We shall use the theory from the previous sections to study the higher-order Lagrange spaces.

9.4 Remarkable Regular Lagrangians of Order k

Let $\mathcal{R}^n = (M, \gamma_{ij}(x))$ be a Riemannian space, $\text{Prol}^k \mathcal{R}^n = (\widetilde{\text{Osc}^k M}, G)$ its prolongation of order k and N the canonical nonlinear connection, with the dual coefficients (1.1). We also consider the Liouville d-vector field $z^{(k)i}$ of N:

$$\begin{aligned} kz^{(k)i} &= ky^{(k)i} + (k-1) \underset{(1)}{M}{}^i{}_j(x, y^{(1)}) y^{(k-1)j} + \\ &+ (k-2) \underset{(2)}{M}{}^i{}_j(x, y^{(1)}, y^{(2)}) y^{(k-2)j} + \cdots + \\ &+ \underset{(k-1)}{M}{}^i{}_j(x, y^{(1)}, ..., y^{(k-1)}) y^{(1)j}. \end{aligned}$$

We can prove:

Theorem 9.4.1 *The function α^2 given by*

$$\alpha^2(x, y^{(1)}, ..., y^{(k)}) = \gamma_{ij}(x) z^{(k)i} z^{(k)j} \tag{4.1}$$

is a regular Lagrangian defined on $\widetilde{\mathrm{Osc}^k M}$.

Proof. Indeed, $\gamma_{ij} \circ \pi$ being a d-tensor field on $\widetilde{\mathrm{Osc}^k M}$ and $z^{(k)i}$ a d–vector field on $\widetilde{\mathrm{Osc}^k M}$, it follows that α^2 is a differentiable Lagrangian. It is regular, since by means of (4.1), we obtain

$$\gamma_{ij} = \frac{1}{2} \frac{\partial^2 \alpha^2}{\partial y^{(k)i} \partial y^{(k)j}} \text{ on } \widetilde{\mathrm{Osc}^k M}. \tag{4.2}$$

q.e.d.

Corollary 9.4.1 *The function*

$$\begin{aligned} L(x, y^{(1)}, ..., y^{(k)}) &= \alpha^2(x, y^{(1)}, ..., y^{(k)}) + \\ &+ b_i(x, y^{(1)}, ..., y^{(k-1)}) z^{(k)i} + a(x, y^{(1)}, ..., y^{(k-1)}), \end{aligned} \tag{4.3}$$

where α^2 is from (4.1), $b_i(x, y^{(1)}, ..., y^{(k-1)})$ is a d–covector field and $a(x, y^{(1)}, ..., y^{(k-1)})$ a function, is a regular Lagrangian.

Of course, if the d–covector b_i and the function a are globally defined on $\mathrm{Osc}^k M$, then the Lagrangian L from (4.3) has the same property.

This is the Lagrangian of the higher-order electrodynamics.

Another remarkable Lagrangian, which is an extension of the Randers function from §1, Ch.5, is as follows

Theorem 9.4.2 *The function*

$$L(x, y^{(1)}, ..., y^{(k)}) = \{\alpha(x, y^{(1)}, ..., y^{(k)}) + \beta(x, y^{(1)}, ..., y^{(k)})\}^2 \tag{4.4}$$

with α from (4.1) and β the following 1–form

$$\beta(x, y^{(1)}, ..., y^{(k)}) = b_i(x, y^{(1)}, ..., y^{(k-1)}) z^{(k)i} \tag{4.5}$$

is a regular Lagrangian, in the points of $\widetilde{\mathrm{Osc}^k M}$ where $\beta > 0$.

Proof. Obviously, L from (4.4) is differentiable on $\widetilde{\mathrm{Osc}^k M}$ because $\alpha = \sqrt{\gamma_{ij} z^{(k)i} z^{(k)j}}$ and β have this property. We prove that L is regular in the points of $\widetilde{\mathrm{Osc}^k M}$ where $\beta > 0$. Indeed, denoting

$$\overset{\circ}{\ell}_i = \frac{1}{\alpha}\gamma_{ij} z^{(k)j},\ \ell_i = \overset{\circ}{\ell}_i + b_i,\ p = \frac{\alpha+\beta}{\alpha} \tag{4.6}$$

and remarking that

$$g_{ij} = \frac{1}{2}\frac{\partial^2(\alpha+\beta)^2}{\partial y^{(k)i}\partial y^{(k)j}} = (\alpha+\beta)\frac{\partial^2 \alpha}{\partial y^{(k)i}\partial y^{(k)j}} + \frac{\partial(\alpha+\beta)}{\partial y^{(k)i}}\frac{\partial(\alpha+\beta)}{\partial y^{(k)j}}$$

we obtain

$$g_{ij} = p[\gamma_{ij} - \overset{\circ}{\ell}_i \overset{\circ}{\ell}_j] + \ell_i \ell_j. \tag{4.7}$$

Applying a known method [183], we can prove that there exists g^{ij} with the property $g_{ij} g^{jm} = \delta_i^{\ m}$ and that

$$\det \|g_{ij}\| = p^{n+1} \det \|\gamma_{ij}\|. \tag{4.8}$$

q.e.d.

Similarly, we can prove:

Theorem 9.4.3 *The function*

$$L(x, y^{(1)}, ..., y^{(k)}) = \left\{ \frac{\alpha^2(x, y^{(1)}, ..., y^{(k)})}{\beta(x, y^{(1)}, ..., y^{(k)})} \right\}^2, \tag{4.9}$$

where α^2 is given by (4.1) and β is given by (4.5) is a regular Lagrangian in the points of $\widetilde{\mathrm{Osc}^k M}$ where $\beta(x, y^{(1)}, ..., y^{(k)}) > 0$.

The previous classes of remarkable Lagrangians can also be considered for the spaces $\mathrm{Prol}^k F^n$. Indeed, if $F^n = (M, F)$ is a Finsler space and $z^{(k)i}$ from (4.1) is constructed with the canonical nonlinear connection (2.5), we can consider the following functions:

$$\alpha^2(x, y^{(1)}, ..., y^{(k)}) = \gamma_{ij}(x, y^{(1)}) z^{(k)} z^{(k)j} \tag{4.10$_1$}$$

$$L(x, y^{(1)}, ..., y^{(k)}) = \alpha^2(x, y^{(1)}, ..., y^{(k)}) + \beta(x, y^{(1)}, ..., y^{(k)}) \tag{4.10$_2$}$$

$$(4.10)_3 \quad L(x,y^{(1)},...,y^{(k)}) = \{\alpha(x,y^{(1)},...,y^{(k)}) + \beta(x,y^{(1)},...,y^{(k)})\}^2$$

$$(4.10)_4 \qquad L(x,y^{(1)},...,y^{(k)}) = \left\{\frac{\alpha^2(x,y^{(1)},...,y^{(k)})}{\beta(x,y^{(1)},...,y^{(k)})}\right\}^2$$

where β is given by (4.4) and $\gamma_{ij}(x,y^{(1)})$ is the fundamental tensor of the Finsler space F^n. Using the previous theory of the present section, we can prove that all functions $(4.10)_1 - (4.10)_4$ are regular Lagrangians in the points of $\widetilde{\mathrm{Osc}^k M}$, where $\beta > 0$. These classes of regular Lagrangians will be used in the next chapter to construct the remarkable classes of the Lagrange space of order $k \geq 2$.

9.5 Problems

1. Write the structure equations for the N-linear connection with the coefficients (1.4) of the space $\mathrm{Prol}^k\mathcal{R}^n$.

2. Determine the tensors of torsion and curvature of the connection (1.4) of the space $\mathrm{Prol}^k\mathcal{R}^n$.

3. Study the geometry of the space $\mathrm{Prol}^k F^n$ in the case when F^n is a Randers space RF^n.

4. For the Lagrangian (4.4):

   1° Determine the contravariant d-tensor g^{ij} with the property $g_{ij}g^{jm} = \delta_i^{\ m}$.

   2° In this case prove the equality (4.8).

   3° Using the canonical nonlinear connection N of the space $\mathrm{Prol}^k\mathcal{R}^n$, where $\mathcal{R}^n = (M,\gamma_{ij}(x))$, determine the N-linear connection $D\Gamma(N) = (L^i{}_{jm}, \underset{(1)}{C}{}^i{}_{jm}, ..., \underset{(k)}{C}{}^i{}_{jm})$ for which:

   (a) $g_{ij|m} = 0,\ g_{ij}\overset{(\beta)}{\Big|}_m = 0,\ (\beta = 1,...,k)$

   (b) $L^i{}_{jm} = L^i{}_{mj},\ \underset{(\beta)}{C}{}^i{}_{jm} = \underset{(\beta)}{C}{}^i{}_{mj},\ (\beta = 1,...,m)$

   (c) g_{ij} is given in (4.7).

Chapter 10

Higher Order Lagrange Spaces

The notion of Lagrange space of order k, $(k \in N^*)$, is an immediate extension of that given in Chapter 6 for the Lagrange space of order 2.

The problems discussed in the mentioned chapter will be treated now for any order $k \geq 2$.

So, a Lagrange space of order k is a pair $L^{(k)n} = (M, L)$ formed by an n-dimensional real manifold M and a regular Lagrangian of order k, $L(x, y^{(1)}, ..., y^{(k)})$, for which its fundamental tensor g_{ij} has a constant signature on $\widetilde{E} = \widetilde{\mathrm{Osc}^k M}$.

We shall study the most important geometrical object fields of the space $L^{(k)n} = (M, L)$ which derive from the fundamental function $L(x, y^{(1)}, ..., y^{(k)})$ and from the d-tensor field $g_{ij}(x, y^{(1)}, ..., y^{(k)})$. Firstly, the existence of the spaces $L^{(k)n}$ will be solved by means of the prolongations to $\widetilde{E}$ of the Riemannian or Finslerian structure, (cf. §4, Ch.9).

The variational problem on the integral of action of the regular Lagrangian $L(x, y^{(1)}, ..., y^{(k)})$, as well as the higher order energies or Nöether theorem of L will be studied as in Chapter 8.

We treat here the canonical nonlinear connection N, the canonical metrical N-linear connection and related problems.

10.1 The Definition of the Space $L^{(k)n}$

Definition 10.1.1 A Lagrange space of order k is a pair $L^{(k)n} = (M, L)$, where

1° M is a real n-dimensional manifold.

2° $L : \mathrm{Osc}^k M \longrightarrow \mathbb{R}$ is a differentiable Lagrangian.

3° The d-tensor field

$$g_{ij}(x, y^{(1)}, ..., y^{(k)}) = \frac{1}{2} \frac{\partial^2 L}{\partial y^{(k)i} \partial y^{(k)j}} \tag{1.1}$$

has the properties:

a. $\mathrm{rank}\, \|g_{ij}\| = n$ on $\widetilde{E} = \widetilde{\mathrm{Osc}^k M}$

b. The quadratic form

$$\psi = g_{ij} \xi^i \xi^j \tag{1.1'}$$

has the constant signature on $\widetilde{E}$.

We continue to say that L is the *fundamental function* and g_{ij} is the *fundamental tensor* field of the space $L^{(k)n}$.

Firstly, we prove:

Theorem 10.1.1 *If the base manifold M is paracompact then there exist the Lagrange spaces of order k, $L^{(k)n} = (M, L)$, for which the fundamental tensor g_{ij} is positively defined.*

Proof. M being a paracompact manifold, there exists at least a Riemannian structure γ_{ij} on M. Theorem 9.1.1, assures the existence of a nonlinear connection with the coefficients $\underset{(1)}{M}{}^i{}_j, ..., \underset{(k)}{M}{}^i{}_j$ from (1.1), Ch.9, which depend only on γ_{ij}. Therefore, the Liouville d-vector field $z^{(k)i}$ from (4.1), Ch.9, depends only on γ_{ij}. The Lagrangian

$$L(x, y^{(1)}, ..., y^{(k)}) = \gamma_{ij}(x) z^{(k)i} z^{(k)j}$$

is defined on $\widetilde{E}$, is a differentiable Lagrangian, and has the fundamental tensor $g_{ij}(x, y^{(1)}, ..., y^{(k)}) = \gamma_{ij}(x)$. Thus, the pair (M, L) is a Lagrange space of order k. **q.e.d.**

Remark. In Section 4, Chapter 9, we have examples of Lagrange space of order k (cf. $(4.10)_1 - (4.10)_4$).

Let us consider the integral of action (1.6), Ch.8, of the Lagrangian L, denoted by $I(c)$. From Theorem 8.1.3, it follows:

Theorem 10.1.2 *For $k > 1$, there are no Lagrange spaces $L^{(k)n} = (M, L)$ for which the fundamental function L should satisfy the Zermelo conditions $I^1(L) = \cdots = I^{k-1}(L) = 0$, $I^k(L) = L$.*

In Chapter 8 we discussed the variational problem for the Lagrangian L. If c_ε are the curves (2.1), Ch.8, then the Theorem 8.2.1 is true:

Theorem 10.1.3 *In order that $I(c)$ be an extremal value for the functionals $I(c_\varepsilon)$, it is necesary that the following Euler-Lagrange equations hold:*

$$(1.2)\quad \begin{aligned} &\overset{\circ}{E}_i(L) := \frac{\partial L}{\partial x^i} - \frac{d}{dt}\frac{\partial L}{\partial y^{(1)i}} + \cdots + (-1)^k \frac{1}{k!}\frac{d^k}{dt^k}\frac{\partial L}{\partial y^{(k)i}} = 0 \\ &y^{(1)} = \frac{dx^i}{dt}, ..., y^{(k)i} = \frac{1}{k!}\frac{d^k x^i}{dt^k}. \end{aligned}$$

Of course, $\overset{\circ}{E}_i(L)$ is a covector field.

Let $\mathcal{E}^k_c(L)$ be the energy of order k. It follows that Theorems 8.5.1 and 8.5.2 are valid.

So, we have

$$(1.3)\quad \frac{d\mathcal{E}^k_c(L)}{dt} = -\overset{\circ}{E}_i(L)\frac{dx^i}{dt}.$$

And we can affirm: *$\mathcal{E}^k_c(L)$ is conserved along the solution curves of the Euler-Lagrange equations* (1.2).

Also, the Nöether Theorems 8.6.2 and 8.6.3, hold for the fundamental function L of the Lagrange space $L^{(k)n}$. Concerning the Jacobi-Ostrogradski momenta, exposed in §7 from the mentioned chapter, we can repeat the expresions (7.3) of $p_{(1)i}, ..., p_{(k)i}$ of these momenta and Theorem 8.7.1 regarding the Hamilton-Jacobi equations.

The fundamental function $L(x, y^{(1)}, ..., y^{(k)})$ is, first of all, a regular Lagrangian. Therefore, it gives rise to a k-spray which depends only on L.

Theorem 8.8.1, assures the existence of a k–spray with the mentioned property. We repeat this teorem for the Lagrange spaces $L^{(k)n}$.

Theorem 10.1.4 (Miron–Atanasiu) *In a Lagrange space* $L^{(k)n} = (M, L)$ *there exists a k-spray on E,*

$$(1.4)\quad S = y^{(1)i}\frac{\partial}{\partial x^i} + 2y^{(2)i}\frac{\partial}{\partial y^{(1)i}} + \cdots + ky^{(k)i}\frac{\partial}{\partial y^{(k-1)i}} - (k+1)G^i\frac{\partial}{\partial y^{(k)i}}$$

with the coefficients

$$(1.5)\qquad (k+1)G^i = \frac{1}{2}g^{ij}\left\{\Gamma\left(\frac{\partial L}{\partial y^{(k)j}}\right) - \frac{\partial L}{\partial y^{(k-1)j}}\right\}$$

which depend only on the fundamental function L.

This k-spray is also called *canonical* for the Lagrange space $L^{(k)n}$.

The nonlinear connection derived from the k–spray S is also called *canonical.*

So, we may repeat Theorem 8.8.2 [197]–[201]:

Theorem 10.1.5 *In a Lagrange space* $L^{(k)n} = (M, L)$ *the canonical nonlinear connection has the dual coefficients*

$$(1.6)\qquad \begin{aligned} \underset{(1)}{M}{}^i{}_j &= \frac{\partial G^i}{\partial y^{(k)j}} \\ \underset{(\alpha)}{M}{}^i{}_j &= \frac{1}{\alpha}\left\{S\ \underset{(\alpha-1)}{M}{}^i{}_j + \underset{(1)}{M}{}^i{}_m\ \underset{(\alpha-1)}{M}{}^m{}_j\right\},\ (\alpha = 2, ..., k). \end{aligned}$$

Of course, the coefficients $\underset{(1)}{N}{}^i{}_j, ..., \underset{(k)}{N}{}^i{}_j$ are expressed by means of dual coefficients, as follows:

$$
(1.7)\qquad
\begin{aligned}
\underset{(1)}{N}{}^i{}_j &= \underset{(1)}{M}{}^i{}_j, \underset{(2)}{N}{}^i{}_j = \underset{(2)}{M}{}^i{}_j - \underset{(1)}{M}{}^m{}_j \underset{(1)}{N}{}^i{}_m, ...,\\
\underset{(k)}{N}{}^i{}_j &= \underset{(k)}{M}{}^i{}_j - \underset{(1)}{M}{}^m{}_j \underset{(k-1)}{N}{}^i{}_m - \cdots -\\
&- \underset{(k-2)}{M}{}^m{}_j \underset{(2)}{N}{}^i{}_m - \underset{(k-1)}{M}{}^m{}_j \underset{(1)}{N}{}^i{}_m.
\end{aligned}
$$

The Liouville d-vector fields $z^{(1)i}, ..., z^{(k)i}$ are given by

$$
(1.8)\qquad
\begin{aligned}
&z^{(1)i} = y^{(1)i}\\
&2z^{(2)i} = 2y^{(2)i} + \underset{(1)}{M}{}^i{}_m y^{(1)m}\\
&\cdots\cdots\cdots\cdots\cdots\cdots\cdots\cdots\cdots\cdots\cdots\cdots\cdots\cdots\cdots\\
&kz^{(k)i} = ky^{(k)i} + (k-1) \underset{(1)}{M}{}^i{}_m y^{(k-1)m} + \cdots + \underset{(k-1)}{M}{}^i{}_m y^{(1)m}.
\end{aligned}
$$

N_0 being the canonical nonlinear connection of the space $L^{(k)n}$, we can determine the J–vertical distributions

$$
(1.9)\qquad N_1 = J(N_0), N_2 = J^2(N_0), ..., N_{k-1} = J^{k-1}(N_0).
$$

One obtains the direct sum of linear spaces:

$$
(1.10)\qquad T_u(E) = N_0(u) \oplus N_1(u) \oplus \cdots \oplus N_{k-1}(u) \oplus V_k(u),\ \forall u \in \widetilde{E}.
$$

The adapted basis to the previous direct decomposition is given by

$$
(1.11)\qquad \left\{ \frac{\delta}{\delta x^i}, \frac{\delta}{\delta y^{(1)i}}, ..., \frac{\delta}{\delta y^{(k)i}} \right\}
$$

where

$$
(1.11)'\qquad
\begin{cases}
\dfrac{\delta}{\delta x^i} = \dfrac{\partial}{\partial x^i} - \underset{(1)}{N}{}^j{}_i \dfrac{\partial}{\partial y^{(1)j}} - \cdots - \underset{(k)}{N}{}^j{}_i \dfrac{\partial}{\partial y^{(k)j}}\\
\dfrac{\delta}{\delta y^{(1)i}} = J\left(\dfrac{\delta}{\delta x^i}\right), ..., \dfrac{\delta}{\delta y^{(k-1)i}} = J^{k-1}\left(\dfrac{\delta}{\delta x^i}\right),\\
\dfrac{\delta}{\delta y^{(k)i}} = \dfrac{\partial}{\partial y^{(1)i}},
\end{cases}
$$

the coefficients $\underset{(\alpha)}{N}{}^{j}{}_{i}$, $(\alpha = 1, ..., k)$ being of the canonical nonlinear connection N.

The cobasis $\delta x^i, \delta y^{(1)i}, ..., \delta y^{(k)i}$, dual of the basis (1.11), is given by

$$\begin{aligned} &\delta x^i = dx^i \\ &\delta y^{(1)i} = dy^{(1)i} + \underset{(1)}{M}{}^{i}{}_{j} dx^j \\ (1.12) \qquad &\dots\dots\dots\dots\dots\dots\dots\dots\dots\dots \\ &\delta y^{(k)i} = dy^{(k)i} + \underset{(1)}{M}{}^{i}{}_{j} dy^{(k-1)i} + \cdots + \\ &\qquad + \underset{(k-1)}{M}{}^{i}{}_{j} dy^{(1)i} + \underset{(k)}{M}{}^{i}{}_{j} dx^j \end{aligned}$$

Following the considerations from Chapter 6 we have:

Theorem 10.1.6 *The horizontal curves of the space $L^{(k)n}$ are characterized by the system of differential equations*

$$\frac{\delta y^{(1)i}}{dt} = \cdots = \frac{\delta y^{(k)i}}{dt} = 0.$$

Theorem 10.1.7 *The autoparallel curves of the canonical nonlinear connection of the space $L^{(k)n}$ are characterized by the system of differential equations*

$$\begin{aligned} &y^{(1)i} = \frac{dx^i}{dt}, ..., y^{(k)i} = \frac{1}{k!}\frac{d^k x^i}{dt^k}, \\ &\frac{\delta y^{(1)i}}{dt} = 0, ..., \frac{\delta y^{(k)i}}{dt} = 0. \end{aligned}$$

Using the results from Chapter 7 we can state:

Theorem 10.1.8 *The canonical nonlinear connection of the space $L^{(k)n}$ is integrable if, and only if, the following equations hold:*

$$\underset{(01)}{R}{}^{i}{}_{jh} = \cdots = \underset{(0k)}{R}{}^{i}{}_{jh} = 0.$$

10.2 Canonical Metrical N–Connections

Let N be the canonical nonlinear connection of the Lagrange space of order k, $L^{(k)n} = (M, L)$. Applying the method used in Chapter 5, we can prove:

Theorem 10.2.1 *The following properties hold:*

1° *There exists a unique N–linear connection D on $\widetilde{E}$ verifying the axioms:*

$$(2.1)\qquad g_{ij|h} = 0,\ g_{ij}\overset{(\alpha)}{|}_{h} = 0,\ (\alpha = 1, ..., k)$$

$$(2.2)\qquad \begin{aligned} &\underset{(0)}{T}{}^{i}{}_{jh} = L^{i}{}_{jh} - L^{i}{}_{hj} = 0,\\ &\underset{(\alpha)}{S}{}^{i}{}_{jh} = \underset{(\alpha)}{C}{}^{i}{}_{jh} - \underset{(\alpha)}{C}{}^{i}{}_{hj} = 0,\ (\alpha = 1, ..., k). \end{aligned}$$

2° *The coefficients $C\Gamma(N) = (L^{i}{}_{jh}, \underset{(1)}{C}{}^{i}{}_{jh}, ..., \underset{(k)}{C}{}^{i}{}_{jh})$ of this connection are given by the generalized Christoffel sumbols:*

$$(2.3)\qquad \begin{aligned} L^{m}{}_{ij} &= \frac{1}{2}g^{ms}\left(\frac{\delta g_{is}}{\delta x^{j}} + \frac{\delta g_{sj}}{\delta x^{i}} - \frac{\delta g_{ij}}{\delta x^{s}}\right),\\ \underset{(\alpha)}{C}{}^{m}{}_{ij} &= \frac{1}{2}g^{ms}\left(\frac{\delta g_{is}}{\delta y^{(\alpha)j}} + \frac{\delta g_{sj}}{\delta y^{(\alpha)i}} - \frac{\delta g_{ij}}{\delta y^{(\alpha)s}}\right)\\ &\qquad (\alpha = 1, ..., k). \end{aligned}$$

3° *This connection depends only on the fundamental function $L(x, y^{(1)}, ..., y^{(k)})$ of the space $L^{(k)n}$.*

The connection D from the previous theorem is the *canonical metrical N-connection* of the space $L^{(k)n}$. Its set of coefficients was denoted by $C\Gamma(N)$.

Taking into account (2.2), we can apply the theory from §6, Ch.6, and obtain:

Theorem 10.2.2 *In a space $L^{(k)n}$, the canonical metrical N-connection satisfies the following Ricci identities:*

$$
(2.4)\quad
\begin{aligned}
&X^i{}_{|j|h} - X^i{}_{|h|j} = X^m R_m{}^i{}_{jh} - \sum_{\gamma=1}^{k} \underset{(0\gamma)}{R}{}^m{}_{jh} X^i \overset{(\alpha)}{\Big|}_m \\
&X^i{}_{|j} \overset{(\alpha)}{\Big|}_h - X^i \overset{(\alpha)}{\Big|}_{h|j} = X^m \underset{(\alpha)}{P}{}_m{}^i{}_{jh} - \underset{(\alpha)}{C}{}^m{}_{jh} X^i{}_{|m} + \\
&\qquad + L^i{}_{jh} X^i \overset{(\alpha)}{\Big|}_m - \sum_{\gamma=1}^{k} \underset{(\alpha\gamma)}{B}{}^m{}_{jh} X^i \overset{(\gamma)}{\Big|}_m \\
&X^i \overset{(\alpha)}{\Big|}_j \overset{(\beta)}{\Big|}_h - X^i \overset{(\beta)}{\Big|}_h \overset{(\alpha)}{\Big|}_j = X^m \underset{(\alpha\beta)}{S}{}_m{}^i{}_{jh} - \underset{(\beta)}{C}{}^m{}_{jh} X^i \overset{(\alpha)}{\Big|}_m - \\
&\qquad - \underset{(\alpha)}{C}{}^m{}_{jh} X^i \overset{(\beta)}{\Big|}_m - \sum_{\gamma=1}^{k} \underset{(\alpha\beta)}{\overset{(\gamma)}{C}}{}^m{}_{jh} X \overset{(\gamma)}{\Big|}_m \\
&\qquad\qquad (\beta \le \alpha,\quad \alpha, \beta = 1, ..., k),
\end{aligned}
$$

where the coefficients from the right hand side are d-tensors of curvature and torsion of $C\Gamma(N)$ (cf. Ch.6).

From here we deduce:

Theorem 10.2.3 *In a space $L^{(k)n}$ endowed with the canonical metrical N-connection the system of partial differential equations:*

$$
(2.5)\qquad X^i{}_{|h} = 0, X^i \overset{(1)}{\Big|}_h = 0, ..., X^i \overset{(k)}{\Big|}_h = 0
$$

is completely integrable if, and only if

$$
R_m{}^i{}_{jh} = 0,\ \underset{(\alpha)}{P}{}_m{}^i{}_{jh} = 0,\ \underset{(\alpha\beta)}{S}{}_m{}^i{}_{jh} = 0 \quad (\beta \le \alpha,\ \alpha, \beta = 1, ..., k).
$$

Applying the Ricci identities to the fundamental d–tensor field g_{ij} of the space $L^{(k)n}$ and using Theorem 10.2.1, we get:

Theorem 10.2.4 *In a Lagrange space of order k, $L^{(k)n}$ endowed with the canonical metrical N-connection $C\Gamma(N)$, the following tensorial equations hold:*

$$
\begin{aligned}
& g_{sj} R_i{}^s{}_{hm} + g_{is} R_j{}^s{}_{hm} = 0, \\
(2.6) \qquad & g_{sj} \underset{(\alpha)}{P}{}_i{}^s{}_{hm} + g_{is} \underset{(\alpha)}{P}{}_j{}^s{}_{hm} = 0, \\
& g_{sj} \underset{(\alpha\beta)}{S}{}_i{}^s{}_{hm} + g_{is} \underset{(\alpha\beta)}{S}{}_j{}^s{}_{hm} = 0.
\end{aligned}
$$

Now, let us consider the $h-$ and v_α–deflection tensors of $C\Gamma(N)$:

$$
(2.7) \qquad \overset{(\alpha)}{D}{}^i{}_j = z^{(\alpha)i}{}_{|j}, \quad \overset{(\beta\alpha)}{d}{}^i{}_j = z^{(\beta)i} \overset{(\alpha)}{\Big|}_j .
$$

They have the expressions (3.12), Ch.7.

Let us consider the covariant $h-$ and v_α–deflection tensors of $C\Gamma(N)$:

$$
(2.7)' \qquad \overset{(\alpha)}{D}_{ij} = g_{is} \overset{(\alpha)}{D}{}^s{}_j, \quad \overset{(\beta\alpha)}{d}{}_{ij} = g_{is} \overset{(\beta\alpha)}{d}{}^s{}_j .
$$

Thus, taking into account the Ricci identities for the Liouville d–vector fields $z^{(\alpha)i}$, $(\alpha = 1, ..., k)$, we have:

Theorem 10.2.5 *In a space $L^{(k)n}$ endowed with the canonical metrical N-connection $C\Gamma(N)$, the covariant deflection tensors satisfy the following identities:*

$$
(2.8) \qquad
\begin{aligned}
& \overset{(\alpha)}{D}{}_{ih|m} - \overset{(\alpha)}{D}{}_{im|h} = z^{(\alpha)s} R_{sihm} - \sum_{\gamma=1}^{k} \underset{(0\gamma)}{R}{}^s{}_{hm} \overset{(\alpha\gamma)}{d}{}_{is} \\
& \overset{(\alpha)}{D}{}_{ih} \overset{(\beta)}{\Big|}_m - \overset{(\alpha\beta)}{d}{}_{im|h} = z^{(\alpha)s} \underset{(\beta)}{P}{}_{sihm} - \underset{(\beta)}{C}{}^s{}_{hm} \overset{(\alpha)}{D}{}_{is} + \\
& \qquad + L^s{}_{mh} \overset{(\alpha\beta)}{d}{}_{is} - \sum_{\gamma=1}^{k} \underset{(\beta\gamma)}{B}{}^s{}_{hm} \overset{(\alpha\beta)}{d}{}_{is} \\
& \overset{(\alpha\beta)}{d}{}_{ih} \overset{(\delta)}{\Big|}_m - \overset{(\alpha\delta)}{d}{}_{im} \overset{(\beta)}{\Big|}_h = z^{(\alpha)s} \underset{(\beta\delta)}{S}{}_{sihm} - \underset{(\delta)}{C}{}^s{}_{hm} \overset{(\alpha\beta)}{d}{}_{is} + \\
& \qquad + \underset{(\beta)}{C}{}^s{}_{mh} \overset{(\alpha\delta)}{d}{}_{is} - \sum_{\gamma=1}^{k} \underset{(\beta\delta)}{\overset{(\gamma)}{C}}{}^s{}_{hm} \overset{(\alpha\gamma)}{d}{}_{is}, \\
& \qquad (\delta \leq \beta, \ \beta, \delta = 1, ..., k),
\end{aligned}
$$

where

$$(2.8)' \qquad R_{ijhm} = g_{js}R_i{}^s{}_{hm}, \quad \underset{(\alpha)}{P}{}_{ijhm} = g_{js}\underset{(\alpha)}{P}{}_i{}^s{}_{hm}, \qquad \underset{(\alpha\beta)}{S}{}_{ijhm} = g_{js}\underset{(\alpha\beta)}{S}{}_i{}^s{}_{hm}.$$

We say that $C\Gamma(N)$ is of Cartan type if

$$(2.9) \qquad \overset{(\alpha)}{D}{}^i{}_j = 0, \quad \overset{(\beta\alpha)}{d}{}^i{}_j = \delta^i{}_j.$$

It follows:

Proposition 10.2.1 *If $C\Gamma(N)$ is of Cartan type, then the following equations hold:*

$$z^{(\alpha)s}R_s{}^i{}_{hm} = \sum_{\gamma=1}^{k} \underset{(0\gamma)}{R}{}^i{}_{hm}$$

$$z^{(\alpha)s}\underset{(\beta)}{P}{}_s{}^i{}_{hm} = \sum_{\gamma=1}^{k} \underset{(\beta\gamma)}{B}{}^i{}_{hm} - L^i{}_{hm}$$

$$z^{(\alpha)s}\underset{(\beta\delta)}{S}{}_s{}^i{}_{hm} = \underset{(\delta)}{C}{}^i{}_{hm} - \underset{(\beta)}{C}{}^i{}_{mh} + \sum_{\gamma=1}^{k} \underset{(\beta\delta)}{\overset{(\gamma)}{C}}{}_{hm},$$

$$(\delta \leq \beta, \ \beta, \delta = 1, ..., k).$$

In the end of this section we remark some specific features of the structure equations of $C\Gamma(N)$.

First, the 1–forms connection of $C\Gamma(N)$ are

$$(2.10) \qquad \omega^i{}_j = L^i{}_{jh}dx^h + \underset{(1)}{C}{}^i{}_{jh}\delta y^{(1)h} + \cdots + \underset{(k)}{C}{}^i{}_{jh}\delta y^{(k)h}.$$

Therefore, we can formulate:

Theorem 10.2.6 *The structure equations of the canonical metrical N-connection $C\Gamma(N)$ of the Lagrange space of order k, are given by*

$$(2.11) \qquad \begin{aligned} &d(dx^i) - dx^m \wedge \omega^i{}_m = -\overset{(0)}{\Omega}{}^i \\ &d(\delta y^{(\alpha)i}) - \delta y^{(\alpha)m} \wedge \omega^i{}_m = -\overset{(\alpha)}{\Omega}{}^i, \ (\alpha = 1, ..., k) \\ &d\omega^i{}_j - \omega^m{}_j \wedge \omega^i{}_m = -\Omega^i{}_j, \end{aligned}$$

where the 2–forms of torsion are

$$(2.12)\quad \begin{aligned} \overset{(0)}{\Omega}{}^i &= \sum_{\gamma=1}^{k} \underset{(\gamma)}{C}{}^i{}_{jh} dx^j \wedge \delta y^{(\gamma)h}, \\ \overset{(\alpha)}{\Omega}{}^i &= \frac{1}{2} \underset{(0\alpha)}{R}{}^i{}_{jh} dx^j \wedge dx^h + \sum_{\gamma=1}^{k} \underset{(\gamma\alpha)}{B}{}^i{}_{jm} dx^j \wedge \delta y^{(\gamma)m} + \\ &+ \sum_{\beta,\gamma=1}^{k} \underset{(\alpha\gamma)}{\overset{(\beta)}{C}}{}^i{}_{jm} \delta y^{(\beta)j} \wedge \delta y^{(\gamma)m} - L^i{}_{jm} dx^j \wedge \delta y^{(\alpha)m} + \\ &+ \sum_{\gamma=1}^{k} \underset{(\gamma)}{C}{}^i{}_{jm} \delta y^{(\gamma)j} \wedge \delta y^{(\alpha)m}, \end{aligned}$$

and where the 2–forms of the curvature are

$$(2.13)\quad \begin{aligned} \Omega^i{}_j &= \frac{1}{2} R_j{}^i{}_{pq} dx^p \wedge dx^q + \sum_{\gamma=1}^{k} \underset{(\gamma)}{P}{}_j{}^i{}_{pq} dx^p \wedge \delta y^{(\gamma)q} + \\ &+ \sum_{\alpha,\gamma=1}^{k} \underset{(\alpha\beta)}{S}{}_j{}^i{}_{pq} \delta y^{(\alpha)p} \wedge \delta y^{(\beta)q}. \end{aligned}$$

If we denote

$$(2.14)\qquad \Omega_{ij} = \Omega^s{}_i g_{sj}$$

then we can formulate:

Proposition 10.2.2 *The following property holds:*

$$(2.15)\qquad \Omega_{ij} + \Omega_{ji} = 0.$$

The Bianchi identities of $C\Gamma(N)$ can be obtained from (2.11) applying the operator d of the exterior differentiation and calculating $d\overset{(0)}{\Omega}{}^i, d\overset{(\alpha)}{\Omega}{}^i, d\Omega^i{}_j$ from (2.12), (2.13), modulo the system of equations (2.11).

10.3 The Riemannian $(k-1)n$–Contact Model of the Space $L^{(k)n}$

According to the theory expounded in Chapter 7, we can introduce the so–called Riemannian $(k-1)n$–contact model of the Lagrange space of order k, $L^{(k)n}$ endowed with the canonical metrical N–connection D, whose coefficients $C\Gamma(N)$ are from (2.3).

The Sasaki N–lift of a metric structure given by the fundamental tensor g_{ij} of the space $L^{(k)n}$ is defined on $\widetilde{E} = \widetilde{\mathrm{Osc}^k M}$, by:

$$(3.1)\qquad G = g_{ij}dx^i \otimes dx^j + g_{ij}\delta y^{(1)i} \otimes \delta y^{(1)j} + \cdots + g_{ij}\delta y^{(k)i} \otimes \delta y^{(k)j}.$$

G is a pseudo–Riemannian structure on the manifold $\widetilde{E}$.

It follows:

Theorem 10.3.1 *The tensor G from (3.1) is absolutely parallel with respect to the canonical metrical N-connection D:*

$$(3.2)\qquad D_X G = 0,\ \forall X \in \mathcal{X}(\widetilde{E}).$$

Proposition 10.3.1 *The distributions $N_0, N_1, ..., N_{k-1}, V_k$ are orthogonal with respect to the structure G.*

Let us consider the almost $(k-1)n$–contact structure $(\mathbb{F}, \underset{(1)a}{\xi}, ..., \underset{(k-1)a}{\xi}, \overset{(1)a}{\eta}, ..., \overset{(k-1)a}{\eta})$ determined by the canonical nonlinear connection N, defined in §7, Ch.6.

Thus, $(\underset{(1)a}{\xi}, ..., \underset{(k-1)a}{\xi})$, $(a = 1, ..., n)$, is a local basis adapted to the direct decomposition

$$N_1 \oplus \cdots \oplus N_{k-1}$$

and $(\overset{(1)a}{\eta}, ..., \overset{(k-1)a}{\eta})$ is its dual cobasis. We have

$$
\begin{aligned}
&\mathbb{F}(\underset{(\alpha)a}{\xi}) = 0, \ \overset{(\alpha)a}{\eta}(\underset{(\beta)b}{\xi}) = \delta^{\alpha}{}_{\beta}\delta^{a}{}_{b}\ (\alpha,\beta = 1,...,k-1);\\
(3.3)\qquad &\mathbb{F}^2(X) = -X + \sum_{a=1}^{n}\sum_{\alpha=1}^{k-1}\{\overset{(\alpha)a}{\eta}(X)\underset{(\alpha)a}{\xi}\},\ \forall X \in \mathcal{X}(E);\\
&\mathbb{F}^3 + \mathbb{F} = 0;\\
&\text{rank}\,\|\mathbb{F}\| = 2n.
\end{aligned}
$$

Theorem 10.3.2 *The following property holds:*

1° *The set* $\{\mathbb{F}, \underset{(1)a}{\xi},..., \underset{(k-1)a}{\xi}, \overset{(1)a}{\eta},..., \overset{(k-1)a}{\eta}, G\}$ *is a Riemannian almost* $(k-1)n$*–contact structure on* $\widetilde{E}$*, determined only by the Lagrange space* $L^{(k)n} = (M, L)$.

2° *The canonical metrical N-connection D is campatible with this structure:*

$$D\mathbb{F} = 0,\ DG = 0.$$

3° J*–vertical distibutions* $N_1, ..., N_{k-1}$ *are parallel with respect to D.*

Proof. Theorem 6.1.1 proves the property 1° and Theorem 6.3.2 gives us $d\mathbb{F} = 0$. The property (3.2), i.e. $DG = 0$, and the fact that D is an N–connection assure 3°. **q.e.d.**

The manifold $\widetilde{\text{Osc}^k M} = \widetilde{E}$ endowed with the previous structure gives us the so–called *Riemannian almost* $(k-1)n$*–contact model* of the space $L^{(k)n}$. We denote it, shortly, by $H^{(k+1)n} = (\widetilde{E}, G, \mathbb{F}, \underset{(\alpha)a}{\xi}, \overset{(\alpha)a}{\eta})$.

Using the Nijenhuis tensor of the structure $\mathbb{F}$:

$$
(3.4)\qquad \begin{aligned}\mathcal{N}_{\mathbb{F}}(X,Y) = [\mathbb{F}X, \mathbb{F}Y] + \mathbb{F}^2[X,Y] - \mathbb{F}[\mathbb{F}X, Y] - \mathbb{F}[X, \mathbb{F}Y]\\ \forall X, Y \in \mathcal{X}(\widetilde{E})\end{aligned}
$$

and the Lemma 7.7.1, we can study the case when the structure $(\mathbb{F}, \underset{(\alpha)a}{\xi}, \overset{(\alpha)a}{\eta})$ is normal.

Indeed, by means of Theorem 6.7.5, it follows:

Theorem 10.3.3 *The almost $(k-1)n$–contact structure of the model $H^{(k+1)n}$ is normal if, and only if:*

$$\begin{aligned}\mathcal{N}_{\mathbb{F}}(X,Y) \quad + & \sum_{i=1}^{n}\sum_{\alpha=1}^{k-1}\{\frac{1}{2}\underset{(0\alpha)}{R}{}^{i}{}_{jm}dx^{m}\wedge dx^{j}+ \\ + & \sum_{\gamma=1}^{k}\underset{(\gamma\alpha)}{B}{}^{i}{}_{jm}\delta y^{(\gamma)m}\wedge dx^{j}+ \\ + & \sum_{\beta,\gamma=1}^{k}\overset{(\beta)}{\underset{(\gamma\alpha)}{C}}{}^{i}{}_{jm}\delta y^{(\gamma)m}\wedge \delta y^{(\beta)j}\}(X,Y)=0,\end{aligned}$$

for

$$\{X,Y\}=\left\{\frac{\delta}{\delta x^{i}},\frac{\delta}{\delta y^{(1)}},...,\frac{\delta}{\delta y^{(k)}}\right\}.$$

Other special cases of the spaces $H^{(k+1)n}$ were studied by many authors [cf. References].

10.4 The Gravitational and Electromagnetic Fields in $L^{(k)n}$

As applications of the Riemannian almost $(k-1)n$–contact model of the space $L^{(k)n}$ endowed with the canonical metrical N–connection D, we shall study the gravitational and electromagnetic fields in $L^{(k)n}$.

In a fixed local coordinates on $\widetilde{E}$, we consider every component of the fundamental tensor g_{ij} of the space $L^{(k)n}$ as the gravitational potentials.

Using the N–Sasaki lift G of g_{ij} and the model $H^{(k+1)n}$, we can give

Definition 10.4.1 The Einstein equations of the space $H^{(k+1)n} = \{\widetilde{E},G,F,\underset{(\alpha)a}{\xi},\overset{(\alpha)a}{\eta}\}$ are called Einstein equations of the Lagrange space of order k, $L^{(k)n}$ endowed with the canonical metrical N–connection D.

Let $I\!R$ be the curvature tensor of D, $Ric\, I\!R$ its Ricci tensor and $\mathcal{R}$ its scalar curvature. Then the Einstein equations of the space $L^{(k)n}$ are expressed by

$$Ric\, I\!R - \frac{1}{2}\mathcal{R}G = \kappa\mathcal{T} \tag{4.1}$$

where κ is a constant and $\mathcal{T}$ is the energy momentum tensor field.

In the adapted basis (1.11), the curvature tensor $I\!R$ has the components $R_i{}^m{}_{jh}$, $\underset{(\alpha)}{P}{}_i{}^m{}_{jh}$, $\underset{(\alpha)}{S}{}_i{}^m{}_{jh} = \underset{(\alpha\alpha)}{S}{}_i{}^m{}_{jh}$ and $\underset{(\alpha\beta)}{S}{}_i{}^m{}_{jh}$ $(\alpha \neq \beta,\ \alpha, \beta = 1, ..., k)$. Therefore, the Ricci tensors are given by

$$(4.2)\quad \begin{cases} R_{ij} = R_i{}^s{}_{js}, & \overset{1}{\underset{(\alpha)}{P}}{}_{ij} = \underset{(\alpha)}{P}{}_i{}^s{}_{js}, \quad \overset{2}{\underset{(\alpha)}{P}}{}_{ij} = \underset{(\alpha)}{P}{}_i{}^s{}_{sj}, \\ \underset{(\alpha)}{S}{}_{ij} = \underset{(\alpha)}{S}{}_i{}^s{}_{js}, & \overset{1}{\underset{(\alpha\beta)}{S}}{}_{ij} = \underset{(\alpha\beta)}{S}{}_i{}^s{}_{js}, \quad \overset{2}{\underset{(\alpha\beta)}{S}}{}_{ij} = \underset{(\alpha\beta)}{S}{}_i{}^s{}_{sj} \end{cases}$$

and the scalar curvature $\mathcal{R}$ is of the form

$$(4.2)' \qquad \mathcal{R} = g^{ij}(R_{ij} + \underset{(1)}{S}{}_{ij} + \cdots + \underset{(k)}{S}{}_{ij}).$$

Finally, we obtain

Theorem 10.4.1 *The Einstein equations of the Lagrange space of order k, $L^{(k)n}$, corresponding to the cannonical metrical N-connection $C\Gamma(N)$, are given by*

$$(4.3)\quad \begin{cases} R_{ij} - \frac{1}{2}\mathcal{R}g_{ij} = \kappa \overset{H}{T}{}_{ij}, \quad \overset{1}{\underset{(\alpha)}{P}}{}_{ij} = \kappa \overset{1}{\underset{(\alpha)}{T}}{}_{ij}, \quad \overset{2}{\underset{(\alpha)}{P}}{}_{ij} = -\kappa \overset{2}{T}{}_{ij}, \\ \underset{(\alpha)}{S}{}_{ij} - \frac{1}{2}\mathcal{R}g_{ij} = \kappa \overset{V}{\underset{(\alpha)}{T}}{}_{ij}, \quad \overset{1}{\underset{(\alpha\beta)}{S}}{}_{ij} = \kappa \overset{1}{\underset{(\alpha\beta)}{T}}{}_{ij}, \\ \overset{2}{\underset{(\alpha\beta)}{S}}{}_{ij} = -\kappa \overset{2}{\underset{(\alpha\beta)}{T}}{}_{ij}, \qquad (\alpha, \beta = 1, ..., k), \end{cases}$$

where $\mathcal{R}$ is in (4.2)′ and $\overset{H}{T}{}_{ij}, ..., \overset{2}{\underset{(\alpha\beta)}{T}}{}_{ij}$ are the components of the energy-momentum tensor field $\mathcal{T}$ in the adapted basis (1.11).

Using the Bianchi identities of $C\Gamma(N)$ we can prove:

Theorem 10.4.2 *The law of conservation in the space $L^{(k)n}$ endowed with the canonical metrical N–connection $C\Gamma(N)$ is given by:*

$$[R^i{}_j - \frac{1}{2}\mathcal{R}\delta^i{}_j]_{|i} + \sum_{\varphi=1}^{k} \overset{1}{\underset{(\varphi)}{P}}{}^i{}_j \overset{(\varphi)}{|}{}_i = 0$$

$$(4.4)\qquad [\underset{(\beta)}{S}{}^i{}_j - \frac{1}{2}\mathcal{R}\delta^i{}_j] \overset{(\beta)}{|}{}_i - \overset{2}{\underset{(\beta)}{P}}{}^i{}_{j|i} - \sum_{\varphi=1}^{\beta-1} \overset{2}{\underset{(\varphi\beta)}{P}}{}^i{}_j \overset{(\varphi)}{|}{}_i + \sum_{\varphi=\beta+1}^{k} \overset{1}{\underset{(\beta\varphi)}{P}}{}^i{}_j \overset{(\varphi)}{|}{}_i = 0,$$

where $R^i{}_j = g^{is}R_{sj}$, etc.

It is difficult to prove the two last theorems. In the case $k = 1$, they can be found in the book [195] (see References).

Now, let us consider the covariant $h-$ and v_α–deflection tensors $\overset{(\alpha)}{D}_{ij}$, $\overset{(\beta\alpha)}{d}{}_{ij}$ of the canonical metrical N–connection. They are given by (2.7)′ and satisfy the identities (2.8). So, these d–tensors fields can be used to describe the electromagnetic field of the Lagrange space $L^{(k)n}$, using the theory from the case $k = 1$.

The Liouville d–vector fields $z^{(1)i}, ..., z^{(k)i}$ from (1.8) depend on the canonical nonlinear connection N of the space $L^{(k)n}$. Then the d–covector fields

$$(4.5)\qquad \overset{(1)}{z_i} = g_{ij}z^{(1)j}, ..., \overset{(k)}{z_i} = g_{ij}z^{(k)j}$$

depend on the fundamental function L of the space $L^{(k)n}$. Therefore, in the preferential local coordinates, $\overset{(1)}{z_i}, ..., \overset{(k)}{z_i}$ will be called the electromagnetic potentials of the space $L^{(k)n}$.

Let us consider the d–tensors

$$(4.6)\qquad F^{(\alpha)}{}_{ji} = \frac{1}{2}\left(\frac{\delta z^{(\alpha)}{}_i}{\delta x^j} - \frac{\delta z^{(\alpha)}{}_j}{\delta x^i}\right), f^{(\alpha\beta)}{}_{ji} = \frac{1}{2}\left(\frac{\delta z^{(\alpha)}{}_i}{\delta y^{(\beta)j}} - \frac{\delta z^{(\alpha)}{}_j}{\delta y^{(\beta)i}}\right),$$
$$(\alpha, \beta = 1, ..., k).$$

They will be called the *electromagnetic tensor fields* of the space $L^{(k)n}$.

The fact that $F^{(\alpha)}_{ji}, f^{(\alpha\beta)}_{ji}$ are d–tensor fields is expressed by the following property:

Proposition 10.4.1 *We have*

$$F^{(\alpha)}_{ji} = \frac{1}{2}(\overset{(\alpha)}{D}_{ij} - \overset{(\alpha)}{D}_{ji}),\ f^{(\alpha\beta)}_{ji} = \frac{1}{2}(\overset{(\alpha\beta)}{d}_{ij} - \overset{(\alpha\beta)}{d}_{ji}). \tag{4.7}$$

An immediate consequence is as follows:

Proposition 10.4.2 *If the canonical metrical N-conection is of Cartan type, then the electromagnetic tensors $F^{(\alpha)}_{ji}, f^{(\alpha\beta)}_{ji}$ of the space $L^{(k)n}$ vanish.*

We shall see that the d–tensor fields $F^{(\alpha)}_{ji},\ f^{(\alpha\beta)}_{ji}$ satisfy some laws of conservation given by the generalized Maxwell equations.

Theorem 10.4.3 *The electromagnetic d–tensor fields of the Lagrange space of order k, $L^{(k)n}$ satisfy the following generalized Maxwell equations:*

$$\begin{aligned}
&F^{(\alpha)}{}_{pi|q} + F^{(\alpha)}{}_{iq|p} + F^{(\alpha)}{}_{qp|i} = \\
&\quad = \frac{1}{2}\underset{(piq)}{\Sigma}\{z^{(\alpha)r}R_{ripq} - \sum_{\varphi=1}^{k} d^{(\alpha\varphi)}{}_{ir}\underset{(0\varphi)}{R}{}^{r}{}_{pq}\} \\
&F^{(\alpha)}{}_{pi}\overset{(\beta)}{|}_{q} + F^{(\alpha)}{}_{iq}\overset{(\beta)}{|}_{p} + F^{(\alpha)}{}_{qp}\overset{(\beta)}{|}_{i} + f^{(\alpha\beta)}{}_{pi|q} + f^{(\alpha\beta)}{}_{iq|p} + \\
&\quad + f^{(\alpha\beta)}{}_{qp|i} = \frac{1}{2}\underset{(piq)}{\Sigma}\{z^{(\alpha)r}(\underset{(\beta)}{P}{}_{ripq} - \underset{(\beta)}{P}{}_{riqp}) + \\
&\quad + \sum_{\varphi=1}^{k} d^{(\alpha\varphi)}{}_{ir}(\underset{(\beta\varphi)}{B}{}^{r}{}_{pq} - \underset{(\beta\varphi)}{B}{}^{r}{}_{qp})\} \\
&f^{(\alpha\beta)}{}_{pi}\overset{(\gamma)}{|}_{q} + f^{(\alpha\beta)}{}_{iq}\overset{(\gamma)}{|}_{p} + f^{(\alpha\beta)}{}_{qp}\overset{(\gamma)}{|}_{i} = \\
&\quad = \frac{1}{2}\underset{(piq)}{\Sigma}\{z^{(\alpha)r}\underset{(\beta r)}{S}{}_{ripq} - \sum_{\varphi=1}^{k} d^{(\alpha\varphi)}{}_{ir}(\underset{(\varphi)}{C}{}^{r}{}_{pq} - \underset{(\varphi)}{C}{}^{r}{}_{qp})\}
\end{aligned} \tag{4.8}$$

$(\gamma \leq \beta,\ \beta,\gamma = 1,...,k; \Sigma$ *means "cyclic sum"*$)$.

Some particular cases can be studied by means of the previous equations.

If $k = 1$, all considerations from the present section reduce to those given in Chapter 1 or in the book [195].

10.5 The Generalized Lagrange Spaces of Order k

The notion of generalized Lagrange spaces of order k is a natural extension of that studied in §12 of Chapter 1 and in §, Chapter 2. It was used by the author together with T. Kawaguchi in a study of the higher order relativistic geometrical optics [205], [206].

Definition 10.5.1 A generalized Lagrange space of order k is a pair $GL^{(k)} = (M, g_{ij}(x, y^{(1)}, ..., y^{(k)}))$ formed by a real differentiable n–dimensional manifold M and a C^∞–covariant of type $(0,2)$, symmetric d–tensor g_{ij} on $\widetilde{E}$, having two properties:

a. g_{ij} has a constant signature on $\widetilde{E}$.

b. $\text{rank}\,\|g_{ij}\| = n$ on $\widetilde{E}$.

We say that g_{ij} is the fundamental tensor field of the space $GL^{(k)n}$.

Of course, any Lagrange space of order k, $L^{(k)n} = (M, L)$ determines a generalized Lagrange space $GL^{(k)n} = (M, g_{ij})$, with

$$g_{ij} = \frac{1}{2}\frac{\partial^2 L}{\partial y^{(k)i}\partial y^{(k)j}}, \tag{5.1}$$

the conditions a and b from the previous definition being verified. But not conversely. If $g_{ij}(x, y^{(1)}, ..., y^{(k)})$ is given, it is possible that the system of differential partial equations (5.1) did not admit any solution $L(x, y^{(1)}, ..., y^{(k)})$.

Let us consider

$$\underset{(k)}{C}{}_{ijh} = \frac{1}{2}\frac{\partial g_{ij}}{\partial y^{(k)h}}. \tag{5.2}$$

Evidently, $\underset{(k)}{C}{}_{ijh}$ is a d–tensor field covariant of order three.

Now, we ca formulate

Proposition 10.5.1 *A necesary condition so that the system of differential partial equations* (5.1) *admit a solution* $L(x, y^{(1)}, ..., y^{(k)})$ *is that the d–tensor field* $\underset{(k)}{C}{}_{ijh}$ *from* (5.2) *be completely symmetric.*

Indeed, assuming the existence of a solution L of the system (5.1), then we have

$$\underset{(k)}{C}{}_{ijh} = \frac{1}{4}\frac{\partial^3 L}{\partial y^{(k)i}\partial y^{(k)j}\partial y^{(k)h}}.$$

Clearly, $\underset{(k)}{C}{}_{ijh}$ is completely symmetric. **q.e.d.**

In case the system (5.1), for a given generalized Lagrange space of order k, $GL^{(k)n} = (M, g_{ij})$ has no solutions, we say that the space $GL^{(k)n}$ *is not reducible* to a Lagrange space of order k.

So, we have:

Theorem 10.5.1 *If the d–tensor field* $\underset{(k)}{C}{}_{ijh}$ *is not totally symmetric, then the generalized Lagrange space* $GL^{(k)n} = (M, g_{ij})$ *is not reducible to a Lagrange space of order k.*

Example 10.5.1 Let $\mathcal{R}^n = (M, \gamma_{ij}(x))$ be a Riemannian space and $\sigma(x, y^{(1)}, ..., y^{(k)})$ a C^∞–function on $E = \text{Osc}^k M$. We assume that σ exists. Then, we can consider the d–tensor field

$$g_{ij}(x, y^{(1)}, ..., y^{(k)}) = e^{2\sigma(x,y^{(1)},...,y^{(k)})}(\gamma_{ij}\circ\pi)(x, y^{(1)}, ..., y^{(k)}), \quad \forall(x, y^{(1)}, ..., y^{(k)}) \in E. \tag{5.3}$$

Of course

$$\sigma_h = \frac{\partial \sigma}{\partial y^{(k)h}} \tag{5.4}$$

is a d–covector field on E.

Theorem 10.5.2 *If the d–covector* σ_h *does not vanish, then the pair* $GL^{(k)n} = (M, g_{ij})$, *with* g_{ij} *from* (5.3), *will be a generalized Lagrange space of order k. It is not reducible to a Lagrange space of order k.*

Proof. Since $\gamma_{ij} \circ \pi(x, y^{(1)}, ..., y^{(k)}) = \gamma_{ij}(x)$, $g_{ij} = e^{2\sigma}\gamma_{ij}$ is clearly a d–tensor field on $\widetilde{E}$, which satisfies the Definition 10.5.1. Therefore, $GL^{(k)n} = (M, g_{ij})$ is a generalized Lagrange space of order k. The d–tensor $\underset{(k)}{C}{}_{ijh}$ in (5.2) is given by

$$\underset{(k)}{C}{}_{ijh} = \sigma_h g_{ij}.$$

Evidently, if $\sigma_h \neq 0$, then it is not totally symmetric. Theorem 10.5.1 shows that $GL^{(k)n}$ is not reducible to a Lagrange space of order k.

q.e.d.

The previous example shows us that, if the base manifold M is paracompact, then there exist the generalized Lagrange spaces $GL^{(k)n} = (M, g_{ij})$.

Example 10.5.2. Let $\mathcal{R}^n = (M, \gamma_{ij}(x))$ be a Riemann space and $\mathrm{Prol}^k\mathcal{R}^n$ be its prolongation to $\mathrm{Osc}^k M$.

We consider the Liouville d–vector field $z^{(k)i}$ of $\mathrm{Prol}^k\mathcal{R}^n$. The expression of $z^{(k)i}$ is in (4.1), Ch.9. We can consider the d–covector

$$z^{(k)}{}_i = \gamma_{ij} z^{(k)j} \tag{5.5}$$

and assume the existence of a function $n(x, y^{(1)}, ..., y^{(k)}) \geq 1$ on $\widetilde{E}$. Thus,

$$g_{ij}(x, y^{(1)}, ..., y^{(k)}) = \gamma_{ij}(x) + \Big(1 - \frac{1}{n^2(x, y^{(1)}, ..., y^{(k)})}\Big) z^{(k)}{}_i z^{(k)}{}_j \tag{5.6}$$

is a d–tensor field, covariant of type $(0,2)$, symmetric and of rank n. It is easy to see that the pair $GL^{(k)n} = (M, g_{ij})$ is a generalized Lagrange space.

In the case $k = 1$, this is the space of the relativistic geometrical optics, and (5.6) gives us the Synge tensor field [205], [206], [284].

For this space we can take the nonlinear connection N of $\mathrm{Prol}^k\mathcal{R}^n$ and determine N-canonical metrical connection $C\Gamma(N) = (L^i{}_{jh}, \underset{(1)}{C}{}^i{}_{jh}, ..., \underset{(k)}{C}{}^i{}_{jh})$ of the space, exactly as in the case of Lagrange spaces of order k, discussed in the present chapter. Of course, this connection can be used for studying the gravitational and electromagnetic field of this space.

Returning to the generalized Lagrange spaces of order k, $GL^{(k)n}$, we remark the difficulty to find a nonlinear connection N derived only from the fundamental tensor field g_{ij} of the space. In the cases from Examples 10.5.1 and 10.5.2, this is possible. But, in general, it is not.

Therefore, we assume that a nonlinear connection N on E is apriori given. Thus, we shall study the pair $(N, GL^{(k)n})$.

Let $\{\frac{\delta}{\delta x^i}, \frac{\delta}{\delta y^{(1)i}}, ..., \frac{\delta}{\delta y^{(k)i}}\}$ the adapted basis to the direct decomposition (1.10) determined by $N_0 = N$, $N_1 = J(N_0), ..., N_{k-1} = J^{k-1}(N_0)$ and V_k.

We can prove a theorem of the form of Theorem 10.2.1:

Theorem 10.5.3

1° *There exists a unique N-linear connection D for which*

$$g_{ij|h} = 0, \qquad g_{ij} \overset{(\alpha)}{|}_h = 0 \qquad (\alpha = 1, ..., k)$$
$$L^i{}_{jh} = L^i{}_{hj}, \quad \underset{(\alpha)}{C}{}^i{}_{jh} = \underset{(\alpha)}{C}{}^i{}_{hj} \quad (\alpha = 1, ..., k).$$

2° *The coefficients of this N-linear connection are given by* (2.3), *in which* $\{\frac{\delta}{\delta x^i}, \frac{\delta}{\delta y^{(1)i}}, ..., \frac{\delta}{\delta y^{(k)i}}\}$ *are constructed by means of the nonlinear connection N.*

Using this theorem, it is not difficult to study the geometry of the generalized Lagrange space of order k, $GL^{(k)n}$.

10.6 Problems

1° Let $GL^{(k)n} = (M, g_{ij})$ be, with g_{ij} given in (5.6). Prove that $GL^{(k)n}$ is a generalized Lagrange space of order k.

2° $GL^{(k)n}$ is not reducible to a Lagrange space of order k.

3° If N is the nonlinear connection of the the $\text{Prol}^k \mathcal{R}^n$ and $\mathcal{R}^n = (M, \gamma_{ij}(x))$ a Riemannian space, determine the N–linear connection D with the properties from Theorem 10.5.3.

4° Write the Einstein equations of $G^{(k)n}$ endowed with the N–connection mentioned to the point 3°.

5° Write the electromagnetic tensors and the corresponding Maxwell equations of the space $G^{(k)n}$ endowed with the N–linear connection D from the point 3°.

Chapter 11

Subspaces in Higher Order Lagrange Spaces

In this chapter we shall study the general principles of the geometry of subspaces in a Lagrange space of order k, $L^{(k)n} = (M, L)$. The main geometrical object fields associated to the space $L^{(k)n}$ determine the corresponding induced geometrical object fields on a submanifold $\check{M}$ of M and $\mathrm{Osc}^k\check{M}$ of $\mathrm{Osc}^k M$. Thus, we obtain the induced nonlinear connection, the induced cannonical metrical connection, and, of course, the Gauss–Codazzi equations.

All this theory is a natural extension of the theory of subspaces in a Lagrange space from case $k = 1$ [194].

11.1 Submanifolds in $\mathrm{Osc}^k M$

Let M be a C^∞, real, n–dimensional manifold and $\check{M}$ be a C^∞, real m–dimensional manifold, immersed in M through the immersion $i : \check{M} \longrightarrow M$. Locally, i can be given in the form

$$(1.1) \qquad x^i = x^i(u^1, ..., u^m), \ \text{rank} \left\| \frac{\partial x^i}{\partial u^\alpha} \right\| = m$$

The indices $i, j, h, r, s, p, q, ...$ run over the set $\{1, ..., n\}$ and $\alpha, \beta, \gamma, ...$ run on the set $\{1, ..., m\}$. We assume $1 < m < n$. If i is an embedding, then we identify $\check{M}$ to $i(\check{M})$ and say that $\check{M}$ is a submanifold of the manifold M. Therefore, (1.1) will be called the parametric equations of the submanifold $\check{M}$ in the manifold M.

The embedding $i : \check{M} \to M$ determines an immersion $\mathrm{Osc}^k i$: $\mathrm{Osc}^k \check{M} \to \mathrm{Osc}^k M$, defined by the covariant functor $\mathrm{Osc}^k : \mathrm{Man} \to \mathrm{Man}$, was introduced in §1, Ch.6.

The mapping $\mathrm{Osc}^k i : \mathrm{Osc}^k \check{M} \longrightarrow \mathrm{Osc}^k M$ has the parametric equations:

$$(1.2)\quad \begin{cases} x^i = x^i(u^1, ..., u^m), \ \mathrm{rank} \left\| \dfrac{\partial x^i}{\partial u^\alpha} \right\| = m \\ y^{(1)i} = \dfrac{\partial x^i}{\partial u^\alpha} v^{(1)\alpha} \\ 2y^{(2)i} = \dfrac{\partial y^{(1)i}}{\partial u^\alpha} v^{(1)\alpha} + 2\dfrac{\partial y^{(1)i}}{\partial v^{(1)\alpha}} v^{(2)\alpha} \\ \cdots\cdots\cdots\cdots\cdots\cdots\cdots\cdots\cdots\cdots \\ ky^{(k)i} = \dfrac{\partial y^{(k-1)i}}{\partial u^\alpha} v^{(1)\alpha} + 2\dfrac{\partial y^{(k-1)i}}{\partial v^{(1)\alpha}} v^{(2)\alpha} + \cdots + \\ \qquad\qquad + k\dfrac{\partial y^{(k-1)i}}{\partial v^{(k-1)\alpha}} v^{(k)\alpha} \end{cases}$$

where

$$(1.3)\quad \begin{cases} \dfrac{\partial x^i}{\partial u^\alpha} = \dfrac{\partial y^{(1)i}}{\partial v^{(1)\alpha}} = \cdots = \dfrac{\partial y^{(k)i}}{\partial v^{(k)\alpha}} \\ \dfrac{\partial y^{(h)i}}{\partial u^\alpha} = \dfrac{\partial y^{(h+1)i}}{\partial v^{(1)\alpha}} = \cdots = \dfrac{\partial y^{(k)i}}{\partial v^{(k-h)\alpha}} \ (h = 1, ..., k-1). \end{cases}$$

The Jacobian matrix of (1.2) is

$$(1.4)\qquad J(\mathrm{Osc}^k i) = \left\| \begin{matrix} \dfrac{\partial x^i}{\partial u^\alpha} & 0 & 0 \dots 0 & \\ \dfrac{\partial y^{(1)i}}{\partial u^\alpha} & \dfrac{\partial x^i}{\partial u^\alpha} & 0 \dots 0 & \\ \vdots & \vdots & \vdots & \\ \dfrac{\partial y^{(k)i}}{\partial u^\alpha} & \dfrac{\partial y^{(k)i}}{\partial v^{(1)\alpha}} & \dfrac{\partial y^{(k)i}}{\partial v^{(2)\alpha}} \cdots & \dfrac{\partial x^i}{\partial u^\alpha} \end{matrix} \right\| .$$

It has the rank equal to $(k+1)m$. So, $\mathrm{Osc}^k i$ is an immersion. The differential i_* of the mapping $\mathrm{Osc}^k i : \mathrm{Osc}^k \check{M} \longrightarrow \mathrm{Osc}^k M$ leads to the relation between the natural basis of the modules $\mathcal{X}(\mathrm{Osc}^k\check{M})$ and $\mathcal{X}(\mathrm{Osc}^k M)$ given by

$$(1.5)\quad i_*\left\|\frac{\partial}{\partial u^\alpha}\ \frac{\partial}{\partial v^{(1)\alpha}}\cdots\frac{\partial}{\partial v^{(k)\alpha}}\right\| = \left\|\frac{\partial}{\partial x^i}\ \frac{\partial}{\partial y^{(1)i}}\cdots\frac{\partial}{\partial y^{(k)i}}\right\| J(\mathrm{Osc}^k i).$$

So, putting $i_*\frac{\partial}{\partial u^\alpha} = \frac{\partial}{\partial u^\alpha}, \dots, i_*\frac{\partial}{\partial v^{(k)\alpha}} = \frac{\partial}{\partial v^{(k)\alpha}}$, we have:

$$(1.6)\quad \begin{aligned} \frac{\partial}{\partial u^\alpha} &= \frac{\partial x^i}{\partial u^\alpha}\frac{\partial}{\partial x^i} + \frac{\partial y^{(1)i}}{\partial u^\alpha}\frac{\partial}{\partial y^{(1)i}} + \cdots + \frac{\partial y^{(k)i}}{\partial u^\alpha}\frac{\partial}{\partial y^{(k)i}} \\ \frac{\partial}{\partial v^{(1)\alpha}} &= \frac{\partial y^{(1)i}}{\partial v^{(1)\alpha}}\frac{\partial}{\partial y^{(1)i}} + \cdots + \frac{\partial y^{(k)i}}{\partial v^{(1)\alpha}}\frac{\partial}{\partial y^{(k)i}} \\ &\vdots \\ \frac{\partial}{\partial v^{(k)\alpha}} &= \frac{\partial y^{(k)i}}{\partial v^{(k)\alpha}}\frac{\partial}{\partial y^{(k)i}} \end{aligned}$$

The distributions: v_1 generated by $\{\frac{\partial}{\partial v^{(1)\alpha}}, \dots, \frac{\partial}{\partial v^{(k)\alpha}}\}$, v_2 generated by $\{\frac{\partial}{\partial v^{(2)\alpha}}, \dots, \frac{\partial}{\partial v^{(k)\alpha}}\}$ and so on, v_k generated by $\{\frac{\partial}{\partial v^{(k)\alpha}}\}$ on $\mathrm{Osc}^k\check{M}$ satisfy the properties

1° $v_1 \supset v_2 \supset \cdots \supset v_k$

2° $\dim v_1 = km, \dim v_2 = (k-1)m, \dots, \dim v_k = m$

3° $v_1 \subset V_1, v_2 \subset V_2, \dots, v_k \subset V_k$.

Let be the Liouville vector fields of $\mathrm{Osc}^k\check{M}$:

$$(1.7)\quad \begin{aligned} \overset{1}{\gamma} &= v^{(1)\alpha}\frac{\partial}{\partial v^{(k)\alpha}} \\ \overset{2}{\gamma} &= v^{(1)\alpha}\frac{\partial}{\partial v^{(k-1)\alpha}} + 2v^{(2)\alpha}\frac{\partial}{\partial v^{(k)\alpha}} \\ &\cdots\cdots\cdots\cdots\cdots\cdots\cdots\cdots \\ \overset{k}{\gamma} &= v^{(1)\alpha}\frac{\partial}{\partial v^{(1)\alpha}} + 2v^{(2)\alpha}\frac{\partial}{\partial v^{(2)\alpha}} + \cdots + kv^{(k)\alpha}\frac{\partial}{\partial v^{(k)\alpha}}. \end{aligned}$$

Of course, $\overset{1}{\gamma}$ belongs to the ditribution v_k, $\overset{2}{\gamma}$, belongs to $v_{k-1}, \dots, \overset{k}{\gamma}$ belongs to the distibution v_1.

It is not difficult to prove

Theorem 11.1.1 *On* $\mathrm{Osc}^k \check{M}$ *we have*

$$i_* \overset{1}{\gamma} = \overset{1}{\Gamma}, \ldots, i_* \overset{k}{\gamma} = \overset{k}{\Gamma} . \tag{1.8}$$

The differential i_* of the immersion $\mathrm{Osc}^k i : \mathrm{Osc}^k \check{M} \longrightarrow \mathrm{Osc}^k M$ maps the cotangent space $T^*_w(\mathrm{Osc}^k M)$ into the cotangent space $T^*_u(\mathrm{Osc}^k \check{M})$ by the rule

$$\begin{aligned}
dx^i &= \frac{\partial x^i}{\partial u^\alpha} du^\alpha \\
dy^{(1)i} &= \frac{\partial y^{(1)i}}{\partial u^\alpha} du^\alpha + \frac{\partial y^{(1)i}}{\partial v^{(1)\alpha}} dv^{(1)\alpha} \\
&\cdots\cdots\cdots\cdots\cdots\cdots\cdots\cdots \\
dy^{(k)i} &= \frac{\partial y^{(k)i}}{\partial u^\alpha} du^\alpha + \frac{\partial y^{(k)i}}{\partial v^{(1)\alpha}} dv^{(1)\alpha} + \cdots + \frac{\partial y^{(k)i}}{\partial v^{(k)\alpha}} dv^{(k)\alpha}.
\end{aligned} \tag{1.9}$$

11.2 Subspaces in the Lagrange Space of Order k, $L^{(k)}M$

Let us consider a Lagrange space of order k, $L^{(k)n} = (M, L(x, y^{(1)}, \ldots, y^{(k)}))$, having $g_{ij} = \dfrac{1}{2} \dfrac{\partial^2 L}{\partial y^{(k)i} \partial y^{(k)j}}$ as fundamental tensor field, positively defined. Therefore, the restriction $\check{L}$ of the Lagrangian L to the manifold $\check{E} = \mathrm{Osc}^k \check{M}$ is as follows

$$\check{L}(u, v^{(1)}, \ldots, v^{(k)}) = L(x(u), y^{(1)}(u, v^{(1)}), \ldots, y^{(k)}(u, v^{(1)}, \ldots, v^{(k)})). \tag{2.1}$$

Theorem 11.2.1 *The pair* $\check{L}^{(k)m} = (\check{M}, \check{L})$ *is a Lagrange space of order* k.

Proof. $\check{M}$ is a real m-dimensional C^∞ manifold. $\check{L} : \check{E} \longrightarrow R$ is a differentiable Lagrangian, since the second hand of (2.1) has this property. Finally,

$$\check{g}_{\alpha\beta} = \frac{1}{2} \frac{\partial^2 \check{L}}{\partial v^{(k)\alpha} \partial v^{(k)\beta}} \tag{2.2}$$

is a positively defined d–tensor field with respect to the transformations of the coordinates on $\check{E}$:

$$(2.3)\qquad \begin{cases} \bar{u}^{\alpha} = \bar{u}^{\alpha}(u^1, ..., u^m), \text{ rank } \left\| \dfrac{\partial \bar{u}^{\alpha}}{\partial u^{\beta}} \right\| = m \\ \bar{v}^{(1)\alpha} = \dfrac{\partial \bar{u}^{\alpha}}{\partial u^{\beta}} v^{(1)\beta} \\ \dots\dots\dots\dots\dots\dots\dots\dots\dots\dots\dots\dots \\ \bar{v}^{(k)\alpha} = \dfrac{\partial \bar{v}^{(k-1)\alpha}}{\partial u^{\beta}} v^{(1)\beta} + \cdots + k \dfrac{\partial \bar{v}^{(k-1)\alpha}}{\partial v^{(k-1)\beta}} v^{(k)\beta} \end{cases}$$

and

$$(2.3)' \qquad \frac{\partial \bar{u}^{\alpha}}{\partial u^{\beta}} = \cdots = \frac{\partial \bar{v}^{(k)\alpha}}{\partial v^{(k)\beta}} \quad \text{etc.}$$

The last statement is proved by means of the equations

$$(2.4)\qquad \check{g}_{\alpha\beta} = \frac{\partial x^i}{\partial u^{\alpha}} \frac{\partial x^j}{\partial u^{\beta}} g_{ij}, \quad \text{on } \widetilde{E}.$$

q.e.d.

$\check{L}^{(k)m} = (\check{M}\check{L})$ will be called the *Lagrange subspaces of order k* of the Lagrange spaces $L^{(k)n} = (M, L)$.

Let us consider

$$(2.5)\qquad B^i{}_{\alpha} = \frac{\partial x^i}{\partial u^{\alpha}}.$$

Thus, $\{B^i{}_1, ..., B^i{}_m\}$ are m–linear independent d–vector fields on $\check{E}$. Also, $\{B^1{}_{\alpha}, ..., B^n{}_{\alpha}\}$ are d–covector fields, with respect to the transformations of coordinates (2.3).

Of course, d–vector fields $\{B^i{}_{\alpha},\ \alpha = 1, ..., m\}$ are tangent to the submanifold $\check{M}$.

We say that a d–vector field $\xi^i(x, y^{(1)}, ..., y^{(k)}))$ is *normal* along $\check{E}$ if $g_{ij}(x(u), y^1(u, v^{(1)}), ..., y^{(k)}(u, v^{(1)}, ..., v^{(k)}))B^i_{\alpha}\xi^j(x(u), ..., y^{(k)}(u, v^{(1)}, ..., v^{(k)})) = 0$. Consequently, locally, there are $n - m$ unit vector fields $B^i_{\bar{\alpha}}$, $(\bar{\alpha} = 1, ..., n - m)$, normal along $\check{E}$, and orthogonal to each other:

$$(2.6)\qquad g_{ij} B^i{}_{\alpha} B^j{}_{\bar{\beta}} = 0,\ g_{ij} B^i{}_{\bar{\alpha}} B^j{}_{\bar{\beta}} = \delta_{\bar{\alpha}\bar{\beta}}\ (\bar{\alpha}, \bar{\beta} = 1, ..., n - m)$$

Of course, the system of d–vectors $B^i_{\bar{\alpha}}$, $(\bar{\alpha} = 1, ..., n - m)$ is determined up to the orthogonal transformations of the form

$$B^i{}_{\bar{\alpha}'} = A^{\bar{\beta}}{}_{\bar{\alpha}'} B^i{}_{\bar{\beta}}, \ \|A^{\bar{\beta}}{}_{\bar{\alpha}'}\| \in O(n - m) \tag{2.7}$$

The indices $\bar{\alpha}, \bar{\beta}, ...$ run over throughout the set $(1, 2, ..., n - m)$.

Let us consider the system of d–covector fields $\{B^\alpha_i(u, v^{(1)}, ..., v^{(k)}), B^{\bar{\alpha}}_i(u, v^{(1)}, ..., v^{(k)})\}$ dual to the system $\{B^i_\alpha(u), B^i_{\bar{\alpha}}(u, v^{(1)}, ..., v^{(k)})\}$. It is defined on an open set $\check{\pi}(\check{U})$ in $\check{E}$, $\check{U}$ being a domain of a local chart on the submanifold $\check{M}$.

The condition of duality between $\{B^i_\alpha, B^i_{\bar{\alpha}}\}$ and $\{B^\alpha_i, B^{\bar{\alpha}}_i\}$ are given by

$$\begin{cases} B^i_\alpha B^\beta_i = \delta^\beta_\alpha, \ B^i_\alpha B^{\bar{\beta}}_i = 0, \ B^i_{\bar{\alpha}} B^\beta_i = 0, \ B^i_{\bar{\alpha}} B^{\bar{\beta}}_i = \delta^{\bar{\beta}}_{\bar{\alpha}}, \\ B^i_\alpha B^\alpha_j + B^i_{\bar{\alpha}} B^{\bar{\alpha}}_j = \delta^i_j. \end{cases} \tag{2.8}$$

Using (2.6), we deduce, along the open set $\check{\pi}(\check{U})$:

$$g_{\alpha\beta} B^\alpha_i = g_{ij} B^j_\beta, \ \delta_{\bar{\alpha}\bar{\beta}} B^{\bar{\beta}}_j = g_{ij} B^i_{\bar{\alpha}}. \tag{2.9}$$

So, we can look to the set $\mathcal{R} = \{w;\ B^i_\alpha(u), B^i_{\bar{\alpha}}(u, v^{(1)}, ..., v^{(k)})\}$, $\forall w = (u, v^{(1)}, ..., v^{(k)}) \in \check{\pi}(\check{U})$ as to a *moving frame*.

It is now obvious that we can represent in $\mathcal{R}$ the d–tensors from E restricted to the open set $\check{\pi}^{-1}(\check{U})$. For instance, we have:

$$\begin{aligned} g_{ij} &= \check{g}_{\alpha\beta} B^\alpha_i B^\beta_j + \delta_{\bar{\alpha}\bar{\beta}} B^{\bar{\alpha}}_i B^{\bar{\beta}}_j, \\ g^{ij} &= \check{g}^{\alpha\beta} B^i_\alpha B^j_\beta + \delta^{\bar{\alpha}\bar{\beta}} B^i_{\bar{\alpha}} B^j_{\bar{\beta}}. \end{aligned} \tag{2.10}$$

In a next sections we shall study the Gauss–Weingarten formulae for the moving frame $\mathcal{R}$.

11.3 Induced Nonlinear Connection

Now, let us consider the canonical nonlinear connection N of the Lagrange space of order k, $L^{(k)n} = (M, L)$ having the dual coefficients

$(\underset{(1)}{M}{}^i{}_j, ..., \underset{(k)}{M}{}^i{}_j)$ given by (1.6), Ch.10. We will prove that the restriction of the nonlinear connection N to $\check{E}$ uniquely determines an induced nonlinear connection $\check{N}$ on $\check{E}$. Of course, $\check{N}$ is well determined by means of its dual coefficients or by means of its adapted coframe $(du^\alpha, \delta v^{(1)\alpha}, ..., \delta v^{(k)\alpha})$.

Definition 11.3.1 A nonlinear connection $\check{N}$ in $\check{L}^{(k)m}$ is called *induced* by the canonical nonlinear connection N if we have

$$\delta v^{(1)\alpha} = B_i^\alpha \delta y^{(1)i}, ..., \delta v^{(k)\alpha} = B_i^\alpha \delta y^{(k)i}. \tag{3.1}$$

Of course (3.1) implies:

$$\delta \bar{v}^{(1)\alpha} = \frac{\partial \bar{u}^\alpha}{\partial u^\beta}\delta v^{(1)\beta}, ..., \delta \bar{v}^{(k)\alpha} = \frac{\partial \bar{u}^\alpha}{\partial u^\beta}\delta v^{(k)\beta}.$$

This fact has as a consecquence that $\{\underset{(1)}{\check{M}}{}^\alpha{}_\beta, ..., \underset{(k)}{\check{M}}{}^\alpha{}_\beta\}$ are the dual coefficients of a nonlinear connection along $\check{E}$.

Theorem 11.3.1 *The dual coefficients of the nonlinear connection $\check{N}$, induced in $\check{L}^{(k)m}$ by the canonical nonlinear connection N of the Lagrange space of order k, $L^{(k)n}$ are given by the following formulae:*

$$\underset{(1)}{\check{M}}{}^\alpha{}_\beta = B_i^\alpha \underset{(1)}{\check{M}}{}^i{}_\beta, ..., \underset{(k)}{\check{M}}{}^\alpha{}_\beta = B_i^\alpha \underset{(k)}{\check{M}}{}^i{}_\beta, \tag{3.2}$$

where

$$\begin{aligned}
\underset{(1)}{\check{M}}{}^i{}_\beta &= \underset{(1)}{M}{}^i{}_j \frac{\partial y^{(1)j}}{\partial v^{(1)\beta}} + \frac{\partial y^{(1)i}}{\partial u^\beta} \\
\underset{(2)}{\check{M}}{}^i{}_\beta &= \underset{(2)}{M}{}^i{}_j \frac{\partial y^{(2)j}}{\partial v^{(2)\beta}} + \underset{(1)}{M}{}^i{}_j \frac{\partial y^{(2)j}}{\partial v^{(1)\beta}} + \frac{\partial y^{(2)i}}{\partial u^\beta} \\
&\cdots\cdots\cdots\cdots\cdots\cdots\cdots\cdots\cdots \\
\underset{(k)}{\check{M}}{}^i{}_\beta &= \underset{(k)}{M}{}^i{}_j \frac{\partial y^{(k)j}}{\partial v^{(k)\beta}} + \underset{(k-1)}{M}{}^i{}_j \frac{\partial y^{(k)j}}{\partial v^{(k-1)\beta}} + \cdots + \\
&\quad + \underset{(1)}{M}{}^i{}_j \frac{\partial y^{(k)j}}{\partial v^{(1)\beta}} + \frac{\partial y^{(k)i}}{\partial u^\beta}.
\end{aligned} \tag{3.3}$$

Proof. The first equation (3.1) leads to

$$dv^{(1)\alpha} + \underset{(1)}{\check{M}}{}^{\alpha}{}_{\beta} du^{\beta} = B^{\alpha}_{i}[B^{i}_{\beta} dv^{(1)\beta} + (\frac{\partial y^{(1)i}}{\partial u^{\beta}} + \underset{(1)}{M}{}^{i}{}_{j} \frac{\partial x^{j}}{\partial u^{\beta}}) du^{\beta}].$$

After the evident reductions we get the first equality (3.3). Similarly, we deduce the other formulae (3.3). **q.e.d.**

In the following lines, it is important to write the components of the coframe $\{dx^i, \delta y^{(1)i}, ..., \delta y^{(k)i}\}$, adapted to the canonical nonlinear connection N and to the vertical distribution V, in the moving frame $\mathcal{R}$.

Proposition 11.3.1 *The cobasis* $\{dx^i, \delta y^{(1)i}, ..., \delta y^{(k)i}\}$ *is uniquely represented in the moving frame* $\mathcal{R}$ *in the form:*

$$(3.4)\quad \begin{aligned} dx^i &= B^i_\beta du^\beta \\ \delta y^{(1)i} &= B^i_\beta \delta v^{(1)\beta} + B^i_{\bar\alpha} \underset{(1)}{H}{}^{\bar\alpha}_{\beta} du^\beta \\ &\cdots\cdots\cdots\cdots\cdots\cdots \\ \delta y^{(k)i} &= B^i_\beta \delta v^{(k)\beta} + B^i_{\bar\alpha}\{\underset{(1)}{H}{}^{\bar\alpha}_{\beta} dv^{(k-1)\beta} + \cdots + \underset{(k)}{H}{}^{\bar\alpha}_{\beta} du^\beta\} \end{aligned}$$

where

$$(3.4)'\quad \underset{(1)}{H}{}^{\bar\alpha}_{\beta} = B^{\bar\alpha}_{i} \underset{(1)}{\check{M}}{}^{i}{}_{\beta}, ..., \underset{(k)}{H}{}^{\bar\alpha}_{\beta} = B^{\bar\alpha}_{i} \underset{(k)}{\check{M}}{}^{i}{}_{\beta}.$$

Proof. The equation

$$\delta y^{(1)i} = B^i_\alpha \delta v^{(1)\alpha} + B^i_{\bar\alpha} B^{\bar\alpha}_{j} (\frac{\partial y^{(1)j}}{\partial v^{(1)\beta}} dv^{(1)\beta} + \frac{\partial y^{(1)j}}{\partial u^\beta} du^\beta + \underset{(1)}{M}{}^{i}_{j} B^j_\beta du^\beta)$$

has as a consequence the second equality (3.4). Similarly, for the other equalities (3.4). **q.e.d.**

The previous expressions, (3.4), are not convenient for us because in the right hands we have the natural cobasis $(du^\alpha, dv^{(1)\alpha}, ..., dv^{(k)\alpha})$. This means that the coefficients have not a geometrical meaning.

We get the following theorem:

Theorem 11.3.2 *The cobasis $\{dx^i, \delta y^{(1)i}, ..., \delta y^{(k)i}\}$ determined by the canonical nonlinear connection N from the Lagrange space of order k, $L^{(k)n}$ is uniquely represented in the moving frame $\mathcal{R}$ in the form*

$$
(3.5)\quad
\begin{aligned}
dx^i &= B^i{}_\beta du^\beta \\
\delta y^{(1)i} &= B^i{}_\beta \delta v^{(1)\beta} + B^i{}_{\bar\alpha} \underset{(1)}{K}{}^{\bar\alpha}{}_\beta du^\beta \\
\delta y^{(2)i} &= B^i{}_\beta \delta v^{(2)\beta} + B^i{}_{\bar\alpha}\{\underset{(1)}{K}{}^{\bar\alpha}{}_\beta \delta v^{(1)\beta} + \underset{(2)}{K}{}^{\bar\alpha}{}_\beta du^\beta\} \\
&\cdots\cdots\cdots\cdots\cdots\cdots\cdots\cdots \\
\delta y^{(k)i} &= B^i{}_\beta \delta v^{(k)\beta} + B^i{}_{\bar\alpha}\{\underset{(1)}{K}{}^{\bar\alpha}{}_\beta \delta v^{(k-1)\beta} + \cdots + \\
&\qquad + \underset{(k)}{K}{}^{\bar\alpha}{}_\beta du^\beta\}
\end{aligned}
$$

where

$$
(3.6)\quad
\begin{aligned}
\underset{(1)}{K}{}^{\bar\alpha}{}_\beta &= \underset{(1)}{H}{}^{\bar\alpha}{}_\beta, \\
\underset{(2)}{K}{}^{\bar\alpha}{}_\beta &= \underset{(2)}{H}{}^{\bar\alpha}{}_\beta - \underset{(1)}{H}{}^{\bar\alpha}{}_\gamma \underset{(1)}{\check N}{}^\gamma{}_\beta \\
&\cdots\cdots\cdots\cdots\cdots\cdots\cdots\cdots \\
\underset{(k)}{K}{}^{\bar\alpha}{}_\beta &= \underset{(k)}{H}{}^{\bar\alpha}{}_\beta - \underset{(k-1)}{H}{}^{\bar\alpha}{}_\gamma \underset{(1)}{\check N}{}^\gamma{}_\beta - \cdots - \underset{(1)}{H}{}^{\bar\alpha}{}_\gamma \underset{(k-1)}{\check N}{}^\gamma{}_\beta .
\end{aligned}
$$

and where $\underset{(1)}{\check N}{}^\alpha_\beta, ..., \underset{(k)}{\check N}{}^\alpha_\beta$ are the coefficients of induced nonlinear connection $\check N$.

Proof. Taking into account (3.4) and using the fact that $du^\alpha = \delta u^\alpha$, $dy^{(1)\alpha} = \delta v^{(1)\alpha} - \underset{(1)}{\check N}{}^\alpha_\beta \delta u^\beta, ..., dv^{(k)\alpha} = \delta v^{(k)\alpha} - \underset{(1)}{\check N}{}^\alpha_\beta \delta v^{(k-1)\beta} - \cdots - \underset{(k)}{\check N}{}^\alpha_\beta \delta u^\beta$, we get first of all: dx^i and $\delta y^{(1)i}$ from (3.5) the same as in (3.4) and after that

$$\delta y^{(2)i} = B^i_\alpha \delta y^{(2)\alpha} + B^i_{\bar\alpha} \underset{(1)}{H}{}^{\bar\alpha}_\beta \delta v^{(1)\beta} + B^i_{\bar\alpha}(\underset{(2)}{H}{}^{\bar\alpha}{}_\beta - \underset{(1)}{H}{}^{\bar\alpha}_\gamma \underset{(k)}{\check N}{}^\gamma_\beta).$$

This is exactly the expression of $\delta y^{(2)i}$ from (3.5), (3.6). Similarly, for all the other equalities (3.5). **q.e.d.**

The last theorem has as a consequence:

Corollary 11.3.1 *With respect to the transformations of coordinates on $\check{E}$ and to the transformations* (2.7), $\underset{(1)}{K}{}^{\bar{\alpha}}_{\beta}, \dots, \underset{(k)}{K}{}^{\bar{\alpha}}_{\beta}$ *are mixed d–tensor fields.*

Generally, a set of functions $T^{i\dots\alpha\dots\bar{\alpha}}_{j\dots\beta\dots\bar{\beta}}(u, v^{(1)}, \dots, v^{(k)})$ which are d–tensors in the index $i, j, \dots$, and d–tensors in the index $\alpha, \beta, \dots$, and tensors with respect to the transformations (2.7) in the index $\bar{\alpha}, \bar{\beta}, \dots$ is called a *mixed* d–tensor field on $\check{E}$.

For instance, $B^i_\alpha, B^i_{\bar{\alpha}}, B^\alpha_i, B^{\bar{\alpha}}_i, g_{\alpha\beta}, g_{ij}, \delta_{\bar{\alpha}\bar{\beta}}, \underset{(1)}{K}{}^{\bar{\alpha}}_{\beta}, \dots, \underset{(k)}{K}{}^{\bar{\alpha}}_{\beta}$ are mixed d–tensor fields.

The previous definition of the mixed d–tensor fields can be extended to any mixed geometric d–object field.

11.4 The Relative Covariant Derivative

Now, we shall construct an operator ∇ of relative covariant derivation in the algebra of mixed d–tensor fields. It is clear that ∇ will be known if its action on functions and on the vector fields of the form

(4.1)
$$X^i(x(u), y^{(1)}(u, v^{(1)}), \dots, y^{(k)}(u, v^{(1)}, \dots, v^{(k)})), X^\alpha(u, v^{(1)}, \dots, v^{(k)}),$$
$$X^{\bar{\alpha}}(u, v^{(1)}, \dots, v^{(k)})$$

are known.

Definition 11.4.1 We call a *coupling* of the canonical metrical N–connection D of the space $L^{(k)n} = (M, L)$ to the induced nonlinear connection $\check{N}$ along $\check{E}$ the operator $\check{D}$ with the property

$$\check{D}X^i = DX^i (\text{modulo}\,(3.5)). \tag{4.2}$$

Here

$$DX^i = dX^i + X^j \omega^i{}_j \tag{4.2$'$}$$

and

$$\omega^i{}_j = L^i{}_{jh} dx^h + \underset{(1)}{C}{}^i{}_{jh} \delta y^{(1)h} + \cdots + \underset{(k)}{C}{}^i{}_{jh} \delta y^{(k)h}. \tag{4.2$''$}$$

Then, let be

$$\check{D}X^i = dX^i + X^j\check{\omega}^i{}_j, \tag{4.3}$$

where $\check{\omega}^i{}_j$ are the connection 1–forms of $\check{D}$.

We have

Theorem 11.4.1 *The connection 1-forms $\check{\omega}^i{}_j$ of $\check{D}$ are given by:*

$$\check{\omega}^i{}_j = \check{L}^i{}_{j\alpha}du^\alpha + \underset{(1)}{\check{C}}{}^i{}_{j\alpha}\delta v^{(1)\alpha} + \cdots + \underset{(k)}{\check{C}}{}^i{}_{j\alpha}\delta v^{(k)\alpha} \tag{4.4}$$

where

$$\begin{cases} \check{L}^i{}_{j\beta} & = L^i{}_{jh}B^h{}_\beta + (\underset{(1)}{C}{}^i{}_{jh}\underset{(1)}{K}{}^{\bar{\alpha}}{}_\beta + \cdots + \\ & + \underset{(k)}{C}{}^i{}_{jh}\underset{(k)}{K}{}^{\bar{\alpha}}{}_\beta)B^h{}_{\bar{\alpha}} \\ \underset{(1)}{\check{C}}{}^i{}_{j\beta} & = \underset{(1)}{C}{}^i{}_{jh}B^h{}_\beta + (\underset{(2)}{C}{}^i{}_{jh}\underset{(1)}{K}{}^{\bar{\alpha}}{}_\beta + \cdots + \\ & + \underset{(k)}{C}{}^i{}_{jh}\underset{(k-1)}{K}{}^{\bar{\alpha}}{}_\beta)B^h{}_{\bar{\alpha}} \\ \vdots & \\ \underset{(k-1)}{\check{C}}{}^i{}_{j\beta} & = \underset{(k-1)}{C}{}^i{}_{jh}B^h{}_\beta + \underset{(k)}{C}{}^i{}_{jh}\underset{(1)}{K}{}^{\bar{\alpha}}{}_\beta B^h{}_{\bar{\alpha}} \\ \underset{(k)}{\check{C}}{}^i{}_{j\beta} & = \underset{(k)}{C}{}^i{}_{jh}B^h{}_\beta \end{cases} \tag{4.5}$$

Evidently, the coupling $\check{D}$ of the canonical metrical N–connection D to $\check{N}$ depends only on the embedding $i : \check{M} \longrightarrow M$ and on the fundamental function L of the space $L^{(k)n}$.

Of course, we can write $\check{D}X^i$ in the form:

$$\check{D}X^i = X^i_{|\alpha}du^\alpha + X^i \overset{(1)}{|}_\alpha \delta v^{(1)\alpha} + \cdots + X^i \overset{(k)}{|}_\alpha \delta v^{(k)\alpha}, \tag{4.6}$$

where

$$\begin{aligned} & X^i_{|\alpha} = \frac{\delta X^i}{\delta u^\alpha} + X^j\check{L}^i{}_{j\alpha},\ X^i \overset{(1)}{|}_\alpha = \frac{\delta X^i}{\delta v^{(1)\alpha}} + X^j \underset{(1)}{\check{C}}{}^i{}_{j\alpha}, \ldots, \\ & X^i \overset{(k)}{|}_\alpha = \frac{\delta X^i}{\delta v^{(k)\alpha}} + X^j \underset{(k)}{\check{C}}{}^i{}_{j\beta}. \end{aligned} \tag{4.6$'$}$$

Definition 11.4.2 We call the induced tangent connection on $\check{E}$ by the canonical metrical N–connection D the operator $D^{\top}$ given by

$$D^{\top}X^{\alpha}=B^{\alpha}_{i}\check{D}X^{i},\quad\text{for}\quad X^{i}=B^{i}_{\alpha}X^{\alpha}.\tag{4.7}$$

Next, we shall set

$$D^{\top}X^{\alpha}=dX^{\alpha}+X^{\beta}\omega^{\alpha}{}_{\beta},\tag{4.7$'$}$$

where $\omega^{\alpha}{}_{\beta}$ are the connection 1–forms of $D^{\top}$.

Theorem 11.4.2 *The connection 1–forms $\omega^{\alpha}{}_{\beta}$ of the induced tangent connection $D^{\top}$ are given by the following formulae:*

$$\omega^{\alpha}{}_{\beta}=L^{\alpha}{}_{\beta\gamma}du^{\gamma}+\underset{(1)}{C}{}^{\alpha}{}_{\beta\gamma}\delta v^{(1)\gamma}+\cdots+\underset{(k)}{C}{}^{\alpha}{}_{\beta\gamma}\delta v^{(k)\gamma},\tag{4.8}$$

where:

$$\begin{cases}L^{\alpha}{}_{\beta\gamma}=B^{\alpha}_{i}(B^{i}_{\beta\gamma}+B^{j}_{\beta}\check{L}^{i}{}_{j\gamma})\\ \underset{(1)}{C}{}^{\alpha}{}_{\beta\gamma}=B^{\alpha}_{i}B^{j}_{\beta}\underset{(1)}{\check{C}}{}^{i}{}_{j\gamma},\ldots,\underset{(k)}{C}{}^{\alpha}{}_{\beta\gamma}=B^{\alpha}_{i}B^{j}_{\beta}\underset{(k)}{\check{C}}{}^{i}{}_{j\gamma}.\end{cases}\tag{4.8$'$}$$

As in the case of $\check{D}$ we can write

$$D^{\top}X^{\alpha}=X^{\alpha}_{|\gamma}du^{\gamma}+X^{\alpha}\overset{(1)}{|}_{\gamma}\delta v^{(1)\gamma}+\cdots+X^{\alpha}\overset{(k)}{|}_{\gamma}\delta v^{(k)\gamma},\tag{4.9}$$

where

$$\begin{aligned}X^{\alpha}{}_{|\gamma}&=\frac{\delta X^{\alpha}}{\delta u^{\gamma}}+X^{\beta}L^{\alpha}{}_{\beta\gamma}\\ X^{\alpha}\overset{(1)}{|}_{\gamma}&=\frac{\delta X^{\alpha}}{\delta v^{(1)\gamma}}+X^{\beta}\underset{(1)}{C}{}^{\alpha}{}_{\beta\gamma},\ldots,X^{\alpha}\overset{(k)}{|}_{\gamma}=\frac{\delta X^{\alpha}}{\delta v^{(k)\gamma}}+X^{\beta}\underset{(k)}{C}{}^{\alpha}{}_{\beta\gamma}.\end{aligned}\tag{4.9$'$}$$

Looking at the formulae (4.8), it follows that $D^{\top}$ depends on the embedding $i:\check{M}\longrightarrow M$ and on the space $L^{(k)n}$.

Definition 11.4.3 We call the *induced normal connection* by the canonical metrical N–connection D the operator $D^{\perp}$ given by

$$D^{\perp}X^{\bar{\alpha}}=B^{\bar{\alpha}}_{i}\check{D}X^{i},\quad\text{for}\quad X^{i}=B^{i}_{\bar{\alpha}}X^{\bar{\alpha}}.\tag{4.10}$$

As before, we set

$$D^{\perp}X^{\bar{\alpha}} = dX^{\bar{\alpha}} + X^{\bar{\beta}}\omega^{\bar{\alpha}}{}_{\bar{\beta}} \tag{4.10$'$}$$

where $\omega^{\bar{\alpha}}{}_{\bar{\beta}}$ is the connection 1–forms of $D^{\perp}$.

By means of (4.3), (4.10) and (4.10)$'$, we deduce:

Theorem 11.4.3 *The connection* 1*–forms* $\omega^{\bar{\alpha}}{}_{\bar{\beta}}$ *of the induced normal connection* $D^{\perp}$ *are as follows*

$$\omega^{\bar{\alpha}}{}_{\bar{\beta}} = L^{\bar{\alpha}}{}_{\bar{\beta}\gamma}du^{\gamma} + \underset{(1)}{C}{}^{\bar{\alpha}}{}_{\bar{\beta}\gamma}\delta v^{(1)\gamma} + \cdots + \underset{(k)}{C}{}^{\bar{\alpha}}{}_{\bar{\beta}\gamma}\delta v^{(k)\gamma}, \tag{4.11}$$

where

$$\begin{aligned} L^{\bar{\alpha}}{}_{\bar{\beta}\gamma} &= B^{\bar{\alpha}}_{i}\left(\frac{\delta B^{i}_{\bar{\beta}}}{\delta u^{\gamma}} + B^{j}_{\bar{\beta}}\check{L}^{i}{}_{j\gamma}\right), \\ \underset{(1)}{C}{}^{\bar{\alpha}}{}_{\bar{\beta}\gamma} &= B^{\bar{\alpha}}_{i}\left(\frac{\delta B^{i}_{\bar{\beta}}}{\delta v^{(1)\gamma}} + B^{j}_{\bar{\beta}}\underset{(1)}{\check{C}}{}^{i}{}_{j\gamma}\right), \\ &\cdots\cdots\cdots\cdots\cdots\cdots \\ \underset{(k)}{C}{}^{\bar{\alpha}}{}_{\bar{\beta}\gamma} &= B^{\bar{\alpha}}_{i}\left(\frac{\delta B^{i}_{\bar{\beta}}}{\delta v^{(k)\gamma}} + B^{j}_{\bar{\beta}}\underset{(k)}{\check{C}}{}^{i}{}_{j\gamma}\right). \end{aligned} \tag{4.12}$$

Now, we may set

$$D^{\perp}X^{\bar{\alpha}} = X^{\bar{\alpha}}{}_{|\beta}du^{\beta} + X^{\bar{\alpha}}\overset{(1)}{|}_{\beta}\delta v^{(1)\beta} + \cdots + X^{\bar{\alpha}}\overset{(k)}{|}_{\beta}\delta v^{(k)\beta}, \tag{4.13}$$

where

$$\begin{aligned} X^{\bar{\alpha}}{}_{|\beta} &= \frac{\delta X^{\bar{\alpha}}}{\delta u^{\beta}} + X^{\bar{\gamma}}L^{\bar{\alpha}}{}_{\bar{\gamma}\beta}, \quad X^{\bar{\alpha}}\overset{(1)}{|}_{\beta} = \frac{\delta X^{\bar{\alpha}}}{\delta v^{(1)\beta}} + X^{\bar{\gamma}}\underset{(1)}{C}{}^{\bar{\alpha}}{}_{\bar{\gamma}\beta}, \ldots, \\ X^{\bar{\alpha}}\overset{(k)}{|}_{\beta} &= \frac{\delta X^{\bar{\alpha}}}{\delta v^{(k)\beta}} + X^{\bar{\gamma}}\underset{(k)}{C}{}^{\bar{\alpha}}{}_{\bar{\gamma}\beta}. \end{aligned} \tag{4.14}$$

Now, we can define the relative (or mixed) covariant derivation ∇ enounced at the beginning of this section.

Definition 11.4.4 A relative (mixed) covariant derivation in the algebra of mixed d–tensor fields is an operator ∇ for which the following properties hold:

$$\begin{aligned} &\nabla f = df, \ \ \forall f \in \mathcal{F}(\check{E}) \\ &\nabla X^i = \check{D}X^i, \ \nabla X^\alpha = D^\top X^\alpha, \ \nabla X^{\bar{\alpha}} = D^\perp X^{\bar{\alpha}}. \end{aligned} \tag{4.15}$$

The connection 1–forms $\check{\omega}^i{}_j, \omega^\alpha{}_\beta$ and $\omega^{\bar{\alpha}}{}_{\bar{\beta}}$ will be called the *connection 1–forms of* ∇.

Of course, we can write the d–tensors of torsion and curvature of the relative covariant derivation ∇ taking into account every component $\check{D}$, $D^\top$ and $D^\perp$ of it.

Then, it is easy to prove

Theorem 11.4.4 *The structure equations of* ∇ *are as follows:*

$$\begin{cases} d(du^\alpha) - du^\beta \wedge \omega^\alpha{}_\beta = -\Omega^\alpha \\ d(\delta v^{(a)\alpha}) - \delta v^{(a)\beta} \wedge \omega^\alpha{}_\beta = -\Omega^{(a)\alpha}, \ (a = 1, ..., k) \end{cases} \tag{4.16}$$

and

$$\begin{cases} d\check{\omega}^i{}_j - \check{\omega}^h{}_j \wedge \check{\omega}^i{}_h = -\check{\Omega}^j{}_h \\ d\omega^\alpha{}_\beta - \omega^\gamma{}_\beta \wedge \omega^\alpha{}_\gamma = -\Omega^\alpha{}_\beta \\ d\omega^{\bar{\alpha}}{}_{\bar{\beta}} - \omega^{\bar{\gamma}}{}_{\bar{\beta}} \wedge \omega^{\bar{\alpha}}{}_{\bar{\gamma}} = -\Omega^{\bar{\alpha}}{}_{\bar{\beta}} \end{cases} \tag{4.17}$$

in which $\Omega^\alpha, \Omega^{(a)\alpha}$ *are the 2–forms of torsion:*

$$\begin{aligned} \Omega^\alpha &= \underset{(1)}{C}{}^\alpha{}_{\beta\gamma} du^\beta \wedge \delta v^{(1)\gamma} + \cdots + \underset{(k)}{C}{}^\alpha{}_{\beta\gamma} du^\beta \wedge \delta v^{(k)\gamma}, \\ \Omega^{(a)\alpha} &= \frac{1}{2} \underset{(0a)}{R}{}^\alpha{}_{\beta\gamma} du^\beta \wedge \delta u^\gamma + \cdots + \sum_{b=1}^{k} \underset{(b)}{C}{}^\alpha{}_{\beta\gamma} \delta v^{(b)} \wedge \delta v^{(a)\gamma}, \end{aligned} \tag{4.18}$$

and where the 2–form of curvatures are given by:

$$
(4.19)\quad
\begin{aligned}
\check{\Omega}^i{}_j &= \frac{1}{2}\check{R}_h{}^i{}_{\alpha\beta}du^\alpha \wedge du^\beta + \sum_{a=1}^{k} \underset{(a)}{\check{P}}{}_h{}^i{}_{\alpha\beta}du^\alpha \wedge \delta v^{(a)\beta} + \\
&+ \sum_{a,b=1}^{k} \underset{(ab)}{\check{S}}{}_h{}^i{}_{\alpha\beta}\delta v^{(a)\alpha} \wedge \delta v^{(b)\beta} \\
\Omega^\alpha{}_\beta &= \frac{1}{2}R_\beta{}^\alpha{}_{\gamma\varphi}du^\gamma \wedge du^\varphi + \sum_{a=1}^{k} \underset{(a)}{P}{}_\beta{}^\alpha{}_{\gamma\varphi}du^\gamma \wedge \delta v^{(a)\varphi} + \\
&+ \sum_{a,b=1}^{k} \underset{(ab)}{S}{}_\beta{}^\alpha{}_{\gamma\varphi}\delta v^{(a)\gamma} \wedge \delta v^{(b)\varphi} \\
\Omega^{\bar\alpha}{}_{\bar\beta} &= \frac{1}{2}R_{\bar\beta}{}^{\bar\alpha}{}_{\gamma\varphi}du^\gamma \wedge du^\varphi + \sum_{a=1}^{k} \underset{(a)}{P}{}_{\bar\beta}{}^{\bar\alpha}{}_{\gamma\varphi}du^\gamma \wedge \delta v^{(a)\varphi} + \\
&+ \sum_{a,b=1}^{k} \underset{(ab)}{S}{}_{\bar\beta}{}^{\bar\alpha}{}_{\gamma\varphi}\delta v^{(a)\gamma} \wedge \delta v^{(b)\varphi}.
\end{aligned}
$$

We shall adopt the notations

$$
(4.20)\qquad \check{\Omega}_{ij} = \check{\Omega}^h{}_i g_{hj},\ \ \Omega_{\alpha\beta} = \Omega^\gamma{}_\alpha \check{g}_{\gamma\beta},\ \ \Omega_{\bar\alpha\bar\beta} = \Omega^{\bar\gamma}{}_{\bar\alpha}\delta_{\bar\gamma\bar\beta},
$$

which will be useful in writing the fundamental equations of the embedding $i : \check{M} \longrightarrow M$.

As a direct consequence of Theorems 11.4.1, 11.4.2 and 11.4.3, we get:

Proposition 11.4.1 *The following equations hold:*

$$
(4.21)\qquad \nabla g_{ij} = 0,\ \ \nabla \delta_{\bar\alpha\bar\beta} = 0.
$$

11.5 The Gauss–Weingarten Formulae

Now, we are interested in the moving equations of the moving frame

$$
\mathcal{R} = \{w, B^i_\alpha(x(u)), B^i_{\bar\alpha}(u, v^{(1)}, ..., v^{(k)})\},\ \forall w \in \check{\pi}^{-1}(\check{U})
$$

along the manifold $\check{E} = \mathrm{Osc}^k\check{M}$. These equations, called the *Gauss–Weingarten formulae*, are obtained when the relative covariant derivatives of the vector fields from $\mathcal{R}$ are expressed again in the frame $\mathcal{R}$.

We obtain:

Theorem 11.5.1 *The following Gauss–Weingarten formulae hold good:*

$$\nabla B^i{}_\alpha = B^i{}_{\bar\beta}\Pi^{\bar\beta}_\alpha,\ \ \nabla B^i{}_{\bar\alpha} = -B^i{}_\beta\Pi^\beta{}_{\bar\alpha}, \tag{5.1}$$

where

$$\begin{cases} \Pi^{\bar\beta}{}_\alpha = \underset{(0)}{H}{}^{\bar\beta}{}_{\alpha\gamma}\delta u^\gamma + \underset{(1)}{H}{}^{\bar\beta}{}_{\alpha\gamma}\delta v^{(1)\gamma} + \cdots + \underset{(k)}{H}{}^{\bar\beta}{}_{\alpha\gamma}\delta v^{(k)\gamma} \\ \Pi^\beta{}_{\bar\alpha} = g^{\beta\gamma}\delta_{\bar\alpha\bar\beta}\Pi^{\bar\beta}{}_\gamma \end{cases} \tag{5.2}$$

and

$$\begin{aligned} &\underset{(0)}{H}{}^{\bar\beta}{}_{\alpha\gamma} = B^{\bar\beta}{}_h(B^h{}_{\alpha\gamma} + B^j{}_\alpha \check{L}^h{}_{j\gamma}), \underset{(1)}{H}{}^{\bar\beta}{}_{\alpha\gamma} = B^{\bar\beta}{}_h B^j{}_\alpha \underset{(1)}{\check{C}}{}^h{}_{j\gamma}, \ldots, \\ &\underset{(k)}{H}{}^{\bar\beta}{}_{\alpha\gamma} = B^{\bar\beta}{}_h B^j{}_\alpha \underset{(k)}{\check{C}}{}^h{}_{j\gamma}. \end{aligned} \tag{5.3}$$

Proof. Indeed, the first equality (5.1) is given by:

$$\begin{aligned} \nabla B^i{}_\alpha &= B^i{}_{\alpha|\beta}du^\beta + \cdots + B^i{}_\alpha \overset{(k)}{|}_\beta \delta v^{(k)\beta} = B^i{}_{\bar\gamma}\{B^{\bar\gamma}{}_h[B^h{}_{\alpha\beta}du^\beta + \\ &+ B^j{}_\alpha(\check{L}^h{}_{j\beta}du^\beta + \underset{(1)}{\check{C}}{}^h{}_{j\beta}\delta v^{(1)\beta} + \cdots + \underset{(k)}{\check{C}}{}^h{}_{j\beta}\delta v^{(k)\beta})]\} = \\ &= B^i{}_{\bar\gamma}\{\underset{(0)}{H}{}^{\bar\gamma}{}_{\alpha\beta}du^\beta + \underset{(1)}{H}{}^{\bar\gamma}{}_{\alpha\beta}\delta v^{(1)\beta} + \cdots + \underset{(k)}{H}{}^{\bar\gamma}{}_{\alpha\beta}\delta v^{(k)\beta}\} = B^i{}_{\bar\gamma}\Pi^{\bar\gamma}{}_\alpha. \end{aligned}$$

The second equality (5.1) is obtained by applying the operator ∇ to the identities $g_{ij}B^i_\alpha B^j_{\bar\beta} = 0$ and taking into account the fact that $\nabla g_{ij} = 0$ and $\nabla B^i_\alpha = B^i_{\bar\beta}\Pi^{\bar\beta}{}_\alpha$. **q.e.d.**

Obviously, $\underset{(0)}{H}{}^{\bar\beta}{}_{\alpha\gamma}, \ldots, \underset{(k)}{H}{}^{\bar\beta}{}_{\alpha\gamma}$ are mixed d–tensor fields. They will be called the *second fundamental tensors* of $\check{L}^{(k)m}$. Also, $\Pi^{\bar\alpha}{}_\beta$ are 1–forms on $\mathrm{Osc}^k\check{M}$.

Now we can prove

Proposition 11.5.1 *The operator ∇ has the property*

$$\nabla \check{g}_{\alpha\beta} = 0. \tag{5.4}$$

Indeed, from $\check{g}_{\alpha\beta} = g_{ij}B^i{}_\alpha B^j{}_\beta$, (4.21) and (5.1) it follows the announced property.

Therefore we get:

Proposition 11.5.2 *We have*

$$\check{\Omega}_{ij} = -\check{\Omega}_{ji},\ \Omega_{\alpha\beta} = -\Omega_{\beta\alpha},\ \Omega_{\bar{\alpha}\bar{\beta}} = -\Omega_{\bar{\beta}\bar{\alpha}}. \tag{5.5}$$

We can find a very simple form for the 1–forms $\Pi^{\bar{\beta}}{}_\alpha$.

In fact, $y^{(1)i}$ and $v^{(1)\alpha}$ are d–vector fields. The second equality (1.3) gives us

$$\nabla y^{(1)i} = B^i_{\bar{\beta}}\Pi^{\bar{\beta}}_\alpha v^{(1)\alpha} + B^i_\alpha \nabla v^{(1)\alpha}. \tag{5.6}$$

Proposition 11.5.3 *The 1–forms $\Pi^{\bar{\beta}}{}_\alpha$ are given by*

$$\Pi^{\bar{\beta}}{}_\alpha v^{(1)\alpha} = B^{\bar{\beta}}_i \nabla y^{(1)i}. \tag{5.7}$$

Proposition 11.5.4 *If the second fundamental tensors $\underset{(0)}{H}{}^{\bar{\beta}}{}_{\alpha\gamma}, \dots, \underset{(k)}{H}{}^{\bar{\beta}}{}_{\alpha\gamma}$ vanish, then we shall have $\nabla y^{(1)i} = B^i_\alpha \nabla v^{(1)\alpha}$.*

Let us consider a smooth curve $\widetilde{c} : I \longrightarrow \mathrm{Osc}^k \check{M}$, locally represented by

$$u^\alpha = u^\alpha(t), v^{(1)\alpha} = v^{(1)\alpha}(t), \dots, v^{(k)\alpha} = v^{(k)\alpha}(t),\ t \in I \tag{5.8}$$

and a vector field

$$X^i = B^i{}_\alpha X^\alpha. \tag{5.9}$$

Along the curve $\widetilde{c}$ we have

$$\frac{\nabla X^i}{dt} = B^i{}_{\bar{\beta}}\frac{\Pi^{\bar{\beta}}{}_\alpha}{dt}X^\alpha + B^i{}_\alpha \frac{\nabla X^\alpha}{dt}. \tag{5.9'}$$

We say that the subspace $\check{L}^{(k)m}$ is totally geodesic in the Lagrange space of order k, $L^{(k)n}$ if for any vector field X^i from (5.9), we have

$$(5.9)'' \qquad \frac{\nabla X^i}{dt} = B^i{}_\alpha \frac{\nabla X^\alpha}{dt}$$

along any curve $\widetilde{c}$.

Using this definition from the equation (5.9), it follows:

Theorem 11.5.2 *The Lagrange subspace $\check{L}^{(k)m}$ in a Lagrange space of order k, $L^{(k)n}$, is totally geodesic if, and only if, the second fundamental tensors $\underset{(0)}{H}{}^{\bar{\beta}}{}_{\alpha\gamma}, \dots, \underset{(k)}{H}{}^{\bar{\beta}}{}_{\alpha\gamma}$ vanish.*

11.6 The Gauss–Codazzi Equations

The Gauss–Codazzi equations of a Lagrange subspace $\check{L}^{(k)m} = (\check{M}, \check{L})$ of a Lagrange space of order k, $L^{(k)n} = (M, L)$, endowed with canonical metrical N–connection D are the integrability conditions of the system of equations (5.1). We obtain these conditions by using the exterior calculus.

Theorem 11.6.1 *The Gauss–Codazzi equations of a Lagrange subspace $\check{L}^{(k)m}$ in the Lagrange space of order k, $L^{(k)n}$, are as follows:*

$$(6.1) \qquad \begin{aligned} &B^i{}_\alpha B^j{}_\beta \check{\Omega}_{ij} - \Omega_{\alpha\beta} = \Pi_{\beta\bar{\gamma}} \wedge \Pi^{\bar{\gamma}}{}_\alpha, \\ &B^i{}_{\bar{\alpha}} B^j{}_{\bar{\beta}} \check{\Omega}_{ij} - \Omega_{\bar{\alpha}\bar{\beta}} = \Pi_{\gamma\bar{\beta}} \wedge \Pi^{\gamma}{}_{\bar{\alpha}}, \\ &-B^i{}_\alpha B^j{}_{\bar{\beta}} \check{\Omega}_{ij} = \delta_{\bar{\alpha}\bar{\gamma}}(d\Pi^{\bar{\gamma}}{}_\alpha + \Pi^{\bar{\gamma}}{}_\varphi \wedge \omega^\varphi{}_\alpha - \Pi^{\bar{\varphi}}{}_\alpha \wedge \omega^{\bar{\gamma}}{}_{\bar{\varphi}}) \end{aligned}$$

where $\Pi_{\alpha\bar{\beta}} = g_{\alpha\gamma}\Pi^\gamma{}_{\bar{\beta}}$.

Proof. The first equation (5.1) can be written in the form

$$(*) \qquad dB^i{}_\alpha + B^p{}_\alpha \check{\omega}^i{}_p - B^i{}_\beta \omega^\beta{}_\alpha = B^i{}_{\bar{\beta}} \Pi^{\bar{\beta}}{}_\alpha.$$

Through exterior differentiation of both sides of this equation we get

$$(**) \qquad \begin{aligned} &dB^p{}_\alpha \wedge \check{\omega}^i{}_p + B^p{}_\alpha d\check{\omega}^i{}_p - dB^i{}_\beta \wedge \omega^\beta{}_\alpha - B^i{}_\beta d\omega^\beta{}_\alpha = \\ &\qquad = dB^i{}_{\bar{\beta}} \wedge \Pi^{\bar{\beta}}{}_\alpha + B^i{}_{\bar{\beta}} d\Pi^{\bar{\beta}}{}_\alpha. \end{aligned}$$

Looking at the equality $(*)$ and at

$$(***) \qquad dB^i{}_{\bar{\alpha}} + B^p{}_{\bar{\alpha}}\omega^i{}_p - B^i{}_{\bar{\beta}}\omega^{\bar{\beta}}{}_{\bar{\alpha}} = -B^i{}_{\beta}\Pi^{\beta}{}_{\bar{\alpha}},$$

the equality $(**)$ becomes

$$(6.2) \qquad \begin{aligned} -B^p{}_{\alpha}\check{\Omega}^i{}_p + B^i{}_p\Omega^{\beta}{}_{\alpha} &= B^i{}_{\varphi}\Pi^{\bar{\gamma}}{}_{\alpha} \wedge \Pi^{\varphi}{}_{\bar{\gamma}} + \\ &+ B^i{}_{\bar{\gamma}}(d\Pi^{\bar{\gamma}}{}_{\alpha} + \Pi^{\bar{\gamma}}{}_{\beta} \wedge \omega^{\beta}{}_{\alpha} - \Pi^{\bar{\beta}}{}_{\alpha} \wedge \omega^{\bar{\gamma}}{}_{\bar{\beta}}). \end{aligned}$$

Now, multiplying (6.2) by $g_{is}B^s{}_{\beta} = \check{g}_{\beta\gamma}B^{\gamma}{}_i$, we obtain the first Gauss–Codazzi equation (6.1). The same operation, taking the factor $g_{ij}B^j{}_{\bar{\beta}} = \delta_{\bar{\beta}\bar{\alpha}}B^{\bar{\alpha}}{}_i$, leads to the third equation (6.1).

Of course, to deduce the second Gauss–Codazzi equation (6.1), we will take the exterior differential of both sides of the equations $(***)$ and apply the same method. **q.e.d.**

In order to obtain the system of all fundamental equations of the subspace $L^{(k)m}$ in $L^{(k)n}$, we must find the relations between the torsion 2–forms $\overset{(0)}{\Omega}{}^i$, $\overset{(a)}{\Omega}{}^i$, $(a = 1, ..., k)$ of the canonical metrical N–connection D of $L^{(k)n}$ and the torsion 2–forms Ω^{α}, $\overset{(a)}{\Omega}{}^{\alpha}$, $(a = 1, ..., k)$ of the relative connection ∇.

Therefore, we obtain:

Theorem 11.6.2 *The fundamental equations of the Lagrange subspace $L^{(k)m}$ in a Lagrange space of order k, $L^{(k)n}$, endowed with the canonical metrical N–connection D are the Gauss–Codazzi equations* (6.1), *as well as the equations*

$$d(dx^i) - dx^j \wedge \check{\omega}^i{}_j = -\overset{(0)}{\Omega}{}^i, \qquad \text{modulo}\,(3.4)$$
$$d(\delta y^{(a)i}) - \delta y^{(a)j} \wedge \check{\omega}^i{}_j = -\overset{(a)}{\Omega}{}^i, \quad \text{modulo}\,(3.4), \ (a = 1, ..., k).$$

We end here the theory of Lagrange subspaces of order k, in a Lagrange space of the same order.

We underline the importance of this theory for applications in the higher–order Lagrangian mechanics.

The particular case $m = n - 1$, of the hypersurfaces $\check{M}$ in M can be obtained from the previous theory without any difficulties.

11.7 Problems

1. For the subspace $\check{L}^{(k)m}$ in $L^{(k)n}$, determine the canonical k–spray and the canonical nonlinear connection.

2. Study the relations between canonical nonlinear connection of $\check{L}^{(k)m}$ in $L^{(k)n}$ and the induced nonlinear connection.

3. Determine the Gauss–Codazzi equations for the hypersurfaces $\check{L}^{(k)n-1}$ in a Lagrange space of order k, $L^{(k)n}$.

Chapter 12

Gauge Theory in the Higher Order Lagrange Spaces

In this chapter we shall study the extension of the gauge theory, expounded in our book [195], to the gauge transformations on the k–osculator bundles, endowed with the Lagrangians of order k. We adopt here Asanov's point of view [19], [20], using its extension due to Gh. Munteanu [225], [226].

In some physical theories, instead of the transformations of coordinates on a differentiable manifold M (for instance, the Lorentz transformations on the space–time), there appear the transformations determined by the action of a Lie group G, generally a group of interior symmetries, which acts locally or globally, on a manifold F. These are the gauge transformations.

The physical approach is based on the Lagrangian formalism. The geometrical framework is provided by the associated fibre bundle to a principal bundle P, [266]. This geometrical model is adequate, especially when the associate bundle is a vector bundle. In this case there is the possibility to express the whole theory in local coordinates, which is convenient with a large category of scientists. The calculation is easier if we introduce a nonlinear connection in the associate fibre bundle and report to it the main gauge object fields defined in the geometrical model.

Following this point of view, one defines the notion of gauge k–osculator bundle, $(G\,\mathrm{Osc}^k M, \pi, M)$ and defines a gauge transformation as a diffeomorphism $\mathcal{F} : G\mathrm{Osc}^k M \longrightarrow G\,\mathrm{Osc}^k M$ which preserves the geometric structure of the manifold $G\,\mathrm{Osc}^k M$. The geometrical object fields, preserved by the gauge transformations, are the *gauge geometrical* objects. And a gauge theory on $\mathrm{Osc}^k M$ is the study of the gauge geometrical object fields.

This theory is also consistent when one considers the gauge infinitesimal transformations of a Lie group G. But we present here only the general approach ending with the Einstein–Yang–Mills equations.

For details we refer to Gh. Munteanu's paper [225].

12.1 Gauge Transformations in Principal Bundles

Let (P, π_P, M, G) be a principal bundle, with base manifold M and structural group a Lie group G, (U, ψ_U^P) a fibered chart:

$$\psi_U^P : \pi_P^{-1}(U) \longrightarrow U \times G, \quad (p \longrightarrow (x = \pi_P(p), g = \varphi_u(p)) \quad \text{and}$$
$$L : G \times F \longrightarrow F, \ [(g, y) \longrightarrow gy]$$

the action of the group G on a manifold F.

An associated fibre bundle to P has as fibre F and is defined by $E = (P \times F)/G$ identifying the elements (p, y) and $(pg, g^{-1}y)$. A fibred chart on E is (U, ψ_U^E), where

$$\psi_U^E : \pi_E^{-1}(U) \longrightarrow U \times F, \ [(p, y) \longrightarrow (\pi_E(p), \varphi_U(p)y)].$$

The transformations of local charts on $U \cap V$ are of the form

$$(x, y) \overset{\psi_V^E \circ (\psi_U^E)^{-1}}{\longrightarrow} (\tilde{x}(x), g_{UV}(x)y) \quad \text{where } g_{UV}(x) \in G.$$

Let us consider $F^P : P \longrightarrow P$ an automorphism (i.e. a diffeomorphism) with

$$F^P(p, g) = F^P(p)g \tag{1.1}$$

and $F^0 : M \longrightarrow M$ a diffeomorphism.

Definition 12.1.1 A gauge transformation in the principal bundle P is a pair (F^P, F^0) with the property:

$$\pi_P \circ F^P = F^0 \circ \pi_P. \tag{1.2}$$

In the case when G is a group of interior symmetries, then F^0 is id_M. The standard principal bundle is $P = M \times G$. In this case the left translation $L_g : P \longrightarrow P$, $L_g(x, g') = (x, gg')$ is a gauge transformation, called *global*, since it does not depend on the point $x \in U \subset M$.

If (F^P, id_M) is a gauge transformation of the form $F^P(x, g') = (x, h(x)g')$, then it is called *local*.

The automorphism F^P induces an automorphism F^E on the total space of the associated bundle E in a natural way: $F^E([p, y])) = [F^P(p)y]$. It satisfies the condition $\pi_E \circ F^E = F^0 \circ \pi_E$.

Assume that $n = \dim M$, $F = R^m$ and that G is a group of matrix. Then, G acts linearly on R^m. In this case, the associated bundle E to (P, π_P, M, G) is called a *gauge vectorial bundle*. The coordinates transformations on the manifold E are of the following form:

$$
\begin{aligned}
&\tilde{x}^i = \tilde{x}^i(x), \qquad \mathrm{rank} \left\| \frac{\partial \tilde{x}^i}{\partial x^j} \right\| = n \quad (i, j, h ... = 1, ..., n),\\
&\tilde{y}^a = M^a_b(x) y^b, \quad \|M^a_b(x)\| \in G \qquad (a, b, c, ... = 1, ..., m).
\end{aligned}
\tag{1.3}
$$

Also, a gauge transformation (F^E, F^0) on E is of the form:

$$
\begin{aligned}
&\bar{x}^i = X^i(x), \qquad \mathrm{rank} \left\| \frac{\partial X^i}{\partial x^j} \right\| = n,\\
&\bar{y}^a = Y^a(x, y), \quad \mathrm{rank} \left\| \frac{\partial Y^a}{\partial y^b} \right\| = m.
\end{aligned}
\tag{1.4}
$$

The condition (1.1) implies the fact that the gauge transformation (F^E, F^0) is defined on E.

12.2 Gauge k–Osculator Bundles

The previous gauge transformations can be extended to the higher order. One obtains what Gh. Munteanu calls the notion of *Gauge k–osculator bundle* [225].

Let us consider G a compact subgroup in $GL(m, R)$ and $G^{(k)}$ its prolongation of order k. The elements of $G^{(k)}$ are the k–jets of diffeomorphisms of $\mathcal{R}^m$ into R^m which preserves the origin and have, in a fixed basis from R^m, its matrix from G. Let $P_G^{(k)}(M)$ be a principal bundle having the base M and the structural group $G^{(k)}$.

We consider $F = R^{mk}$. The action of $G^{(k)}$ on F is given by

$$gy = (g^a_{b_1} y^{(1)b_1}; g^a_{b_1} y^{(2)b_1} + g^a_{b_1 b_2} y^{(1)b_1} y^{(1)b_2}; ...) \tag{2.1}$$

where $g \in G^{(k)}$ has the form

$$g = (g^a_{b_1}, g^a_{b_1 b_2}, ..., g^a_{b_1 b_2 ... b_k}). \tag{2.1$'$}$$

The fibre bundle $(E_k, \pi^k, M, F, G^{(k)})$ associated to $P_G^{(k)}(M)$ has the local transformation of coordinates on E_k given by

$$\begin{aligned}
\tilde{x}^i &= \tilde{x}^i(x)\\
\tilde{y}^{(1)i} &= g^a_{b_1}(x) y^{(1)b}\\
\tilde{y}^{(2)i} &= g^a_{b_1}(x) y^{(2)b_1} + g^a_{b_1 b_2}(x) y^{(1)b_1} y^{(1)b_2}\\
&\vdots\\
\tilde{y}^{(k)i} &= g^a_{b_1}(x) y^{(k)b_1} + \cdots + g^a_{b_1 ... b_k}(x) y^{(1)b_1} ... y^{(1)b_k}
\end{aligned} \tag{2.2}$$

In the particular case, $m = n$ and $G \subset GL(n, R)$, $P_G^{(k)}(M)$ can be considered as a G–structure of order k. This is a reduction of the group $GL^{(k)}(n, R)$ of the principal bundle of frames of order k to $G^{(k)}$.

In the case $G = GL(n, R)$, taking the element y of the form $y =$ $= (y^{(1)i}, 2! y^{(2)i}, ..., k! y^{(k)i}) \in F$, and $\tilde{x}^i = \tilde{x}^i(x)$, we obtain

$$g^a_{j_1} = \frac{\partial \tilde{x}^i}{\partial x^{j_1}}, ..., g^i_{j_1 ... j_k} = \frac{\partial^k \tilde{x}^i}{\partial x^{j_1} ... \partial x^{j_k}}.$$

So that, the transformations of coordinates on E_k are given by

$$\begin{aligned}
\tilde{x}^i &= \tilde{x}^i(x), \det \left\| \frac{\partial \tilde{x}^i}{\partial x^j} \right\| \neq 0\\
\tilde{y}^{(1)i} &= \frac{\partial \tilde{x}^i}{\partial x^j} y^{(1)j}\\
&\vdots\\
k\tilde{y}^{(k)i} &= k \frac{\partial \tilde{y}^{(k-1)i}}{\partial y^{(k-1)i}} y^{(k)i} + \cdots + \frac{\partial \tilde{y}^{(k-1)i}}{\partial x^j} y^{(1)j},
\end{aligned} \tag{2.3}$$

the conditions (1.5)$'$, Ch.6, being verified.

Definition 12.2.1 A gauge k-osculator bundle, $G\,\mathrm{Osc}^k M$, is a G structure of order k of the principal bundle $P_G^{(k)}(M)$.

In a point $u = (x, y^{(1)}, ..., y^{(k)}) \in G\,\mathrm{Osc}^k M$, the coordinate transformations are (2.3) restricted to the group $G^{(k)}$, that is $\left\| \dfrac{\partial \tilde{x}^i}{\partial x^j} \right\| \in G$. Therefore, the geometrical theory of gauge k–osculator bundle $G\,\mathrm{Osc}^k M$ is the geometrical theory of the k–osculator bundle $\mathrm{Osc}^k M$ restricted to the group $G^{(k)}$.

Evidently, the fibered bundle $(G\,\mathrm{Osc}^k M, \pi^k, M)$ is a fibered bundle over $(G\,\mathrm{Osc}^h M, \pi^h, M)$, $(1 \leq h < k)$. Namely, we have the bundle $(G\,\mathrm{Osc}^k M, \pi^k_h, \mathrm{Osc}^h M)$, with $\pi^k_h(x, y^{(1)}, ..., y^{(k)}) = (x, y^{(1)}, ..., y^{(h)})$ and $\pi^k_h = \pi^h \circ \pi^k_h$.

As we know from Ch. 6, Osc^k : Man$\longrightarrow$Man is a covariant functor. It follows that $f^{(k)} = \mathrm{Osc}^k f^{(0)} : G\,\mathrm{Osc}^k M \longrightarrow G\,\mathrm{Osc}^k M$ restricted to $G^{(k)}$ is a diffeomorphism, if $f^{(0)} : M \longrightarrow M$ has the same property. It will be called the extension of order k of the diffeomorphism $f^{(0)} : M \longrightarrow M$.

Consequently, *the set of k–extensions of the diffeomorphisms $f \in \in \mathrm{Diff}^{(0)} M$ is a group, denoted by* $\mathrm{Diff}^{(k)} M$, *with respect to the operation* $f^{(k)} \circ g^{(k)} = (f \circ g)^k$.

The following diagram is commutative:

$$\begin{array}{ccc} G\,\mathrm{Osc}^k M & \xrightarrow{f^{(k)}} & G\,\mathrm{Osc}^k M \\ \pi^k_h \downarrow & & \downarrow \pi^k_h \\ G\,\mathrm{Osc}^h M & \xrightarrow{f^{(h)}} & G\,\mathrm{Osc}^h M \end{array}$$

for $h = 0, ..., k$, $\pi^k_k = \mathrm{id}$, $\pi_0 = \pi^k$ and $G\,\mathrm{Osc}^0 M = M$.

Consequently, *the set of diffeomorphisms $\{f^{(0)}, f^{(1)}, ..., f^{(k)}\}$ preserves the geometrical structure of the total space $G\,\mathrm{Osc}^k M$.*

Now, we can formulate the following definition:

Definition 12.2.2 A set of diffeomorphisms $\mathcal{F} = \{F^{(0)}. ..., F^{(k)}\}$, $F^{(h)} \in \in \mathrm{Diff}\, G\,\mathrm{Osc}^h M$, $h = 0, ..., k$, is a gauge transformation in $G\,\mathrm{Osc}^k M$ if the following conditions are verified:

$$\pi^k_h \circ F^{(k)} = F^{(h)} \circ \pi^k_h, \quad (h = 0, ..., k-1). \tag{2.4}$$

Evidently, from the previous equality, it follows that

$$(2.4)' \qquad \begin{cases} \pi^\alpha_{\alpha-1} \circ F^{(\alpha-1)} = F^{(\alpha)} \circ \pi^\alpha_{\alpha-1} \ \text{ and } \ \pi^\alpha_\beta \circ F^{(\beta)} = F^{(\alpha)} \circ \pi^\alpha_\beta \\ \forall \alpha = 1, ..., k; \quad \alpha > \beta \geq 0. \end{cases}$$

So, we have

Proposition 12.2.1 *The set of diffeomorphisms $\mathcal{F}$ has the property:*

$$\begin{aligned} &\pi^k_h(F^{(k)}(x, y^{(1)}, ..., y^{(k)})) = F^{(h)}(x, y^{(1)}, ..., y^{(h)}),\\ &(h = 0, 1, ...k-1),\ \forall (x, y^{(1)}, ..., y^{(k)}) \in G\,\mathrm{Osc}^k M. \end{aligned} \tag{2.5}$$

Therefore, we deduce:

Proposition 12.2.2 *The values of diffeomorphisms $F^{(h)}$, $h = 0, ..., k$, are well determined by those of $F^{(k)} \in \mathrm{Diff}\, G\,\mathrm{Osc}^k M$.*

Clearly, the composition of two gauge transformations is a gauge transformation and the set $\{\mathrm{id}\, G\,\mathrm{Osc}^{(0)} M, ..., \mathrm{id}\, G\,\mathrm{Osc}^{(k)} M\}$ is a gauge transformation. That leads to:

Theorem 12.2.1 *The set $\mathrm{Gau}\,(G\,\mathrm{Osc}^k M)$ of all gauge transformations on $G\,\mathrm{Osc}^k M$, together with the operation of composition, is a group.*

The group $\mathrm{Gau}\,(G\,\mathrm{Osc}^k M)$ is sufficiently general. In many cases it is possible to reduce it to identity only.

Therefore, we consider here those gauge transformations $\mathcal{F}$ which operate on an open set $(\pi^{(k)})^{-1}(U) \subset G\,\mathrm{Osc}^k M$, where U is an open set on the manifold M.

12.3 The Local Representation of Gauge Transformations

It is convenient to represent in local coordinates $(x^i, y^{(1)i}, ..., y^{(k)i})$ on $G\,\mathrm{Osc}^k M$ the elements of the group $\mathrm{Gau}\,(G\,\mathrm{Osc}^k M)$.

Let $\mathcal{F} = (F^{(0)}, ..., F^{(k)}) \in \mathrm{Gau}\,(G\,\mathrm{Osc}^k M)$ be a local gauge transformation and assume that $\mathcal{F}$ operates on the open set $(\pi^k)^{-1}(U)$. That is $\mathcal{F} : u \longrightarrow \bar{u}$, $u = (x, y^{(1)}, ..., y^{(k)})$, $\bar{u} = (\bar{x}, \bar{y}^{(1)}, ..., \bar{y}^{(k)})$ and $u \in$ $\in (\pi^k)^{-1}(U_\alpha \cap U)$, $\bar{u} \in (\pi^k)^{-1}(U_\beta \cap U)$, U_α and U_β being two domains of charts in the base manifold M. We can see, without any difficulties, that the following theorem holds.

Theorem 12.3.1 *In local coordinates on the manifold $G\,\mathrm{Osc}^k M$, a gauge transformation $\mathcal{F} = (F^{(0)}, ..., F^{(k)})$ can be represented by equations of the form*

$$
(3.1)\qquad
\begin{aligned}
\bar{x}^i &= X^i(x)\\
\bar{y}^{(1)i} &= Y^{(1)i}(x, y^{(1)})\\
&\vdots\\
\bar{y}^{(k)i} &= Y^{(k)}(x, y^{(1)}, ..., y^{(k)}),
\end{aligned}
$$

where

$$
(3.1)'\qquad \delta = \det\left\|\frac{\partial X^i}{\partial x^j}\right\| \cdot \det\left\|\frac{\partial Y^{(1)i}}{\partial y^{(1)j}}\right\| \cdots \det\left\|\frac{\partial Y^{(k)i}}{\partial y^{(k)j}}\right\| \neq 0.
$$

Of course, the representation (3.1) of the gauge transformation $\mathcal{F}$ depends on the choice of the local charts on $G\,\mathrm{Osc}^k M$ in the points $u = (x, y^{(1)}, ..., y^{(k)})$ and $\bar{u} = (\bar{x}, \bar{y}^{(1)}, ..., \bar{y}^{(k)})$. For other local chart in the point $u \in (\pi^k)^{-1}(\widetilde{U}_{\alpha'} \cap U)$, the new coordinates $(\tilde{x}, \tilde{y}^{(1)}, ..., \tilde{y}^{(k)})$ are related to the old coordinates $(x, y^{(1)}, ..., y^{(k)})$ by the equations (2.3). Independently, we can have a changing of local coordinates in the point $\bar{u} = \mathcal{F}(u)$, $\bar{u} = (\bar{x}, \bar{y}^{(1)}, ..., \bar{y}^{(k)})$. It will be of the form

$$
(3.2)\qquad
\begin{cases}
\tilde{\bar{x}}^i = \tilde{\bar{x}}^i(\bar{x}), & \left[\left\|\dfrac{\partial \tilde{\bar{x}}^i}{\partial \bar{x}^j}(\bar{x})\right\| \in G\right]\\[2ex]
\tilde{\bar{y}}^{(1)i} = \dfrac{\partial \tilde{\bar{x}}^i}{\partial \bar{x}^j}\bar{y}^{(1)j} & \\
\vdots & \\
k\,\tilde{\bar{y}}^i(k) = \dfrac{\partial \tilde{\bar{y}}^{(k-1)i}}{\partial \bar{x}^m}\bar{y}^{(1)m} + \cdots + k\dfrac{\partial \tilde{\bar{y}}^{(k-1)i}}{\partial \bar{y}^{(k-1)j}}\bar{y}^{(k)j} &
\end{cases}
$$

and

$$
(3.2)'\qquad \frac{\partial \tilde{\bar{y}}^{(\alpha)i}}{\partial \bar{x}^m} = \cdots = \frac{\partial \tilde{\bar{y}}^{(k)i}}{\partial \bar{y}^{(k-\alpha)m}} \quad (\alpha = 0, ..., k-1, \bar{y}^{(0)} = \bar{x}) \quad \text{etc.}
$$

Therefore, we get:

Proposition 12.3.1 *With respect to* (3.2), *the equations* (3.1) *are transformed as follows:*

$$(3.3)\qquad \begin{aligned} \tilde{\bar{x}}^i &= \widetilde{X}^i(x) \\ \tilde{\bar{y}}^{(1)i} &= \widetilde{Y}^{(1)i}(x, y^{(1)}) \\ &\vdots \\ \tilde{\bar{y}}^{(k)i} &= \widetilde{Y}^{(k)i}(x, y^{(1)}, ..., y^{(k)}). \end{aligned}$$

Clearly, (3.2) and (3.3) imply:

$$(3.4)\qquad \begin{aligned} \widetilde{X}^i &= \tilde{\bar{x}}^i\,(X(x)) \\ \widetilde{Y}^{(1))i}(x, y^{(1)}) &= \frac{\partial\,\tilde{\bar{x}}^i}{\partial \bar{x}^j} Y^{(1)j}(x, y^{(1)}) \\ &\vdots \\ k\widetilde{Y}^{(1))i}(x, y^{(1)}, ..., y^{(k)}) &= \frac{\partial\,\tilde{\bar{y}}^{(k-1)}}{\partial \bar{x}^j} Y^{(1)j}(x, y^{(1)}) + \cdots + \\ &\quad + k\frac{\partial\,\tilde{\bar{y}}^{(k-1)i}}{\partial \bar{y}^{(k-1)j}} Y^{(k)j}(x, y^{(1)}, ..., y^{(k)}). \end{aligned}$$

Also, from the fact that the gauge transformation $\mathcal{F}$ is well defined on $(\pi^k)^{-1}(U)$ we get:

$$(3.5)\qquad \begin{aligned} \bar{x}^i &= \widetilde{X}^i(\tilde{x}) \\ \bar{y}^{(1)i} &= \widetilde{Y}^{(1)i}(\tilde{x}, \tilde{y}^{(1)}) \\ &\vdots \\ \bar{y}^{(k)i} &= \widetilde{Y}^{(k)i}(\tilde{x}, \tilde{y}^{(1)}, ..., \tilde{y}^{(k)}) \end{aligned}$$

where

$$(3.6)\qquad \begin{aligned} &\widetilde{X}^i(\tilde{x}(x)) = X^i(x) \\ &\widetilde{Y}^{(1)i}(\tilde{x}(x), \tilde{y}^{(1)}(x, y^{(1)}) = Y^{(1)i}(x, y^{(1)}) \\ &\vdots \\ &\widetilde{Y}^{(k)i}(\tilde{x}(x), \tilde{y}^{(1)}(x, y^{(1)}), ..., \tilde{y}^{(k)}(x, y^{(1)}, ..., y^{(k)})) = \\ &\qquad = Y^{(k)}(x, y^{(1)}, ..., y^{(k)}). \end{aligned}$$

Let us consider the Jacobi matrix of the transformation of local coordinates (2.3):

$$(3.7) \qquad J(\tilde{x}, \tilde{y}^{(1)}, ..., \tilde{y}^{(k)}) = \begin{Vmatrix} \dfrac{\partial \tilde{x}^i}{\partial x^j} & 0 & \cdots & 0 \\ \dfrac{\partial \tilde{y}^{(1)i}}{\partial x^j} & \dfrac{\partial \tilde{y}^{(1)i}}{\partial y^{(1)j}} & \cdots & 0 \\ \cdots & \cdots & \cdots & \cdots \\ \dfrac{\partial \tilde{y}^{(k)i}}{\partial x^j} & \dfrac{\partial \tilde{y}^{(k)i}}{\partial y^{(1)j}} & \cdots & \dfrac{\partial \tilde{y}^{(k)i}}{\partial y^{(k)j}} \end{Vmatrix}$$

and the Jacobi matrix of the gauge transformation (3.1):

$$(3.8) \qquad J(X, Y^{(1)}, ..., Y^{(k)}) = \begin{Vmatrix} \dfrac{\partial X^i}{\partial x^j} & 0 & \cdots & 0 \\ \dfrac{\partial Y^{(1)i}}{\partial x^j} & \dfrac{\partial Y^{(1)i}}{\partial y^{(1)j}} & \cdots & 0 \\ \cdots & \cdots & \cdots & \cdots \\ \dfrac{\partial Y^{(k)i}}{\partial x^j} & \dfrac{\partial Y^{(k)i}}{\partial y^{(1)j}} & \cdots & \dfrac{\partial Y^{(k)i}}{\partial y^{(k)j}} \end{Vmatrix}$$

Proposition 12.3.2 *The following property holds:*

$$(3.9) \qquad J(X, Y^{(1)}, ..., Y^{(k)}) = J(\widetilde{X}, \widetilde{Y}^{(1)}, ..., \widetilde{Y}^{(k)}) \cdot J(\tilde{x}, \tilde{y}^{(1)}, ..., \tilde{y}^{(k)}).$$

The proof is immediate by using (3.6).

The inverse $\mathcal{F}^{-1}$ of the gauge transformation $\mathcal{F}$ can be expressed, in local coordinates, in the form:

$$(3.10) \qquad \begin{aligned} x^i &= \overline{X}^i(\bar{x}) \\ y^{(1)i} &= \overline{Y}^{(1)i}(\bar{x}, \bar{y}^{(1)}) \\ \cdots & \cdots \cdots \\ y^{(k)i} &= \overline{Y}^{(k)i}(\bar{x}, \bar{y}^{(1)}, ..., \bar{y}^{(k)}). \end{aligned}$$

It follows

$$(3.11) \qquad J(\overline{X}, \overline{Y}^{(1)}, ..., \overline{Y}^{(1)k}) = [J(X, Y^{(1)}, ..., Y^{(1)k})]^{-1}.$$

Denoting

$$(3.12)\qquad \begin{aligned} X^i{}_m &= \frac{\partial X^i}{\partial x^m},\ Y^{(1)i}{}_m = \frac{\partial Y^{(1)i}}{\partial y^{(1)m}}, ..., Y^{(k)i}{}_m = \frac{\partial Y^{(k)i}}{\partial y^{(k)m}},\\ \overline{X}^i{}_m &= \frac{\partial \overline{X}^i}{\partial \bar{x}^m},\ \overline{Y}^{(1)i}{}_m = \frac{\partial \overline{Y}^{(1)i}}{\partial \bar{y}^{(1)m}}, ..., \overline{Y}^{(k)i}{}_m = \frac{\partial \overline{Y}^{(k)i}}{\partial \bar{y}^{(k)m}}, \end{aligned}$$

from (3.11) we obtain

$$X^i{}_m \overline{X}^m{}_j = Y^{(1)i}{}_m \overline{Y}^{(1)m}{}_j = \cdots = Y^{(k)i}{}_m \overline{Y}^{(k)m}{}_j = \delta^i{}_j.$$

We need these considerations in the following sections of the present chapter.

Examples.

1. A gauge transformation $\mathcal{F}$ in the fibre bundle $G\,\mathrm{Osc}^k M$ is locally expressed by (3.1). Assuming $F^{(0)} = \mathrm{id}$, we have

$$(3.13)\qquad \bar{x}^i = x^i,\ \bar{y}^{(\alpha)} = Y^{(\alpha)}(x, y^{(1)}, ..., y^{(\alpha)}),\ (\alpha = 1, ..., k).$$

In this case, $\mathcal{F}$ preserves the fibres of the bundle $G\,\mathrm{Osc}^k M$. For $k = 1$ we get the gauge transformations studied by G. Asanov [19].

2. The gauge transformations $\mathcal{F}$ from (3.1), useful in applications, are as follows:

$$(3.14)\qquad \bar{x}^i = X^i(x),\ \bar{y}^{(\alpha)i} = Y^{(\alpha)i}{}_j(x) y^{(\alpha)j},\ (\alpha = 1, ..., k).$$

3. The gauge transformations (3.14) are particular forms of the following gauge transformations

$$(3.15)\qquad \begin{aligned} \bar{x}^i &= X^i(x),\\ \bar{y}^{(\alpha)i} &= g^i{}_{j_1}(x) y^{(\alpha)j_1} + \cdots g^i{}_{j_1 ... j_\alpha}(x) y^{(1)j_1} ... y^{(1)j_\alpha},\\ &\qquad (\alpha = 1, ..., k). \end{aligned}$$

where $g(x) = ({g^i}_{j_1}(x), ..., {g^i}_{j_1...j_\alpha}(x)) \in G^{(k)}$ are the extensions of order k of the Lie group G.

From (3.15), it follows

$$(3.16)\qquad \begin{cases} \dfrac{\partial \bar{y}^{(1)i}}{\partial y^{(1)j}} = \cdots = \dfrac{\partial \bar{y}^{(k)i}}{\partial y^{(k)j}} \\ \dfrac{\partial \bar{y}^{(\alpha)i}}{\partial x^j} = \dfrac{\partial \bar{y}^{(\alpha+1)i}}{\partial y^{(1)j}} = \cdots = \dfrac{\partial \bar{y}^{(k)i}}{\partial y^{(k-\alpha)j}}, \\ \qquad\qquad (\alpha = 1, ..., k-1). \end{cases}$$

Generally, we have $\dfrac{\partial \bar{y}^{(1)j}}{\partial y^{(1)j}} \neq \dfrac{\partial \bar{x}^i}{\partial x^j}$.

In particular, if the $F^{(0)} \in \mathrm{Diff}\,(M)$ has the local representation $\bar{x}^i = g^i(x)$, then the condition $\dfrac{\partial \bar{y}^{(1)j}}{\partial y^{(1)j}} = \dfrac{\partial \bar{x}^i}{\partial x^j}$ is verified. In this case, $F^{(0)}$ determines the gauge transformation $\mathcal{F}_{F^{(0)}}$. So, the gauge transformation $\mathcal{F}_{F^{(0)}}$ determines transformations in the atlases of the associated bundle $G\,\mathrm{Osc}^k M$.

12.4 Gauge d–Tensor Fields. Gauge Nonlinear Connections

Based on the previous theory, we can define the notion of gauge tensor fields.

Definition 12.4.1 A gauge d–tensor field of type (r, s) on $(\pi^k)^{-1}(U) \subset \subset G\,\mathrm{Osc}^k M$ is an ordered system of functions $T^{i_1...i_r}_{j_1...j_s}(x, y^{(1)}, ..., y^{(k)})$ which has the following properties:

1. With respect to the gauge transformations (3.1), it is transformed by the rule:

$$(4.1)\qquad \begin{aligned} &\overline{T}^{i_1...i_r}_{j_1...j_s}(\bar{x}, \bar{y}^{(1)}, ..., \bar{y}^{(k)}) = \\ &= X^{i_1}_{h_1} ... X^{i_r}_{h_r} \overline{X}^{\ell_1}_{j_1} ... \overline{X}^{\ell_s}_{j_s} T^{h_1...h_r}_{\ell_1...\ell_s}(x, y^{(1)}, ..., y^{(k)}). \end{aligned}$$

2. With respect to (2.3), its transformation rule is that of the d–tensor field of type (r,s), i.e.:

$$\begin{aligned}&\widetilde{T}^{i_1\dots i_r}_{j_1\dots j_s}(\tilde{x},\tilde{y}^{(1)},\dots,\tilde{y}^{(k)})=\\ &=\frac{\partial\tilde{x}^{i_1}}{\partial x^{h_1}}\cdots\frac{\partial\tilde{x}^{i_r}}{\partial x^{h_r}}\frac{\partial x^{\ell_1}}{\partial\tilde{x}^{j_1}}\cdots\frac{\partial x^{\ell_s}}{\partial\tilde{x}^{j_s}}T^{h_1\dots h_r}_{\ell_1\dots\ell_s}(x,y^{(1)},\dots,y^{(k)}).\end{aligned}\tag{4.2}$$

3. In the point $\bar{u}=(\bar{x},\bar{y}^{(1)},\dots,\bar{y}^{(k)})$, with a changing of local coordinates, $T^{i_1\dots i_r}_{j_1\dots j_s}$ has the same rule (4.2) of transformation.

For instance, a gauge vector field $U^i(x,y^{(1)},\dots,y^{(k)})$ has the rules of transformations

$$\overline{U}^i=X^i{}_mU^m;\ \widetilde{U}^i=\frac{\partial x^i}{\partial x^j}U^j;\ \stackrel{\approx}{U}^i=\frac{\partial\stackrel{\approx}{x}^i}{\partial\bar{x}^j}\overline{U}^j.\tag{4.3}$$

A Lagrangian of order k, $L(x,y^{(1)},\dots,y^{(k)})$ has the rule of transformations

$$\begin{aligned}&\overline{L}(\bar{x},\bar{y}^{(1)},\dots,\bar{y}^{(k)})=L(x,y^{(1)},\dots,y^{(k)})\\ &\widetilde{L}(\tilde{x},\tilde{y}^{(1)},\dots,\tilde{y}^{(k)})=L(x,y^{(1)},\dots,y^{(k)})\\ &\stackrel{\approx}{L}(\stackrel{\approx}{x},\stackrel{\approx}{y}^{(1)},\dots,\stackrel{\approx}{Y}^{(k)})=\overline{L}(\bar{x},\bar{y}^{(1)},\dots,\bar{y}^{(k)})\end{aligned}\tag{4.4}$$

The set of all gauge d–tensor fields, together with the addition and tensor product form an algebra over the ring of gauge, functions.

Let N be a nonlinear connection on $G\,\mathrm{Osc}^kM$ with the local coefficients

$$\underset{(1)}{N}{}^i{}_j,\dots,\underset{(k)}{N}{}^i{}_j.\tag{4.5}$$

As usually, we set: $N_0=N, N_1=J(N_0),\dots,N_{k-1}=J(N_{k-2})$ and $V_k=J(N_{k-1})$, the J–vertical distributions having the adapted basis:

$$\begin{aligned}&\frac{\delta}{\delta x^i}=\frac{\partial}{\partial x^i}-\underset{(1)}{N}{}^j{}_i\frac{\partial}{\partial y^{(1)j}}-\dots-\underset{(k)}{N}{}^j{}_i\frac{\partial}{\partial y^{(k)j}}\\ &\frac{\delta}{\delta y^{(1)j}}=J\left(\frac{\delta}{\delta x^i}\right),\dots,\frac{\partial}{\partial y^{(k)i}}=J\left(\frac{\delta}{\delta y^{(k-1)j}}\right).\end{aligned}\tag{4.6}$$

Let us consider $\mathcal{F} = \{F^{(0)}, ..., F^{(k)}\}$ a local gauge transformation, $\mathcal{F} : u \longrightarrow \bar{u}$, expressed in (3.1).

The tangent map $\mathcal{F}_* : T_u(G\,\mathrm{Osc}^k M) \longrightarrow T_{\bar{u}}(G\,\mathrm{Osc}^k M)$ is well defined.

Definition 12.4.2 A nonlinear connection N is called a *gauge nonlinear* connection if the tangent mapping $\mathcal{F}_*$ preserves $N_0 = N$ and the J-vertical distributions $N_1, ..., N_{k-1}, V_k$.

Let $\left(\dfrac{\delta}{\delta \bar{x}^i}, \dfrac{\delta}{\delta \bar{y}^{(1)i}}, ..., \dfrac{\delta}{\delta \bar{y}^{(k)i}}\right)$ be the adapted basis to a gauge nonlinear connection in the point $\bar{u} = (\bar{x}, \bar{y}^{(1)}, ..., \bar{y}^{(k)})$, i.e.:

$$(4.6)' \qquad \begin{aligned} &\frac{\delta}{\delta \bar{x}^i} = \frac{\partial}{\partial \bar{x}^i} - \underset{(1)}{\bar{N}^j}{}_i \frac{\partial}{\partial \bar{y}^{(1)j}} - \cdots - \underset{(k)}{\bar{N}^j}{}_i \frac{\partial}{\partial \bar{y}^{(k)j}} \\ &\frac{\delta}{\delta \bar{y}^{(1)i}} = J\left(\frac{\delta}{\delta \bar{x}^i}\right), ..., \frac{\delta}{\delta \bar{y}^{(k)i}} = J\left(\frac{\delta}{\delta \bar{y}^{(k-1)i}}\right). \end{aligned}$$

Then, by means of (3.1) and of the tangent map $\mathcal{F}_*$, we get:

$$(4.7) \qquad \begin{aligned} &\frac{\delta}{\delta x^i} = X_i^m \frac{\delta}{\delta \bar{x}^m}, \frac{\delta}{\delta y^{(1)i}} = Y^{(1)m}{}_i \frac{\delta}{\delta \bar{y}^{(1)m}}, ..., \\ &\frac{\partial}{\partial y^{(k)i}} = Y^{(k)m}{}_i \frac{\partial}{\partial \bar{y}^{(k)m}}. \end{aligned}$$

So, we have

Proposition 12.4.1 *In order that N be a gauge nonlinear connection it is necessary that*

$$(4.8) \qquad X^i{}_m = Y^{(1)i}{}_m = \cdots = Y^{(k)i}{}_m.$$

But, if we take into consideration only the gauge transformations (3.1), which verify (4.8) and the following conditions

$$(4.8)' \qquad \frac{\partial Y^{(\alpha)i}}{\partial x^j} = \frac{\partial Y^{(\alpha+1)i}}{\partial y^{(1)j}} = \cdots = \frac{\partial Y^{(k)i}}{\partial y^{(k-\alpha)j}}, \ (\alpha = 1, ..., k-1),$$

we deduce that $\left(\dfrac{\delta}{\delta x^i}, \dfrac{\delta}{\delta y^{(1)i}}, ..., \dfrac{\delta}{\delta y^{(k)i}}\right)$ are the gauge d-vector fields.

Theorem 12.4.1

1. *With respect to the gauge transformations* (3.1), (4.8) *and* (4.8)′, *the coefficients of a nonlinear connection* N *is transformed by the rules:*

$$
(4.9)\qquad
\begin{aligned}
&\underset{(1)}{\overline{N}}{}^{i}{}_{m}X^{m}{}_{j} = \underset{(1)}{N}{}^{m}{}_{j}X^{i}{}_{m} - \frac{\partial Y^{(1)i}}{\partial x^{j}}\\
&\underset{(2)}{\overline{N}}{}^{i}{}_{m}X^{m}{}_{j} = \underset{(2)}{N}{}^{m}{}_{j}X^{i}{}_{m} + \underset{(1)}{N}{}^{m}{}_{j}\frac{\partial Y^{(1)i}}{\partial x^{m}} - \frac{\partial Y^{(2)i}}{\partial x^{j}}\\
&\dots\dots\dots\dots\dots\dots\dots\dots\dots\dots\\
&\underset{(k)}{\overline{N}}{}^{i}{}_{m}X^{m}{}_{j} = \underset{(k)}{N}{}^{m}{}_{j}X^{i}{}_{m} + \underset{(k-1)}{N}{}^{m}{}_{j}\frac{\partial Y^{(1)i}}{\partial x^{m}} + \cdots +\\
&\qquad + \underset{(1)}{N}{}^{m}{}_{j}\frac{\partial Y^{(k-1)i}}{\partial x^{m}} - \frac{\partial Y^{(k)i}}{\partial x^{j}}.
\end{aligned}
$$

2. *The dual coefficients* $\{\underset{(1)}{M}{}^{i}{}_{j}, ..., \underset{(k)}{M}{}^{i}{}_{j}\}$ *of* N *are transformed as follows:*

$$
(4.10)\qquad
\begin{aligned}
&\underset{(1)}{M}{}^{m}{}_{j}X^{i}{}_{m} = \underset{(1)}{\overline{M}}{}^{i}{}_{m}X^{m}{}_{j} + \frac{\partial Y^{(1)i}}{\partial x^{i}}\\
&\underset{(2)}{M}{}^{m}{}_{j}X^{i}{}_{m} = \underset{(2)}{\overline{M}}{}^{i}{}_{m}X^{m}{}_{j} + \underset{(1)}{\overline{M}}{}^{i}{}_{m}\frac{\partial Y^{(1)m}}{\partial x^{j}} + \frac{\partial Y^{(2)m}}{\partial x^{i}}\\
&\dots\dots\dots\dots\dots\dots\dots\dots\dots\dots\\
&\underset{(k)}{M}{}^{m}{}_{j}X^{i}{}_{m} = \underset{(k)}{\overline{M}}{}^{i}{}_{m}X^{m}{}_{j} + \underset{(k-1)}{\overline{M}}{}^{i}{}_{m}\frac{\partial Y^{(1)m}}{\partial x^{j}} + \cdots +\\
&\qquad + \underset{(1)}{\overline{M}}{}^{i}{}_{m}\frac{\partial Y^{(k-1)m}}{\partial x^{j}} + \frac{\partial Y^{(k)i}}{\partial x^{j}}.
\end{aligned}
$$

Of course, the dual basis $(\delta x^{i}, \delta y^{(1)i}, ..., \delta y^{(k)i})$ are transformed by (3.1), (4.8) and (4.8)′ by the formula:

$$
(4.10)' \qquad \delta\bar{x}^{i} = X^{i}{}_{m}\delta x^{i}, ..., \delta\bar{y}^{(k)i} = X^{i}{}_{m}\delta y^{(k)i}.
$$

Using the previous theory, we can show that a regular gauge Lagrangian of order k, determines a gauge nonlinear connection (cf.Ch.10).

12.5 Gauge N-Linear Connections and Gauge h– and v^α–Covariant Derivatives

Let N be a gauge nonlinear connection and (4.6) the adapted basis to the distributions $N_0, N_1, ..., V_k$. An N–linear connection D with the coefficients $D\Gamma(N) = (L^i{}_{jm}, \underset{(1)}{C}{}^i{}_{jm}, ..., \underset{(k)}{C}{}^i{}_{jm})$ is a gauge N–linear connection if for any gauge vector field U^i, the h– and v^α-covariant derivatives

$$(5.1)\qquad \begin{cases} U^i_{|m} = \dfrac{\delta U^i}{\delta x^m} + U^s L^i{}_{sm} \\ U^i \overset{(\alpha)}{|}_m = \dfrac{\delta U^i}{\delta y^{(\alpha)m}} + U^s \underset{(\alpha)}{C}{}^i{}_{sm} \ (\alpha = 1, ..., k) \end{cases}$$

are gauge d–tensor fields.

Theorem 12.5.1 *An N–linear connection D with the coefficients $D\Gamma(N) = (L^i{}_{jm}, \underset{(\alpha)}{C}{}^i{}_{jm}, (\alpha = 1, ..., k)$ is a gauge N–linear connection if, and only if, with respect to (3.1), we have:*

$$(5.2)\qquad \begin{aligned} \bar{L}^i_{pq} X^p{}_j X^q{}_m &= X^i{}_m L^m{}_{pq} - \frac{\partial X^i{}_j}{\partial x^m} \\ \underset{(\alpha)}{\bar{C}}{}^i{}_{pq} X^p{}_j X^q{}_m &= X^i{}_m \underset{(\alpha)}{C}{}^m{}_{pq} \ (\alpha = 1, ..., k) \end{aligned}$$

So, we can give

Definition 12.5.1 An N–linear connection, whose coefficients satisfy (5.2), is called a gauge N–linear connection.

It is not difficult to prove:

Theorem 12.5.2 *If N is a gauge nonlinear connection, then $B\Gamma(N) = = (\overset{B}{L}{}^i{}_{jh}, \underset{(\alpha)}{\overset{B}{C}}{}^i{}_{jh})$, where*

$$(5.3)\qquad \overset{B}{L}{}^i{}_{jh} = \frac{\delta \underset{(1)}{N}{}^i{}_j}{\delta y^{(1)h}}, \quad \underset{(\alpha)}{\overset{B}{C}}{}^i{}_{jh} = 0, \ \alpha = 1, ..., k$$

is a gauge (Berwald) N-linear connection.

We denote by $d^h_m T^{i_1...i_r}_{j_1...j_s}$ and $d^{v_\alpha}_m T^{i_1...i_r}_{j_1...j_s}$ the $h-$ and $v_\alpha-$ $(\alpha = 1, ..., k)$ the gauge $h-$ and v^α–covariant derivatives of a gauge d–tensor field.

So, we have

$$(5.4)\quad \begin{aligned} d^h_m T^{i_1...i_r}_{j_1...j_s} &= \frac{\delta T^{i_1...i_r}_{j_1...j_s}}{\delta x^m} + L^{i_1}_{hm} T^{h...i_r}_{j_1...j_s} + \cdots - L^h_{j_s m} T^{i_1...i_r}_{j_1...h}, \\ d^{v_\alpha}_m T^{i_1...i_r}_{j_1...j_s} &= \frac{\delta T^{i_1...i_r}_{j_1...j_s}}{\delta y^{(\alpha)m}} + \underset{(\alpha)}{C}{}^{i_1}{}_{hm} T^{h...i_r}_{j_1...j_s} + \cdots - \underset{(\alpha)}{C}{}^{h}{}_{j_s m} T^{i_1...i_r}_{j_1...h}. \end{aligned}$$

Let us consider g_{ij} a gauge metric tensor field. A gauge N–linear connection $D\Gamma(N) = (L^i{}_{jh}, \underset{(\alpha)}{C}{}^i{}_{jh})$ is called metric with respect to g_{ij} if

$$d^h{}_m g_{ij} = 0,\ d^{v_\alpha}{}_m g_{ij} = 0,\ (\alpha = 1, ..., k).$$

Theorem 12.5.3 *If g_{ij} is a gauge metric tensor field then its generalized Christoffel symbols are the coefficients* (2.3), *Ch.*10, *of a gauge N–linear connection, which is metric with respect to g_{ij}.*

The properties of the curvatures and torsions of a gauge N–linear connection can be studied without any difficulties.

Let us consider the fibre bundle $\mathcal{N}$ of nonlinear connections on $G\,\mathrm{Osc}^k M$, $\pi_{\mathcal{N}} : \mathcal{N} \longrightarrow G\,\mathrm{Osc}^k M$, in which in a local chart a point $\eta \in \mathcal{N}$, $\eta = (x, y^{(1)}, ..., y^{(k)}, \underset{(1)}{N}, ..., \underset{(k)}{N})$ has the coordinates $(x^i, y^{(1)i}, ..., y^{(k)i}, \underset{(1)}{N}{}^i{}_j, ..., \underset{(k)}{N}{}^i{}_j)$ and $\pi_{\mathcal{N}}(x, y^{(1)}, ..., y^{(k)}, \underset{(1)}{N}, ..., \underset{(k)}{N}) = (x, y^{(1)}, ..., y^{(k)})$.

The transformations of the local coordinates are given by the transformations $(x^i, y^{(1)i}, ..., y^{(k)i}, \underset{(1)}{N}{}^i{}_j, ..., \underset{(k)}{N}{}^i{}_j) \longrightarrow (\tilde{x}^i, \tilde{y}^{(1)i}, ..., \tilde{y}^{(k)i}, \underset{(1)}{\widetilde{N}}{}^i{}_j, ..., \underset{(k)}{\widetilde{N}}{}^i{}_j)$ from (1.5), (4.10), Ch.6. Then, the group $\mathrm{Gau}\,(G\,\mathrm{Osc}^k M)$ acts on $\mathcal{N}$ by the rules (2.3) and (4.9).

Analogously, on the fibre bundle $\mathcal{M}$, $\pi_{\mathcal{M}} : \mathcal{M} \to G\,\mathrm{Osc}^k M$ of the N–linear connections, whose local coordinates are $(x^i, y^{(1)i}, ..., y^{(k)i}, \underset{(1)}{N}{}^i{}_j, ..., \underset{(k)}{N}{}^i{}_j, L^i{}_{jh}, \underset{(1)}{C}{}^i{}_{jh}, ..., \underset{(k)}{C}{}^i{}_{jh})$ acts on the group $\mathrm{Gau}\,(G\,\mathrm{Osc}^k M)$ by the rule (1.5), (4.10), Ch.6 and (5.2). We use these fibre bundles in the next section.

12.6 Einstein–Yang–Mills Equations

Let $L_0(x^i, y^{(1)i}, ..., y^{(k)i})$ be a Lagrangian defined on a domain $\Omega \subset R^{n(k+1)}$, $N = (\underset{(1)}{N}{}^i{}_j, ..., \underset{(k)}{N}{}^i{}_j)$ a gauge nonlinear connection and g_{ij} a gauge metric structure on $G\,\mathrm{Osc}^k M$. We assume that L_0 is a function of $(x^i, y^{(1)i}, ..., y^{(k)i})$ of the form:

$$L_0(x^i, y^{(1)i}, ..., y^{(k)i}) = L\left(\Phi^A, \frac{\delta\Phi^A}{\delta x^i}, \frac{\delta\Phi^A}{\delta y^{(1)i}}, ..., \frac{\delta\Phi^A}{\delta y^{(k)i}}\right), \tag{6.1}$$

where Φ^A, $(A = 1, ..., p)$ are p gauge scalar fields, which in applications will be scalar fields on the total space $\mathcal{M}$ of the fibre bundle of gauge N–linear connections (the functions of waves).

In the following lines we denote Φ^A by Φ.

For a change of local coordinates on $G\,\mathrm{Osc}^k M$, the density of Lagrangian

$$\mathcal{L}(x^i, y^{(1)i}, ..., y^{(k)i}) = \sqrt{g^{k+1}} L_0(x^i, y^{(1)i}, ..., y^{(k)i}),\ g = \det\|g_{ij}\| \tag{6.2}$$

is transformed by the rule

$$\mathcal{L}(x^i, y^{(1)i}, ..., y^{(k)i}) = \widetilde{\mathcal{L}}(\tilde{x}^i, \tilde{y}^{(1)i}, ..., \tilde{y}^{(k)i}) \det J(\tilde{x}^i, \tilde{y}^{(1)i}, ..., \tilde{y}^{(k)i}). \tag{6.3}$$

Therefore, the integral action of the Lagrangian L_0 can be given in the form

$$I(\Phi) = \int_\Omega \mathcal{L}(x^i, y^{(1)i}, ..., y^{(k)i}) d\omega \tag{6.4}$$

with the volume element $d\omega = dx^1 \wedge \cdots \wedge dx^n \wedge dy^{(1)1} \wedge \cdots \wedge dy^{(k)n}$.

According to (6.3), it follows that the action $I(\Phi)$ does not depend on the choice of local coordinates on $G\,\mathrm{Osc}^k M$.

From the variational principle we have $\delta I(\Phi) = 0$, where

$$\delta I(\Phi) = \int_\Omega \left(\frac{\partial\mathcal{L}}{\partial\Phi}\delta\Phi + \frac{\partial\mathcal{L}}{\partial(\frac{\delta\Phi}{\delta x^i})}\delta\left(\frac{\delta\Phi}{\delta x^i}\right) + \cdots + \frac{\partial\mathcal{L}}{\partial(\frac{\delta\Phi}{\delta y^{(k)i}})}\delta\left(\frac{\delta\Phi}{\delta y^{(k)i}}\right)\right) d\omega. \tag{6.5}$$

Assuming that the operators δ and ∂ commute, we obtain, from (6.5), the *Euler–Lagrange equations* for the gauge scalar fields Φ^A:

$$\frac{\partial \mathcal{L}}{\partial \Phi} - \frac{\partial}{\partial x^i}\left(\frac{\partial \mathcal{L}}{\partial(\frac{\partial \Phi}{\partial x^i})}\right) - \frac{\partial}{\partial y^{(1)i}}\left(\frac{\partial \Phi}{\partial(\frac{\partial \Phi}{\partial y^{(1)i}})}\right) - \cdots - \tag{6.6}$$

$$- \frac{\partial}{\partial y^{(k)i}}\left(\frac{\partial \mathcal{L}}{\partial(\frac{\partial \Phi}{\partial y^{(k)i}})}\right) = 0.$$

In order to prove the geometrical invariance of the Euler–Lagrange equation (6.6), it is convenient to write it with respect to the gauge adapted basis $\left(\frac{\delta}{\delta x^i}, \ldots, \frac{\delta}{\delta y^{(k)i}}\right)$, taking into account the formulae (6.8), Ch.6:

$$\begin{cases} \dfrac{\partial}{\partial x^i} = \dfrac{\delta}{\delta x^i} + \underset{(1)}{M}{}^m{}_i \dfrac{\delta}{\delta y^{(1)m}} + \cdots + \underset{(k)}{M}{}^m{}_i \dfrac{\partial}{\partial y^{(k)i}} \\ \dfrac{\partial}{\partial y^{(1)i}} = J\left(\dfrac{\partial}{\partial x^i}\right), \ldots, \dfrac{\partial}{\partial y^{(k)i}} = J\left(\dfrac{\delta}{\delta y^{(k-1)i}}\right). \end{cases} \tag{6.7}$$

Remarking that

$$\frac{\partial L}{\partial\left(\frac{\partial \Phi}{\partial x^i}\right)} = \frac{\partial L}{\partial\left(\frac{\delta \Phi}{\delta x^i}\right)}$$

$$\frac{\partial L}{\partial\left(\frac{\partial \Phi}{\partial y^{(1)i}}\right)} = \frac{\partial L}{\partial\left(\frac{\delta \Phi}{\delta x^j}\right)}\left(-\underset{(1)}{N}{}^i{}_j\right) + \frac{\partial L}{\partial\left(\frac{\delta \Phi}{\delta y^{(1)i}}\right)\Big| \frac{\delta \Phi}{\delta x^j} = c_1 = const.}$$

....

and that $\sqrt{g}$ does not depend on Φ, the equation (6.6) is equivalent to:

$$
(6.8)\quad \begin{aligned}
&\sqrt{g^{k+1}}\left\{\left[\frac{\partial L}{\partial\Phi}-\frac{\delta}{\delta x^i\left(\dfrac{\partial L}{\partial\left(\dfrac{\delta\Phi}{\delta x^i}\right)}\right)}-\frac{\delta}{\delta y^{(1)i}}\left(\frac{\partial L}{\partial\left(\dfrac{\delta\Phi}{\delta y^{(1)i}}\right)}\right)\Bigg|_{c_1}-\cdots-\right.\right.\\
&\left.-\frac{\delta}{\delta y^{(k)i}}\left(\frac{\partial L}{\partial\left(\dfrac{\delta\Phi}{\delta y^{(k)i}}\right)}\right)\Bigg|_{c_1\ldots c_k}\right]+\left[\frac{\delta}{\delta y^{(1)i}}\left(\underset{(1)}{N^i}{}_j\right)\frac{\partial L}{\partial\left(\dfrac{\delta\Phi}{\delta x^i}\right)}\right]+\\
&+\left[\frac{\delta}{\delta y^{(2)i}}\left(\underset{(1)}{N^i}{}_j\right)\frac{\partial L}{\partial\left(\dfrac{\delta\Phi}{\delta y^{(1)j}}\right)}+\frac{\delta}{\delta y^{(2)i}}\left(\underset{(2)}{N^i}{}_j\right)\frac{\partial L}{\partial\left(\dfrac{\delta\Phi}{\delta x^j}\right)}\right]+\cdots+\\
&\left.+\left[\frac{\delta}{\delta y^{(k)i}}\left(\underset{(1)}{N^i}{}_j\right)\frac{\partial L}{\partial\left(\dfrac{\delta\Phi}{\delta y^{(k-1)j}}\right)}+\cdots+\frac{\delta}{\delta y^{(k)i}}\left(\underset{(k)}{N^i}{}_j\right)\frac{\partial L}{\partial\left(\dfrac{\delta\Phi}{\delta x^j}\right)}\right]\right\}-\\
&-\left\{\frac{\delta\sqrt{g^{k+1}}}{\delta x^i}\frac{\partial L}{\partial\left(\dfrac{\delta\Phi}{\delta x^i}\right)}+\frac{\delta\sqrt{g^{k+1}}}{\delta y^{(1)i}}\frac{\partial L}{\partial\left(\dfrac{\delta\Phi}{\delta y^{(1)i}}\right)}\Bigg|_{c_1}+\cdots+\right.\\
&\left.+\frac{\delta\sqrt{g^{k+1}}}{\delta y^{(k)i}}\frac{\partial L}{\partial\left(\dfrac{\delta\Phi}{\delta y^{(k)i}}\right)}\Bigg|_{c_1\ldots c_{k-1}}\right\}=0,
\end{aligned}
$$

where $c_1=\dfrac{\delta\Phi}{\delta x^j},\ldots,c_{k-1}=\dfrac{\delta\Phi}{\delta y^{(k-1)j}}$ are constants.

Now, it is not difficult to prove:

Theorem 12.6.1 *The Euler–Lagrange equation* (6.6) *is invariant with respect to the transformations of local coordinates on* $G\,\mathrm{Osc}^kM$. *If the Lagrangian* $\mathcal{L}$ *is a gauge invariant, then the equation* (6.6) *has the same property.*

If s is a solution of the equation (6.6), it follows that $\bar{s} = g \cdot s$, for every $g \in G^{(k)}$ is a solution of this equation, too.

Remark 12.6.1 An important problem consists in the study of the gauge invariance of the Lagrange $\mathcal{L}$, from (6.2). In the case $k = 1$, it was solved by Utiama's Theorem [307].

The gauge invariance of $\mathcal{L}$, to the infinitesimal transformation of the Lie group G, using the same method given in the Chapter 8 for the Nöther theorem, will be treated in the next section.

12.7 Gauge Invariance of the Lagrangians of Order k

Now, we assume that the manifold $\mathrm{Osc}^k M$ is a homogeneous space, with the fundamental group G having the parameters $\{a^\lambda\}$, $\lambda = 1, ..., m$.

The gauge transformations on $\mathrm{Osc}^k M$ will be the infinitesimal transformations of G. The gauge invariance of the Lagrangian L from (6.1), to the infinitesimal transformations will be studied as in the classical case [266]. In this respect it is useful to write the equation (6.8) in a simplified form.

So, we denote:

$$(7.1) \qquad \overset{h}{\Phi}{}^i{}_A = \frac{\partial L}{\partial\left(\dfrac{\delta \Phi^A}{\delta x^i}\right)}, \quad \overset{v_\alpha}{\Phi}{}^i{}_A = \frac{\partial L}{\partial\left(\dfrac{\delta \Phi^A}{\delta y^{(\alpha)i}}\right)_{|c_1 \dots c_{\alpha-1}}}, \quad (\alpha = 1, ..., k)$$

and consider $D\Gamma(N) = (L^i{}_{jk}, \underset{(\alpha)}{C}{}^i{}_{jk})$ a fixed N-linear connection, as for instance the Berwald N-linear connection $B\Gamma(N)$.

With these data, the equation (6.8) takes the form

$$(7.2) \qquad \frac{\partial L}{\partial \Phi^A} - \overset{h}{\Phi}{}^i{}_{A|i} - \overset{v_1}{\Phi}{}^i{}_A \overset{v_1}{\Big|}_i - \cdots - \overset{v_k}{\Phi}{}^i{}_A \overset{v_k}{\Big|}_i - E_A = 0, \text{ where}$$

$$E_A = \frac{1}{\sqrt{g^{k+1}}}\left\{ \sqrt{g^{k+1}}_{|i} \overset{h}{\Phi}{}^i{}_A + \sqrt{g^{k+1}} \overset{(v_1)}{|}_i \overset{v_1}{\Phi}{}^i{}_A + \cdots + \right.$$

$$(7.3) \quad \left. + \sqrt{g^{k+1}} \overset{(v_k)}{|}_i \overset{v_k}{\Phi}{}^i{}_A \right\} - \left(L^i{}_{ji} + \frac{\delta \underset{(1)}{N}{}^i{}_j}{\delta y^{(1)i}} + \cdots + \frac{\delta \underset{(k)}{N}{}^i{}_j}{\delta y^{(k)i}} \right) \overset{h}{\Phi}{}^j{}_A -$$

$$- \left(\underset{(1)}{C}{}^i{}_{ji} + \frac{\delta \underset{(1)}{N}{}^i{}_j}{\delta y^{(2)i}} + \cdots + \frac{\delta \underset{(k-1)}{N}{}^i{}_j}{\delta y^{(k)i}} \right) \overset{v_1}{\Phi}{}^j{}_A - \cdots - \underset{(k)}{C}{}^i{}_{ji} \overset{v_k}{\Phi}{}^j{}_A.$$

From (7.2) it follows that E_A are the scalar fields. Assuming that $\underset{(\alpha)}{N}{}^i{}_j$ are the functions of the form $\underset{(\alpha)}{N}(x, y^{(1)}, ..., y^{(\alpha)})$ $(\alpha = 1, ..., k)$, then $\dfrac{\delta \underset{(\alpha)}{N}{}^i{}_j}{\delta y^{(\beta)h}} = 0$ for $\alpha < \beta$, and $\dfrac{\delta \underset{(\alpha)}{N}{}^i{}_j}{\delta y^{(k)h}} = \underset{(\alpha)}{L}{}^i{}_{jh}$ have, with respect to a changing of local coordinates, the same rule of transformation as the coefficients of a linear connection on the base manifold M.

Denoting

$$(7.4) \quad \begin{aligned} &\overset{B}{L}{}^i{}_{jh} = \frac{1}{k+1}\left\{ L^i{}_{jh} + \underset{(1)}{L}{}^i{}_{jh} + \cdots + \underset{(k)}{L}{}^i{}_{jh} \right\}, \\ &\underset{(\alpha)}{\overset{B}{C}}{}^i{}_{jh} = 0 \qquad\qquad (\alpha = 1, ..., k) \end{aligned}$$

we obtain $\overset{B}{D}\ \Gamma(N) = (\overset{B}{L}{}^i{}_{jh}, \underset{(\alpha)}{\overset{B}{C}}{}^i{}_{jh})$, an N-linear connection of the Berwald type.

Therefore, the scalar fields E_A take an interesting form:

$$(7.3)' \quad \begin{aligned} E_A = \frac{1}{\sqrt{g^{k+1}}}\left\{ \sqrt{g^{k+1}}_{|i} \overset{h}{\Phi}{}^i{}_A + \sqrt{g^{k+1}} \overset{(v_1)}{|}_i \overset{v_1}{\Phi}{}^i{}_A + \cdots + \right. \\ \left. + \sqrt{g^{k+1}} \overset{(k)}{|}_i \overset{v_k}{\Phi}{}^i{}_A \right\} - (k+1) \overset{B}{L}{}^i{}_{ji} \overset{h}{\Phi}{}^j{}_A. \end{aligned}$$

Let us consider a Lie group G, $\dim G = m$, of transformations on $\mathrm{Osc}^k M$, given by

$$\bar{u} = \mathcal{F}(u, a),\ u = \mathcal{F}(u, 0), \tag{7.5}$$

where $u = (x, y^{(1)}, ..., y^{(k)}) \in \mathrm{Osc}^k M$ and $a = (a^1, ..., a^m)$ are the parameters of G.

An infinitesimal transformation of G is as follows:

$$\bar{u} = u + \xi_\lambda \varepsilon^\lambda,\ \varepsilon_\lambda(u) = \left.\frac{\partial \mathcal{F}(u, a)}{\partial a^\lambda}\right|_{a^\lambda = 0}, \tag{7.6}$$

where $\varepsilon^\lambda = \delta a^\lambda$ are the infinitesimal variations of the parameters.

Of course, (7.6) are expressed in the form:

$$\begin{cases} \bar{x}^i = x^i + \xi^{(0)i}{}_\lambda \varepsilon^\lambda \\ \bar{y}^{(\alpha)i} = y^{(\alpha)i} + \xi^{(\alpha)i}{}_\lambda \varepsilon^\lambda,\ \xi^{(\alpha)i}{}_\lambda = \left.\dfrac{\partial F^{(\alpha)i}}{\partial a^\lambda}\right|_{a^\lambda = 0} \quad (\alpha = 1, ..., m), \end{cases} \tag{7.6$'$}$$

The generators of the group of transformations are given by:

$$X_\lambda = \xi^{(0)i}{}_\lambda \frac{\partial}{\partial x^i} + \xi^{(1)i}{}_\lambda \frac{\partial}{\partial y^{(1)i}} + \cdots + \xi^{(k)i}{}_\lambda \frac{\partial}{\partial y^{(k)i}}. \tag{7.7}$$

and its structure constants $C^\nu{}_{\lambda\mu}$, are obtained from:

$$[X_\lambda, X_\mu] = C^\nu{}_{\lambda\mu} X_\nu. \tag{7.7$'$}$$

Of course, an infinitesimal transformation (7.6) acts on the functions $\Phi(u)$ as follows:

$$\begin{aligned} \overline{\Phi} - \Phi = \delta\Phi = \Big\{ & \xi^{(0)i}{}_\lambda \frac{\partial \Phi}{\partial x^i} + \xi^{(1)i}{}_\lambda \frac{\partial \Phi}{\partial y^{(1)i}} + \cdots + \\ & + \xi^{(k)i}{}_\lambda \frac{\partial \Phi}{\partial y^{(k)i}} \Big\} \varepsilon^\lambda \stackrel{def}{=} X_\lambda \Phi \varepsilon^\lambda. \end{aligned} \tag{7.8}$$

In an adapted basis, the operators X_λ will be written in the form

$$X_\lambda = \chi^{(0)i}{}_\lambda \frac{\delta}{\delta x^i} + \cdots + \chi^{(k)i}{}_\lambda \frac{\delta}{\delta y^{(k)i}}. \tag{7.7$''$}$$

The relationship between the coefficients $\xi^{(\alpha)i}{}_{\lambda}$ and $\chi^{(\alpha)i}{}_{\lambda}$ are not difficult to be found.

Let us consider the function of wave Φ^{A} and a representation of the generators X_{λ} given by the matrix of the form

$$(X_{\lambda})^{A}_{B},\ (A, B = 1, ..., p;\ \lambda = 1, ..., m)$$

An infinitesimal transformation (7.8) of the functions Φ^{A} can be writen:

$$\overline{\Phi}^{A} = \Phi^{A} + \delta\Phi^{A},\ \delta\Phi^{A} = \varepsilon^{\lambda}(X_{\lambda})^{A}_{B}\Phi^{B}. \tag{7.9}$$

The transformation (7.6) in which ε^{λ} are constants is called a *global gauge transformation* on the manifold $\mathrm{Osc}^{k}M$.

In this section we study only the case of the global gauge transformations.

Remark 12.7.1 A global gauge transformation (7.6) satisfies the Definition 12.2.2 if the functions $\xi^{(\alpha)i}$ depend only on $(x, y^{(1)}, ..., y^{(\alpha)})$, $(\alpha = 0, 1, ..., k)$.

The following variations hold:

$$\begin{aligned} &\delta\left(\frac{\delta\Phi^{A}}{\delta x^{i}}\right) = \frac{\delta}{\delta x^{i}}(\delta\Phi^{A}) = \varepsilon^{\lambda}(X_{\lambda})^{A}_{B}\frac{\delta\Phi^{B}}{\delta x^{i}} \\ &\dots\dots\dots\dots\dots\dots\dots\dots\dots\dots\dots \\ &\delta\left(\frac{\delta\Phi^{A}}{\delta y^{(k)i}}\right) = \frac{\delta}{\delta y^{(k)i}}(\delta\Phi^{A}) = \varepsilon^{\lambda}(X_{\lambda})^{A}_{B}\frac{\delta\Phi^{B}}{\delta y^{(k)i}} \end{aligned} \tag{7.10}$$

The invariance of the Lagrangian L to the infinitesimal transformations implies

$$\frac{\partial L}{\partial\Phi^{A}}\delta\Phi^{A} + \frac{\partial L}{\partial(\dfrac{\delta\Phi^{A}}{\delta x^{i}})}\delta\left(\frac{\delta\Phi^{A}}{\delta x^{i}}\right) + \cdots + \frac{\partial L}{\partial(\dfrac{\delta\Phi^{A}}{\delta y^{(k)i}})}\delta\left(\frac{\delta\Phi^{A}}{\delta y^{(k)i}}\right) = 0.$$

Taking into account (7.9) and (7.10) we get

$$\left\{\frac{\partial L}{\partial\Phi^{A}}\Phi^{B} + \overset{h}{\Phi}{}^{i}{}_{A}\frac{\delta\Phi^{B}}{\delta x^{i}} + \overset{v_1}{\Phi}{}^{i}{}_{A}\frac{\delta\Phi^{B}}{\delta y^{(1)i}} + \cdots + \overset{v_k}{\Phi}{}^{i}{}_{A}\frac{\delta\Phi^{B}}{\delta y^{(k)i}}\right\}(\chi_{\lambda})^{A}_{B} = 0\ . \tag{7.11}$$

Combining (7.2) with (7.11), we have:

Theorem 12.7.1 (The law of global conservation). *The condition of the gauge global invariance of the Lagrangian L from (6.1) is given by the equations:*

$$(7.12)\qquad \overset{(h)}{J}{}^{i}{}_{\lambda|i} + \overset{(v_1)}{J}{}^{i}{}_{\lambda}\overset{(v_1)}{\Big|}{}_{i} + \cdots + \overset{(v_k)}{J}{}^{i}{}_{\lambda}\overset{(v_k)}{\Big|}{}_{i} = E_A(X_\lambda)^A_B\Phi^B$$

where

$$(7.12)'\qquad \overset{(h)}{J}{}^{i}{}_{\lambda} = -\overset{h}{\Phi}{}^{i}{}_{A}(X_\lambda)^A_B\Phi^B,\ \overset{(v_\alpha)}{J}{}^{i}{}_{\lambda} = -\overset{(v_\alpha)}{\Phi}{}^{i}{}_{A}(X_\lambda)^A_B\Phi^B \qquad (\alpha = 1,...,k)$$

are the so-called h- and v_α-currents.

In the case when the parameters ε^λ depend on the point $u \in \mathrm{Osc}^k M$, the transformation (7.6) is called *local.*

Consequently, the infinitesimal transformations (7.9) are

$$(7.13)\qquad \overline{\Phi}^a = \Phi^A + \delta\Phi^A,\ \delta\Phi^A = \varepsilon^\lambda(u)(X_\lambda)^A_B\Phi^B$$

and the formulae (7.10) are modified as follows

$$(7.14)\qquad \begin{aligned} \delta\left(\frac{\delta\Phi^A}{\delta x^i}\right) &= \frac{\delta}{\delta x^i}(\delta\Phi^A) = \varepsilon^\lambda(X_\lambda)^A_B\Phi^B + (X_\lambda)^A_B\Phi^B\frac{\delta\varepsilon^\lambda}{\delta x^i} \\ \delta\left(\frac{\delta\Phi^A}{\delta y^{(\alpha)i}}\right) &= \frac{\delta}{\delta y^{(\alpha)i}}(\delta\Phi^A) = \varepsilon^\lambda(X_\lambda)^A_B\Phi^B + (X_\lambda)^A_B\Phi^B\frac{\delta\varepsilon^\lambda}{\delta y^{(\alpha)i}} \end{aligned} \qquad (\alpha = 1,...,k).$$

The previous theory can be extended step by step to this general case. For details, see Munteanu's paper [225].

We end here the introduction into the higher order gauge theory.

References

1. ABRAHAM, R. and MARSDEN, J.E., *Foundations of Mechanics*, The Benjamin–Cummings Publ. Co. 1978.
2. AGUIRRE–DABÁN, E. and SÁNCHEZ–RODRIGUEZ, I., *On Structure Equations for Second Order Connections*, Diff. Geom. and Its Appl. Proc. Int. Conf. Opava, 1992, Silesian Univ. Opava 1993, 257–264.
3. AIKOU, T., *Differential Geometry of Finslerian Vector Bundles*, Ph.D. Thesis, Iaşi, Romania, 1991.
4. ALBU, I.D. et OPRIŞ, D., *Densités Lagrangiennes invariantes par rapport aux automorphismes d'une variété fibrée,* Proc. Int. Workshop on Diff. Geom. Sci. Bul. "Politehnica" Univ. Of Bucharest, S. A, vol. 55, no. 3–4, 1993, 1–7.
5. AMICI, O. and CASCIARO, B., *An Example of Lagrangian Theory of Relativity Satisfying the Conservation Law,* Proc. Nat. Sem. on Finsler and Lagrange Spaces, Braşov, Romania, 1986, 37–42.
6. ANASTASIEI, M., *The Geometry of Time–Dependent Lagrangians*, Math. Comput. Modelling, vol. 20, no. 4/5, 1994, 67–81.
7. ANASTASIEI, M., *On Deflection Tensor Field*, In the volume: Lagrange and Finsler Geometry, Kluwer Acad. Publ. FTPH no. 76, 1996, 1–14.
8. ANASTASIEI, M., *Geometry of Higher Order Sprays* (to appear).
9. ANASTASIEI, M., *Some Tensorial Finsler structures on the Tangent Bundle*, An. şt. Univ. "Al.I.Cuza", Iaşi, XXVII, s. I-a, 1981, 9–16.
10. ANASTASIEI, M. and ANTONELLI, P.L., *The Differential Geometry of Lagrangians which Generate Sprays,* in vol. of Kluwer Acad. Publ. FTPH, no. 76, 1996, 15–34.
11. ANASTASIEI, M. and KAWAGUCHI, H., *A Geometrical Theory of Time Dependent Lagrangians,* I: *Non-linear Connections*, Tensor N.S. vol. 48, 273–282.
12. ANASTASIEI, M. and KAWAGUCHI, H., *A Geometrical Theory of Time Dependent Lagrangians,* II: *M–Connections*, Tensor N.S. vol. 48, 283–293.
13. ANDRÉS, L.C. DE, LÉON, M. DE and RODRIGUES, P.R., *Connections on Tangent Bundle of Higher Order Associated to Regular Lagrangians,* Geometriae Dedicata 39, 17–28, 1991, 17–28.

14. ANTONELLI, P.L. and HRIMIUC, D., *A New Class of Spray-Generating Lagrangians,* in vol. of Kluwer Acad. Publ. FTPH, no. 76, 1996, 81–92.

15. ANTONELLI, P.L., INGARDEN, R.S. and MATSUMOTO, M., *The Theory of Sprays and Finsler Spaces with Applications in Physics and Biology,* Kluwer Acad. Publ. FTPH no. 58, 1993.

16. ANTONELLI, P.L. and . MIRON, R., *Lagrange and Finsler Geometry. Applications to Physics and Biology,* Kluwer Acad. Publ. FTPH no. 76, 1996.

17. ANTONELLI, P.L. and ZASTAWNEAK, T.J., *Lagrange Geometry, Finsler Spaces and Noise Applied in Biology and Physics,* Math. Comp. Mod. 20, no. 4–5, 1994.

18. ARINGAZIN, A.K. and ASANOV, G.S., *Problems of Finslerian Theory of Gauge Fields and Gravitation,* Rep. Math. Phys. 25, 1988, 35–93.

19. ASANOV, G.S., *Finsler Geometry, Relativity and Gauge Theories,* D. Reidel Publ. Co., Dordrecht, 1985.

20. ASANOV, G.S., *Jet Extension of Finslerian Approach,* Fortschr. Phys. 38, 1990, 8, pp 571–610.

21. ASANOV, G.S. and KAWAGUCHI, T., *A Post-Newtonian Estimation for the Metric* $\gamma_{ij}(x) + \frac{\alpha}{c^2} y_i y_j$. Tensor N.S. vol. 49 (1990), 99–102.

22. ASANOV, G.S. and PONOMARENKO, S.F., *Finsler Bundles over Space-Time: Associated Gauge Fields and Connections,* Nauka Ed. Chishinew, 1989.

23. ASANOV, G.S. and STAVRINOS, P.C., *Finslerian Deviations of Geodesics over Tangent Bundle*, Reports on Mathematical Physics, vol. 30 (1991), 63–69.

24. ATANASIU, GH., *Structures Finsler presque horsymplectiques*, An. şt. Univ. "Al.I.Cuza", Iaşi, XXX, (1984), 15–18.

25. BÁCSÓ, S., *Randers and Kropina Spaces in Geodesic Correspondence,* in vol. of Kluwer Acad. Publ. FTPH, no. 76, 1996, 61–64.

26. BALAN, V., *On the Generalized Einstein–Yang–Mills Equations,* Publ. Math. Debrecen 43, 1993, no. 1–2, 1–11.

27. BALAN, V. and STAVRINOS, P.C., *Deviations of Geodesics in the Fibered Finslerian Approach,* In vol. of Kluwer Acad. Publ. FTPH no. 76, 1996, 65–74.

28. BAO, D. and CHERN, S.S., *On a Notable Connection in Finsler Geometry*, Houston J. Math. 19, no. 1, 1993, 135–180.

29. BARBU, V., *Analysis and Control of Nonlinear Infinite Dimensional Systems*, Academic Press, New York, Boston, 1993.

30. BARBU, V., *Mathematical Methods in Optimization of Differential Systems*, Kluwer Acad. Publ. 1994.

31. BARTHEL, W., *Nichtlineare Zusammenhange und Deren Holonomie Gruppen*, J. Reine Angew. Math. 212, 1963, 120–149.

32. BEIL, R.G., *Electrodynamics from a Metric,* Int. Jour. Theor. Phys. 26, 1987, 189–197.

33. BEIL, R.G., *Finsler Gauge Transformations and General Relativity,* Int. Jour. Theor. Phys. vol. 31, 1992, 1025–1044.

34. BEIL, R.G., *New Class of Finsler Metric,* Int. Jour. of Theor. Phys. vol. 28, 1989, 659–667.

35. BLAIR, D.E., *Contact Manifolds in Riemannian Geometry,* Springer–Verlag, 1976.

36. BLEECKER, D., *Gauge Theory and variational principles*, Addison–Wesley Publ. Comp. Inc. 1981.

37. BORŞ, C.I., *Two Body Problem in a Non–Inertial Frame of Reference and the Advance of Perihelion*, Tensor N.S. vol. 53, 1993, 174–178.

38. BORŞ, C.I., *Two Body Problem in a Non–Inertial Frame of Reference and the Theory of Relativity*, An. şt. Univ. "Al.I.Cuza", Iaşi, 1995.

39. BOSKOFF, W., *Hyperbolic Geometry and Barbilian Spaces*, Hadronic Press, 1996, H.P. Inc. Fl. USA.

40. BOURBAKI, N., *Variétés Différentielles et Analytiques. Fascicule des résultats*, Mir Moscow 1975 (Russian).

41. BRANDT, H.E., *Structure of Spacetime Tangent Bundle*, Foundations of Physics Letters, vol. 4, no. 6, 1991, 523–536.

42. BRÂNZEI, D., *Sur une généralisation de la théorie d'homotopie,* C. R. Acad. Sci. Paris, t. 263, 1966, 10–12.

43. BREVIC, V., Nuovo cimento, B. 105, 1990, p. 717.

44. BRÉZIS, H., *Monotonicity Methods in Hilbert Spaces and Some Applications to Nonlinear Partial Differential Equations. Contributions to Nonliner Functional Analysis,* Academic Press, 1971.

45. Brézis, H., *Opératuers maximaux monotones et semigroupes de contractions dans les espaces de Hilbert,* Math. Studies, vol. 5, North–Holland, Amsterdam 1975.

46. Bucătaru, I., *Connection Map in the Higher Order Geometry* (in print).

47. Buchner, K., *Kräftige Lösung der Einsteinschen Gleichungen*, Tensor N.S. 38, 1982, 65–68.

48. Buchner, K. and Roşca, R., *On Real Left Kählerian Space–Time*, Abh. Math. Sem. Univ. Hamburg 51, 1981, 5–12.

49. Burghelea, D., Hangan, Th., Moscovici, H., Verona, A., *Introducere în Topologia Diferenţială*, Ed. Ştiinţifică, Bucureşti, 1973.

50. Cantrijn, F., Crampin, M. and Sarlet, W., *Higher Order Differential Equations and Higher Order Lagrangian Mechanics*, Proc. Camb. Phil. Soc. 99, 1986, 565–587.

51. Capursi, M. and Palombella, A., *On the Almost Hermitian structures Associated to a Generalized Lagrange Space*, Mem. Sect. St. Acad. Române, VIII, 1985, 57–64.

52. Carathéodory, C., *Variationsrechnung und partielle Differentialgleichungen erste Ordnung*, Teubner, Leipzig, Berlin, 1935.

53. Carmo, M.P. do, *Riemannian Geometry*, Birkhäuser, Boston, 1992.

54. Cartan, E., *Les espaces de Finsler,* Hermann, Paris, 1934.

55. Chen, B.Y., *Differential Geometry of Real Submanifolds in a Kähler Manifold*, Monot. für Math. 91, 1981, 257–274.

56. Chern, S.S., *Local Equivalence and Euclidean Connections in Finsler Spaces,* Nat. Tsing. Hua Univ. A5, 1948, 95–121, or Selected Papers, vol. II, 194–212.

57. Chern, S.S., *On Finsler Geometry,* C.R. Acad. Sci. Paris, t. 314, 1992, 757–761.

58. Chinea, D., Léon, M. de and Morero, J.C., *The Constrained Algorithms for Time–Dependent Lagrangians*, Preprint UM–1991.

59. Choquet–Bruhat, Y. and De Witt–Morette, *Analysis, Manifolds and Physics,* North–Holland 1982.

60. Čomic, I., *Curvature Theory of Generalized Miron's d–Connections*, Diff. Geom. and Appl. Proc. Conf. Brno, 1989, 17–26.

61. Čomic, I., *Generalized Second Order Gauge Connections*, Indian J. of Pure & Appl. Math. (1996) (to appear).

62. CRAIG, H.V., *An Extensor Generalization of a Simple Calculus of Variations Problem*, Tensor N.S. 17, 1966, 313–320.

63. CRAIG, H.V., *On a Generalized Tangent Vector,* Amer. Jour. of Math., vol. LVII, 1935, 457–462.

64. CRAIOVEANU, M. and PUTA, M., *Geometric Quantization of Some Constrained M‹chanical Systems*, Proc. of the 22nd Conference on Diff. Geom. and Topology, Bucharest, 1991, 87–91.

65. CRAIOVEANU, M., *On a Class of Non-lLnear Differential Operators* (to appear).

66. CRAMPIN, M., *Jet Bundle Techniques in Analytical Mechanics,* Quaderni del Cons. Naz. delle Ricerche, Gruppo Naz. di Fisica Matematica no. 47, 1995.

67. CRAMPIN, M. and PIRANI, F.A.E., *Applicable Differential Geometry,* LMS Lecture Note S. 59, Cambridge Univ. Press, 1986.

68. CRAMPIN, M., SARLET, W. and CANTRIJN, F., *Higher-Order differential Equations and Higher-Order Lagrangian Mechanics*, Math. Proc. Camb. Phil. Soc. 1986, 99, 565–587.

69. CRUCEANU, V., *Sur la théorie de sous-fibrés vectoriels*, C.R. Acad. Sci. Paris, 302, 1986, 705–708.

70. CRUCEANU, V., *Sur certaines morphismes des structures géométriques*, Rendiconti di Mat. ser. VIII, V.G.N., 1986, 321–332.

71. DEKRÉT, A., *Natural Connections of* 1*-Forms on Tangent Bundles,* Diff. Geom. and its Appl. Proc. Int. Conf. Opava, 1992, Silesian Univ. Opava 1993, 265–272.

72. DIEUDONNÉ, J.A. and . CARRELL, J.B., *Invariant Theory, Old and New*, Academic Press, 1971.

73. DODSON, C.T.J., *Categories, Bundles and Space Time Topology*, Shiva Publ. England, 1980.

74. DODSON, C.T.J. and RADIVOIOVICI, M., *Tangent and Frame Bundle of Order Two,* An. şt. Univ. "Al.I.Cuza", Iaşi, s. I-a, 28, 1982.

75. DOMBROWSKI, P., *On the Geometry of Tangent Bundle*, J. Reine und Angewande Math. 210 (1962), p. 73–88.

76. DRAGOŞ, L, *Principiile Mecanicii Analitice*, Ed. Tehnică, Bucureşti, 1969.

77. DRAGOŞ, L., *Principiile Mecanicii Mediilor Continue,* Ed. Tehnică, Bucureşti, 1981.

78. DRAGOŞ, P, *Nonstandard Differentiable Manifolds* (in Romanian), Ph.D. Thesis, Univ. Timişoara, 1995.

79. DRECHSLER, W. and MAYER, M.E., *Fibre Bundle Techniques in Gauge Theories,* Lecture Note in Physics 67, 1977, Springer–Verlag.

80. DUC, T.V., *Sur la géométrie différentielle des fibrés vectoriels*, Kodai Math. Sem. Rep. 26, 1975, 349–408.

81. ECK, D.J., *Gauge–natural Bundles and Generalized Gauge Theories*, Memoirs of the AMS vol. 33, no. 247, 1981.

82. EHRESMANN, CH., *Introduction à la théorie des structures infinitésimales et des pseudogroupes de Lie*, Coll. de Géom. Diff. de Strasbourg, SNRS, Paris 1953, 97–110.

83. EHRESMANN, CH., *Les connexions infinitésimales d'ordre supérieur*, Atti del V Congresso dell'Unione Matematica Italiana, Pavia–Torino, 1956, 344–346.

84. EHRESMANN, CH., *Les prolongements d'un espace fibré différentiable*, Compte Rend. de l'Acad. Sci. Paris 240, 1955, 1755–1757.

85. ELIOPOULOS, H.A., *A Generalized Metric Space for Electromagnetic Theory*, Bull. Cl. Sci. (5) 51, 1965, 986–995.

86. ELIOPOULOS, H.A., *Subspaces of a Generalized Metric Space*, Canad. J. Math. 11, 1959, 235–255.

87. ENDO, H., *Submanifolds in Contact Metric Manifolds and in Almost Cosymplectic Manifolds,* Ph.D. Thesis, "Al.I. Cuza" Univ. of Iaşi, Romania, 1993.

88. FAVA, FR., *Transformazioni Classiche e fibratti di Getti*, Atti del Convegno Matematico in celebrazione di G. Fubini e F. Severi, Torino, 1982, 187–204.

89. GARCIA, X., PONS, J.M. and ROMÁN–ROY, N., *Higher Order Lagrangian Systems: Geometrie Structures, Dynamics and Constraints*, J. Math. Phys. 32, 10, 1991, 2744–2763.

90. GANCARZEWICZ, J., MIKULSKI, W. and POGODA, Z., *Natural Bundles and Natural Liftings Prolongations of Geometric Structures*, Diff. Geom. and Its Appl. Proc. Int. Conf. Opava, 1992, Silesian Univ. Opava 1993, 281–320.

91. GHEORGHIEV, GH., *Sur les espaces de Riemann–Eisenhart*, Tensor N.S. 46, 1987, 271–277.

92. GHEORGHIEV, GH., MIRON, R., PAPUC, D., *Geometrie Analitică şi Diferenţială,* vol. I, II, Ed. Didactică şi Pedagogică, Bucureşti, 1968 şi 1969.

93. GHINEA, I., *Conexiuni Finsler compatibile cu anumite structuri geometrice*, Teză de doctorat, Univ. Cluj, 1978.

94. GOTTLIEB, I. and VACARU, S., *A Moor's Tensorial Integration in Generalized Lagrange Spaces*, In vol. of Kluwer Acad. Publ. FTPH no. 76, 1996, 209–216.

95. GREUB, W., HALPERIN, S. & VANSTONE, R., *Connections and Cohomology,* vol. I, Academic Press, 1972.

96. GRIFONE, J., *Structures Presque-Tangentes et Connexions*, I, II, An. Inst. Fourier Grenoble, 22, 1972, 287–334; 22, 3, 1972, 291–338.

97. GRIGORE, R.D., *A Generalized Lagrangian Formalism in Particle Mechanics and Classical Field Theory*, Fortschritte der Physik, vol. 41, no. 7, 1993, 569–617.

98. GRIGORE, R.D., *Generalized Lagrangian Dynamics and Noetherian Symmetries*, Int. Jour. Math. Phys. A. 7, 1992, 7153–7168.

99. GRIGORE, R.D., *The Derivation of Einstein Equations from Invariance PrincipleS,* Class Quant. Gravity 9(1992), 1255–1571.

100. GRIGORE, R.D. and POPP, O.T., *Evolution Space and Elementary Systems*, Rev. Roum. Phys. 37, 1992, 447–463.

101. GROSU, M., *Teorii geometrice ale câmpurilor gravitaţional şi electromagnetic,* Teză de doctorat, Inst. de Mat. al Academiei, Bucureşti, 1994.

102. GUILLEMIN, V. and STERNBERG, S., *Symplectic Techniques in Physics,* Cambridge Univ. Press, Cambridge 1984.

103. HAIMOVICI, A., *Sur les espaces à connexion non linéaire à parallelisme absolue,* Rev. Roum. Math. Pures Appl., 10, 1965, 1121–1128.

104. HAIMOVICI, M., *Les formules fondamentales de la théorie des hypersurfaces d'un espace général*, Ann. Sci. Univ. Jassy, 20, 1935, 39–58.

105. HAIMOVICI, M., *Variétés totalement extrémales et variétés totalement géodésiques dans les espaces de Finsler,* Ann. Sci. Univ. Jassy, 29, 1939, 559–644.

106. HANGAN, T., *On Totally Geodesic Distribution of Planes,* Coll. Math. Soc. Janos Bolyai 46, North Holland, 1988, 519–530.

107. HASEGAWA, I., *Sasakian Manifolds and Their Submanifolds*, Ph.D. Thesis, Inst. Math. Acad. Bucureşti, 1995.

108. HASHEGAWA, I., YAMAUCHI, K. and SHIMADA, H., *Sasakian structures of Finsler Manifolds*, In vol. of Kluwer Acad. Publ. FTPH no. 76, 1996, 75–80.

109. HASHIGUCHI, M., *On Conformal Transformations of Finsler Metrics*, J. Math. Kyoto Univ. 16, 1976, 25–50.

110. HASHIGUCHI, M., *On the Generalized Finsler Spaces*, An. şt. Univ. "Al.I.Cuza", Iaşi, XXX, 1984, 69–73.

111. HERMANN, R., *Differential Geometry and the Calculus of Variations*, Acad. Press 1970.

112. HERMANN, R., *Vector Bundles in Mathematical Physics*, vol.I, Benjamin, 1970.

113. HORVARTH, J.I., *New Geometrical Methods of the Theory of Physical Fields*, Nuovo Cimento, 10, 9 (1958, 444-496.

114. HRIMIUC, D., *On the Cartan Connection for a Class of Generalized Lagrange Spaces*, Open system & Information Dynamics (to appear).

115. HRIMIUC, D. and SHIMADA, H., *On the $\mathcal{L}$-Duality between Lagrange and Hamilton Manifolds* (to appear).

116. HUSEMOLLER, D., *Fibre Bundles*, Springer, Berlin 1975.

117. IACOB, C., *Introducere Matematică în Mecanica Fluidelor*, Ed. Acad., Bucureşti, 1953.

118. IACOB, C., *Theoretical Mechanics* (in Romanian), Ed. Didactică şi Pedagogică, Bucureşti, 1980.

119. IANUŞ, S., *Submanifolds of Almost Hermitian Manifolds*, Riv. Mat. Univ. Parma (5)3, 1994, 123–142.

120. IANUŞ, S., *Differential Geometry and Applications to Relativity* (in Romanian), Ed. Acad. Române, Bucureşti, 1983.

121. ICHIJYŌ, Y., *On Some G-Structures Defined on Tangent Bundles*, Tensor N.S. 42, 1985, 179–190.

122. ICHIJYO, Y., MIRON, R., *On Some Structures Defined on Dominant Vector Bundles*, J. Math. Tokushima Univ. 20, 1986, 13–26.

123. IEŞAN, D., *A Theory of Thermoelastic Materials with Voids*, Acta Mechanica 60, 1986, 67–89.

124. IEŞAN, D. *On Saint-Venant's Problem*, Archive for Rational Mechanics and Analysis, vol. 91, nr. 4, 1986, 363–375.

125. IGARASHI, T., *Cartan Spaces and Applications,* Ph.D. Thesis, "Al.I. Cuza" Univ. of Iaşi, Romania, 1994.

126. INGARDEN, R.S., *Differential Geometry and Physics,* Tensor N.S., vol. 30, 1976, 201–209.

127. INGARDEN, R.S., *On Physical Interpretations of Finsler and Kawaguchi Geometries*, Tensor N.S. 46, 1987, 354–360.

128. IONESCU–PALLAS, N., *General Relativity and Cosmology* (in Romanian), Ed. Ştiinţifică, Bucureşti, 1981.

129. IONESCU–PALLAS, I. and SOFONEA, L., *The Lagrange Formalism in the Modelling of "Finite Range" Gravity*, In vol. of Kluwer Acad. Publ. FTPH no. 76, 1996, 217–232.

130. ISHIKAWA, H., *Einstein Equations in Lifted Finsler Spaces*, Nuovo Cimento 56 B, 1980, 252–262.

131. ISU, M., *Notes on Integral Inequality for Riemannian Manifolds Admitting an Infinitesimal Conformal Transformation*, I, II, Tensor N.S. vol. 44, 45, 1987, 189–194, 195–201.

132. ISU, M., *On the Notion of Spray on Tangent Bundle*, Tensor N.S. (to appear).

133. IZUMI, H., YOSHIDA, M., *On the Geometry of Generalized Metric Spaces*, I, II, Publ. Math. Debrecen 39, 1991, 113–134, 185–197.

134. JANYŠKA, J., *Geometrical Properties of Prolongation Functors*, Časopis Pěst. Mat. 110, 1985, 77–86.

135. KADEISVILI, J.V., *Santilli's Isotopies of Contemporary Algebras, Geometries and Relativities*, Hadronic Press, 1994.

136. KAWAGUCHI, M., *An Introduction to the Theory of Higher Order Space,* I: *The Theory of Kawaguchi Spaces*, RAAG Memoirs, vol. 3, 1962.

137. KAWAGUCHI, M., *An Introduction to the Theory of Higher Order Space,* II: *Higher Order Spaces in Multiple Parameters*, RAAG Memoirs, vol. 4, 1968.

138. KAWAGUCHI, M., *Une généralisation du calcul des jets et quelques prolongements généralisés de variétés différentielles*, Tôhoku Math. J., vol. 13, 329–370.

139. KAWAGUCHI, T. and MIRON, R., *A Lagrangian Model for Gravitation and Electromagnetism*, Tensor N.S. vol. 48, no. 2, 1989, 153–168.

140. KAWAGUCHI, A., *On the Vector of Higher Order and the Extended Affine Connections.* Ann. di Matem. Pura ed Appl. IV, 55, 1961, 105–118.

141. KIRKOVITS, M., *Variational Calculus and Lagrange Spaces*, Ph.D. Thesis, "Al.I.Cuza" Univ. of Iaşi, Romania, 1989.

142. KLEIN, J., *Espaces Variationnels et Mécaniques*, Ann. Inst. Fourier (Grenoble), 12, 1962, 1–124.

143. KLEIN, J., *On variational Second Order Differential Equations. Polynomial Case,* Diff. Geom. and its Appl. Proc. Int. Conf. Opava, 1992, Silesian Univ. Opava 1993, 449–459.

144. KLEPP, F., *Remarkable Finsler Structures and the Finsler Geometry of Vector Bundles* (in Romanian), Ph.D. Thesis, "Al.I.Cuza" University of Iaşi, 1982.

145. KOBAYASHI, S., *Transformation Groups in Differential Geometry.* Springer Verlag 1972.

146. KOBAYASHI, S. and NOMIZU, K., *Foundations of Differential Geometry,* vol. I, II, Intersci. Publ., New York, 1963, 1973.

147. KOLÁR, I., *Canonical Forms on the Prolongations of Principal Fibre Bundles*, Rev. Roum. Math. Pures et Appl. 16, 1971, 1091–1106.

148. KOLÁR, I., *On the Prolongations of Geometric Object Fields*, An. şt. Univ. "Al.I.Cuza", Iaşi, S.I-a, Mat. 17, 1971, 437–446.

149. KOLÁR, I., *Some Higher Order Operations with Connections*, Czechoslovak Math. J. 24 (99), 1974, 311–330.

150. KONDO, K., *Construction of Kawaguchi Space–Time by Statistical Observation of Modads and the Origin of Quantum. Physical Fields and Particles*, RAAG Memoirs 3, E-VII, 1962, 263–306.

151. KONDO, K., *Epistemological Foundations of Quasi–Microscopic Phenomena from the Standpoint of Finsler's and Kawaguchi's Higher Order Geometry,* Post RAAG Reports no. 241, 242, 243; 1991.

152. KONDO, K., *On the Physical Meanings of the Zermelo Conditions of Kawaguchi Space*, Tensor N.S. 14, 1963, 191–215.

153. KOSZUL, J.L., *Lecture on Fibre Bundles and Differential Geometry*, Tata Inst. of Fundam. Research. Bombay 1960.

154. KOZMA, L., *On the Classical Conditions for Finsler Type Connections,* Proc. Nat. Seminar on Finsler and Lagrange Spaces, Braşov, Romania, 1988, 207–216.

155. KRUPKA, D., *Lepagean Forms in Higher-Order Variational Theory,* In Proc. of IUTAM-ISIMM Symp. on Modern Developments in Analytical Mechanics, Bologna, 1983, 197–238.

156. KRUPKA, D., *Local Invariants of a Linear Connection,* Coll. Math. Soc. Janos Bolyai 31, Diff. Geom. Budapest, 1979, North Holland 1982, 349–369.

157. KRUPKA, D. and JANYŠKA, J., *Lectures on Differential Invariants,* Univ. Brno, 1990.

158. KRUPKOVA, O., *Variational Analysis on Fibered Manifolds over One-Dimensional Bases,* Ph.D. Thesis, Silesian Univ. Opava, 1992.

159. KUPERSCHMIDT, B.A., *Geometry of Jet Bundles and the Structure of Lagrangian and Hamiltonian Formalism,* Lecture Notes in Math. 775, Geometric Methods in Mathematical Physics, Springer, Berlin 1980, 162–218.

160. LANDAU, L.D., LIFŞIŢ, E.M., *Teoria câmpului,* Ed. Tehnică, Bucureşti, 1963 (Traducere din l. rusă).

161. LÉON, M. DE, OUBIÑA, J.A., RODRIGUES, P. and SALGADO, M., *Almost S-Tangent Manifolds of Higher-Order,* Pacif. J. of Math. 154, 2, 1992, 201–213.

162. LÉON, M. DE and RODRIGUES, P., *Almost Tangent Geometry and Higher-Order Mechanical Systems,* Proc. of Conf. Diff. Geom. Brno 1986, Reidel, 1987, p. 179–195.

163. LÉON, M. DE and RODRIGUES, P., *Dynamical Connections and Non-Autonomous Lagrangian Systems,* Ann. Fac. Sci. Toulouse IX, 2, 1988, 171–181.

164. LÉON, M. DE and RODRIGUES, P., *Generalized Classical Mechanics and Fields Theory,* North Holland, 1985.

165. LÉON, M. DE and RODRIGUES, P., *Methods of Differential Geometry in Analytical Mechanics,* North Holland, 1989.

166. LICHNEROWICZ, A., *Quelques théorèmes de géométrie différentielle globale,* Comm. Math. Helv. 22 (1949), 271–301.

167. LIBERMANN, P., *Sur le problème d'équivalence de certains structures infinitésimales,* Thèse de Doctorat à la Fac. de Sci. Univ. Strasbourg, 1953.

168. LIBERMANN, P. et MARLE, CH.M., *Symplectic Geometry and Analytical Mechanics,* D. Reidel Publ. Comp. 1987.

169. LOVELOCK, D. and RUND, H., *Tensors, Differential Forms and Variational Principles* John Wiley & Sons, New York, 1975.

170. LOVELOK, D. and RUND, H., *Variational Principles in the General Theory of Relativity,* Über Deutsch. Math.-Verein. 74, 1972, 1–65.

171. MANGERONE, D. & IRIMICIUC, N., *Mecanica rigidelor cu aplicaţii în inginerie*, vol. 1, Ed. Tehnică, Bucureşti, 1981.

172. MANGIAROTTI, L. and MODUGNO, M., *Connections and Dif-ferential Calculus on Fibred Manifolds: Application to Field Theory.* Bibliopolis, Naples, 1991.

173. MANGIAROTTI, L., MODUGNO, M., *Fibred Spaces, Jet Spaces and Connections for Field Theories,* Proc. of the Meeting "Geometry and Physics", (Bologna) 1982, 135–165.

174. MANGIAROTTI, L. and MODUGNO, M., *Graded Lie Algebras and Connections on a Fibred Space*, J. Math. Pures Appl. 63, 1984, 111–120.

175. MARTIN, M., *Hermitian Geometry and Involutive Algebras*, Math. Z. 188, 1985, 359–382.

176. MATSUMOTO, K., *Differential Geometry of Locally Conformal Kähler Manifolds and Their Submanifolds*, Ph.D. Thesis, "Al.I.Cuza" University of Iaşi, 1994.

177. MATSUMOTO, K. and MIHAI, I., *On Certain Transformation in a Lorentzian Para-Sasakian Manifold*, Tensor N.S. vol. 47, 1988, 189–197.

178. MATSUMOTO, K., MIHAI, I. and ROŞCA R., *Certain Real Hypersurfaces of a Locally Conformal Kähler Space Form*, Bul. Yamagata Univ. Natural Science, vol. 13, no. 1, 1992, 1–11.

179. MATSUMOTO, M., *Foundations of Finsler Geometry and Special Finsler Spaces,* Kaisheisha Press, Otsu, 1986.

180. MAYER, O., *Opera Matematică,* vol. I, Ed. Acad., Bucureşti, 1974.

181. MIHAI, I., *Capitole speciale de geometria varietăţilor complexe*, Ed. Univ., Bucureşti, 1995.

182. MIHAI, I., *Generic Submanifolds of Some Almost Paracompact Riemannian Manifolds*, Coll. Math. Soc. Janos Bolyai 46, Debrecen, 1984, 777–790.

183. MIRON, R., *General Randers Spaces,* in vol. of Kluwer Acad. Publ. FTPH, no. 76, 1996, 75–80.

184. Miron, R., *Geometry of Vector Subbundles in a Vector Bundle,* Tensor N.S., vol. 53, 1993, 126–139.

185. Miron, R., *Hamilton Geometry,* An. şt. Univ. "Al.I.Cuza", Iaşi, s. I-a Mat., 35, 1989, 33–85.

186. Miron, R., *Hamilton Geometry,* Univ. Timişoara, Sem. Mecanică, no. 3, 1987.

187. Miron, R., *Noether Theorems in Higher-Order Lagrangian Mechanics,* Int. Journ. of Theoretical Phys. vol. 34, no. 7, 1995, 1123–1146.

188. Miron, R., *On the Finslerian Theory of Relativity,* Tensor N.S., 44, 1987, 63–81.

189. Miron, R., *On the Inverse Problem in the Higher-Order Analytical Mechanics,* Proc. of the Congress of the Istituto per la Ricerca di Base (IRB) – in Monteroduni, Italy, July 1995 – Hadronic Press (to appear).

190. Miron, R., *Some Problems in the Geometries of Finsler Type,* Contemporary Mathematics. American Mathematical Society, 1996.

191. Miron, R., *Spaces with Higher Order Metric Structures,* Tensor N.S., vol. 53, 1993, 1–23.

192. Miron, R., *Technics of Finsler Geometry in the Theory of Vector Bundles,* Acta Sci. Math. 49, 1985, 119–129.

193. Miron, R., *The Geometry of Cartan Spaces,* Progress of Math. vol. 22, 1988, 1–38.

194. Miron, R., *The Higher-Order Lagrange Spaces: Theory of Subspaces,* The Proc. of Workshop in Diff. Geometry, Dec. 1995, "Aristoteles" Univ. Thessaloniki.

195. Miron, R. and Anastasiei, M., *The Geometry of Lagrange Spaces: Theory and Applications,* Kluwer Acad. Publ. 1994, FTPH, no. 59.

196. Miron, R. and Anastasiei, M., *Vector Bundles. Lagrange spaces. Applications to the Theory of Relativity* (in Romanian), Ed. Acad. Române, Bucureşti, 1987.

197. Miron, R. and Atanasiu, Gh., *Compendium on the Higher-Order Lagrange Spaces: The Geometry of k-Osculator Bundles. Prolongation of the Riemannian, Finslerian and Lagrangian Structures. Lagrange Spaces,* Tensor N.S. 53, 1993, 39–57.

198. MIRON, R. et ATANASIU, GH., *Compendium sur les espaces Lagrange d'ordre supérieur: La géométrie du fibré k-osculateur; Le prolongement des structures Riemanniennes, Finsleriennes et Lagrangiennes; Les espaces Lagrange* $L^{(k)n}$, Univ. Timişoara, Seminarul de Mecanică, no. 40, 1994.

199. MIRON, R. and ATANASIU, GH., *Differential Geometry of k-Osculator Bundles,* Rev. Roum. de Math. Pures et Appl. 1996 (to appear).

200. MIRON, R. and ATANASIU, GH., *Geometrical Theory of Gravitational and Electromagnetic Fields in the Higher-Order Lagrange Spaces,* Journ. of Tsukuba (to appear).

201. MIRON, R. and ATANASIU, GH., *Higher-Order Lagrange Spaces,* Rev. Roum. de Math. Pures et Appl. 1996 (to appear).

202. MIRON, R. and ATANASIU, GH., *Lagrange Geometry of Second Order,* Math. Comput. Modelling vol. 20, no. 4, 1994, 41–56.

203. MIRON, R., IANUS, S. and ANASTASIEI, M., *The Geometry of the Dual of a Vector Bundle,* Publ. de l'Inst. Math. 46(60), 1989, 145–162.

204. MIRON, R. and ATANASIU, GH., *Prolongation of the Riemannian, Finslerian and Lagrangian Structures,* Rev. Roum. de Math. Pures et Appl. 1996 (to appear).

205. MIRON, R. et KAWAGUCHI, T., *Higher-Order Relativistic Geometrical Optics,* Tensor N.S., vol. 53, 1993.

206. MIRON, R. et KAWAGUCHI, T., *Relativistic Geometrical Optics,* Int. Journ. of Theor. Phys. vol. 30, no. 11, 1991, 1521–1543.

207. MIRON, R. and KAWAGUCHI, T., *Sur la Théorie géométrique de l'optique relativiste,* C.R. Acad. Sci. Paris, t. 312, s. II, (1991), 593–598.

208. MIRON, R., KAWAGUCHI, T. and KAWAGUCHI, M., *On the Electromagnetic Field in Relativistic Geometrical Optics,* Progress of Math., vol. 25, (182), 1991, 1–13.

209. MIRON, R., KIRKOWITS, M. and ANASTASIEI, M., *A Geometrical Model for Variational Problem of Multiple Integrals,* Proc. Int. Conf. Diff. Geom. Appl. Dubrovnik, 1988, 8–25.

210. MIRON, R. and POP, I., *Topologie Algebrică, Omologie, Omotopie, Spaţii de acoperire,* Ed. Academiei, Bucureşti, 1974.

211. MIRON, R. and RADIVOIOVICI–TATOIU, *A Lagrangian Theory of electromagnetism,* Rep. Math. Phys. 27, 1989, 49–84.

212. MIRON, R., ROŞCA, R., ANASTASIEI, M. and BUCHNER, K., *New aspects of Lagrangian Relativity,* Found of Physics Letters, 5, no. 2, 1992, 141–171.

213. MIRON, R. and TAVAKOL, R., *Geometry of Space–Time and Generalized Lagrange Spaces,* Publ. Math. Debrecen, no. 4, 1994.

214. MIRON, R., TAVAKOL, R., BALAN, V., ROXBURG, I., *Geometry of Space–time and Generalized Lagrange Gauge Theory,* Publ. Math. Debrecen 42, 3–4, 1993, 215–224.

215. MIRON, R. and ZET, GH., *Post-Newtonian Approximation in Relativistic Optics,* Tensor N.S. vol. 53, 1993, 92–95.

216. MIRON, R. and ZET, GH., *Relativistic Optics of Nondispersive Media,* Foundations of Physics, vol. 25, no. 9, 1995, 1371–1382.

217. MIRON, R., WATANABE, S. and IZUMI, H., *Geometrical Theory of the Gravitational and Electromagnetic Fields in the Spaces with the Metric* $\exp^{2\sigma}(x,y)\gamma_{ij}(x)$. Mem. Acad. Române.

218. MIRON, R., WATANABE, S. and IKEDA, S., *Some Connections on Tangent Bundle and Their Applications to the General Relativity,* Tensor N.S. 46, 1987, 8–22.

219. MISHRA, R.S., *A Differentiable Manifold with f–Structure of Rank r,* Tensor N.S. 27 no. 3, 1973, 369–378.

220. MIMURA, F. and NÔNO, T., *A Geometric Approach to a Second–Order System of Partial Differential Equations within the Context of the Calculus on Vector–Valued Differential Forms,* Tensor N.S. 51, 1992, 283–300.

221. MOISIL, GR., *Opera Matematică,* vol. I, Ed. Academiei, Bucureşti, 1976.

222. MOOR, A., *Begründung einer affinen Geometrie der Bahnen dritter Ordnung,* Tensor N.S. 16, 1965, 37–55.

223. MORIMOTO, A., *Prolongations of Geometric Structures,* Lecture Notes, Math. Inst. Nagoya Univ., 1969.

224. MOTREANU, D. and POPA, C., *Hamilton–Jacobi Equations on Infinit–Dimensional Riemannian Manifolds,* Nonlinear Analysis Methods & Applications, vol. 9, nr. 7, 1985, 739–761.

225. MUNTEANU, GH., *Lagrangian Gauge Theory of Superior Order,* Proc. Nat. Seminar, Univ. Braşov, 1996 (to appear).

226. Munteanu, Gh., *Techniques of Higher Order Osculator Bundle in Generalized Gauge Theory,* Conf. on diff. Geom. and Appl. Aug. 28–30, 1995, Brno.

227. Munteanu, Gh. and Atanasiu, Gh., *On Miron Connections on Lagrange Space of Second Order,* Tensor N.S. vol. 50, 1991, 241–247.

228. Munteanu, Gh. and Ikeda, S., *On the Gauge Theory of Second Order,* Tensor N.S. vol. (to appear).

229. Myller, Al., *Scrieri Matematice,* Ed. Acad., Bucureşti, 1959.

230. Neagu, A. and Borcea, V., *Some Considerations about Regular Mappings of Finsler Spaces,* Rev. Roum. Math. Pures et Appl. T. XL, no. 2, 1995, 195–211.

231. Neagu, A. and Borcea, V., *Lebesgue Measure and Regular Mappings in Finsler Spaces,* in vol.: P.L. Antonelli and R. Miron, Lagrange and Finsler Geometry. Kluwer Acad. Publ. FTPH, no. 76, 1996, 195–205.

232. Obădeanu, V., *Le problème directe de la mécanique newtonienne,* Studii şi Cerc. Mat. T. 42, no. 3, 1990, 247–249.

233. Obădeanu, V., şi Marinca, V., *Problema inversă în mecanica analitică,* Monografii Matematice no. 44, Univ. Timişoara, 1992.

234. Okumura, M., *On Some Remarks on Special Kawaguchi Spaces,* Tensor N.S. 11, 1961, 154–160.

235. Onicescu, O., *Mecanica,* Ed. Tehnică, Bucureşti, 1969.

236. Opriş, D., *Higher-Order Lagrangian and Hamiltonian Mechanics,* Univ. Timişoara, Seminarul de Mecanică, no. 26, 1990.

237. Opriş, D. and Albu, I.D., *Geometrical Aspects of the Covariant Dynamics of Higher–Order* (Czech. J. of Math. to appear).

238. Otsuki, T., *Theory of Affine Connections of the Space of Tangent Directions of a Differentiable Manifold,* I, II. Math. J. Okayama Univ. 7, 1957, 1–74, 95–122.

239. Palais, R.S., *Natural Equations on Differential Forms,* Trans. Amer. Math. Soc. 92, 1959, 125–141.

240. Papaghiuc, N., *Semi–Invariant Products in Sasakian Manifolds,* An. şt. Univ. "Al.I.Cuza", Iaşi, sect. I-a, Mat. 30, 1984, 69–78.

241. Papuc, I.D., *About a New Metric Geometry of a Time–Oriented Lorentzian Manifold,* Univ. Timişoara, Seminarul de Mecanică, no. 44, 1995, p. 11.

242. PAPUC, D., *Geometrie Diferențială*, Ed. Didactică şi Pedagogică, Bucureşti, 1982.

243. PHAM MAU QUAN, *Induction éctromagnetique en relativité générale*, Cahier de Physique no. 96, 1958, 297–307.

244. PITIŞ, GH., *Foliations and Submanifolds of a Class of Contact Manifolds*, C.R. Acad. Sci. Paris sér. I math. 310, 1990, 197–202.

245. PITIŞ, G., *Trace Forms and Chern Forms of a Class of Submanifolds in a Complex Space Form*, Tensor N.S. vol. 52, 1993, 130–135.

246. POHL, W.F., *Connection in Differential Geometry of Higher Order*, Applied Math. and Statistics Laboratories, Stanford Univ. 1963, 1–23.

247. POHL, W.F., *Differential Geometry of Higher Order* Topology, 1, 1962, 169–211.

248. POMMARET, J.F., *Explicit Calculations of Certain Differential Identities Used in Mathematical Physics.* Proc. Conf. Diff. Geom. Brno 1986, D. Reidel 1987, 271–278.

249. POMMARET, J.F., *Systems of Partial Differential Equations and Lie Pseudogroups,* Gordon and Breach, New York, 1978.

250. POP, I., *On Some Homotopy Lifting and Extension Properties for Sequence of Continuous Maps,* An. şt. Univ. "Al.I.Cuza", Iaşi, s. I. Mat. 19, 1973, 103–111.

251. POPESCU, P., *Structuri geometrice pe fibrate vectoriale*, Teză de doctorat, Univ. "Al.I.Cuza" Iaşi, 1996.

252. POPOVICI, I., ANASTASIEI, M., *Sur les bases de géométrie finsleri-enne*, C.R. Acad. Sci. Paris 290, S. A, 808–810.

253. POPOVICI, I., IORDĂNESCU, R., TURTOI, A., *Graduări simple Jordan şi Lie considerate în geometria diferențială,* Ed. Academiei, Bucureşti, 1971.

254. PRIPOAE, G., *Propriétés de rigidité concernant la courbure des métriques indéfinies,* JGP vol.7, nr.1, 1990.

255. PUTA, M. *Hamiltonian Mechanical Systems and Geometric Quantization*, Kluwer Acad. Publ. 260, 1993.

256. RADIVOIOVICI, M.A., *Geometria Finsler a fibratelor vectoriale. Aplicații în relativitate,* Ph.D. Thesis, "Al.I.Cuza" University of Iaşi, 1990.

257. RĂILEANU, L., *Topological and Differential Manifolds*, Ed. Academiei, Bucureşti (to appear).

258. ROŞCA, R., *Espace-temps possédant la propriété géodésique*, C.R. Acad. Sci. Paris, A 285, 1977, p. 305.

259. ROŞCA, R., *On pseudo-Sasakian Manifolds,* Rend. Matematica, fasc. III, 4, 1984, 167–174.

260. RUND, H., *Adjoint Connections and Group Manifolds and Gauge Transformations*, The Math. Heritage of C.F. Gauss, World Sci. Singapore, 1991, 621–644.

261. RUND, H., *The Differential Geometry of Finsler Spaces*, Springer–Verlag, Berlin, 1959.

262. SAKAGUCHI, T., *Subspaces in Lagrange Spaces,* Ph.D. Thesis, "Al.I. Cuza" University of Iaşi, 1987.

263. SANTILLI, M.R., *Foundations of Theoretical Mechanics*, I: *The inverse problem in Newtonian Mechanics*, Springer–Verlag, 1978.

264. SANTILLI, M.R., *Foundations of Theoretical Mechanics*, II: *Birkhoffian Generalization of Hamiltonian Mechanics*, Springer–Verlag, 1981.

265. SANTILLI, M.R., *Isotopies of Differential Calculus and Its Applications to Mechanics and Geometries*, In Proc. of Workwhop on Diff. Geom., Thessaloniki 1994, Hadronic Press, 1995.

266. SARDANASHVILY, G. and ZAKHAROV, O., *Gauge Gravitation Theory*, World Scientific, Singapore, New Jersey, London, Hong Kong, 1991.

267. SASAKI, S., *Almost Contact Manifolds*, Tôhoku Univ. Sendai, 1, 1965; 2, 1967; 3, 1968.

268. SASAKI, S., *On the Finslerian Theory of Relativity*, Tensor N.S. 44, 1987, 63–81.

269. SASAKI, S., *On the Differential Geometry of Tangent Bundles of Riemannian Manifolds*, Tôhoku Math. J., 10, 1958, 338–354.

270. SATOSHI, IKEDA, *Advanced Studies in APplied Geometry,* Seizansha, Sagamihara, 1995.

271. SAUNDERS, D., *The Geometry of Jet Bundles,* Cambridge Univ.Press, New York, London, 1989.

272. SHEN, Z., *On a Connection in Finsler Geometry* (Houston Jour. Math. 20, 1994, 591–602.

273. SINGH, S.K., *An h-Metrical d-Connection of Special Miron Space*, Indian J. of Pure & Appl. Math., 1995.

274. SINHA, B.B. and SINGH, S.K., *On h-Curvature Tensors of Proper Miron Connections of Finsler Spaces,* Pure and Applied Mathematics Sciences, vol. XXII, no. 1–2, 1985, 71–77.

275. SMARANDA, D., *Immersion d'une modele d'univers avec champ magnetique dans une espace pseudo-euclidean,* Bull. Soc. Sci. Liège, 1967, 23–30.

276. SOFONEA, L., *Representative Geometries and Physical Theories* (in Romanian), Ed. Dacia, Cluj-Napoca, Romania, 1984.

277. SOOS, G., *Uber Gruppen von Affinitäten und Bewegungen in Finslerschen Räumen,* Acta Math. Sci. Hungar., 5, 1954, 73–84.

278. SOURIAU, J.M., *Structure des Systèmes Dynamiques*, Dunod, Paris, 1970.

279. SRIVASTAVA, T.N., *A Few Remarks on Special Kawaguchi Spaces,* Tensor N.S., 15, 1964, 12–19.

280. STAVRE, P., *The d-Linear Connections,* Proc. Nat. Sem. on Finsler and Lagrange Spaces, Braşov, Romania, 1988, 375–382.

281. STAVRINOS, P.C. BALAN, V., PREZAS, N. and MANOUSELIS, P., *Spinor Bundle of Order Two on the Internal Deformed System,* An. şt. Univ. "Al.I.Cuza", Iaşi, 1996 (to appear).

282. STAVRINOS, P.C. and MANOUSELIS, P., *The non-localized Field Theory: Poincaré gravity and Yang-Mills Fields,* Rep. Math. Phys. vol. 36, 1995.

283. STOICA, E. *A Geometrical Characterization of Normal Finsler Connections,* An. şt. Univ. "Al.I.Cuza", Iaşi, XXX, mat. 1984-1.

284. SYNGE, J.L., *Relativity: General Theory,* North-Holland, 1966.

285. SYNGE, J.L., *Some Intrinsic and Derived Vectors in a Kawaguchi Space,* Amer. Jour. of Mathematics, vol. LVII, 1935, 679–691.

286. SZABO, Z.I., *Positively Definite Berwald Spaces,* Tensor N.S., 35, 1981, 25–39.

287. SZENTHE, J., *On a Basic Property of Lagrangians,* Publ. Math. Debrecen 42/3-4, 1993, 247–251.

288. SZILASI, J., KOZMA, L., *Remarks on Finsler-Type Connections,* Proc. Nat. Sem. on Finsler and Lagrange spaces, Braşov, Romania, 1984, 181–197.

289. TAKANO, Y., *On the Theory of Fields in Finsler Spaces,* Proc. Int. Symp. on Relativity and Unified Field Theory, 1975–76, 17–26.

290. ŞANDRU, O., *Hamilton–Lagrange structures Associated to Some Systems of Partial Differential Equations*, Ph.D. Thesis, "Al.I.Cuza" University of Iaşi, 1993.

291. TAMASSY, L., *Affine Connections Inducing a Finsler Connection*, Coll. Math. Soc. Janos Bolyai 46, North–Holland, 1988, 1185–1193.

292. TAMASSY, L. and MATSUMOTO, M., *Direct Method to Characterize Conformally Minkowski Finsler Spaces*, Tensor N.S. 33, 1979, 380–384.

293. TAMIA–DIMOPOULOU, P., *A Relationship between CR-Structures and f-Structures Satisfying $f^3 + f = 0$.* Tensor N.S. 49, 1990, 250–252.

294. TEODORESCU, N., *Géométrie finslerienne et propagation des ondes*, Bull. Sci. Acad. Roumanie, 23, 1942, 138–144.

295. TEODORESCU, P.P., *Sisteme mecanice: modele clasice*, Ed. Tehnică, Bucureşti, 1984.

296. TRAUTMAN, D., *Differential Geometry for Physicists Bibliopolis,* Naples, 1984.

297. TONNELAT, M.A., *Les théories unitaires de l'électromagnetisme et de la gravitation*, Gauthier–Villars, Paris, 1965.

298. TSAGAS, GR., *Special Structures on a Manifold*, Journ. Inst. Math. and Com. Sci. Math. Ser. vol. 8, 1 (1955), 11–17.

299. TSAGAS, GR., *The Spectra of the Laplace Operator on Conformally Flat Manifolds*, Journal of the Tensor Soc. of India, vol. 10, 1994, 76–89.

300. TSAGAS, GR., and SOURLAS, D., *Isomanifolds and Their Isotensor Fields. Isomappings between Isomanifolds*, Algebras, Groups and Geometries vol. 12, no. 1, 1995, 1–89.

301. TULCZYJEW, W.M., *The Euler–Lagrange Resolution,* In Lecture Notes in Math. 836, Diff. Geom. Methods in Math.–Phys., Springer, Berlin 1980, 22–48.

302. ŢARINĂ, M., *Invariant Finsler Connections on Vector Bundles,* An. şt. Univ. "Al.I.Cuza", Iaşi, s. I-a, Mat. 33, 1987, 87–94.

303. ŢIŢEICA, GH., *Œuvres,* Tom I, Imprimeria Naţională, 1941.

304. ŢIŢEICA, Ş., *Mecanica Cuantică*, Ed. Acad., Bucureşti, 1984.

305. UDRIŞTE, C., *Convex Functions and Optimization Methods on Riemannian Manifolds,* Mathematics and Its Applications, 297, Kluwer Academic Publishers, 1994, p. 348.

306. UDRIŞTE, C., *Linii de câmp,* Ed. Tehnică, Bucureşti, 1988, p. 207.

307. UTIAMA, R. *Invariant Theoretical Interpretation of Interaction*, Phys. Rev. 101, 1956, p. 1597–1607.

308. VACARU, S., *Spinor Structures and Nonlinear Connections in Vector Bundles, Generalized Lagrange and Finsler Spaces,* J. Math. Phys. 36(10), 1995, 1-16.

309. VACARU, Ş. and GONCHARENKO, V., *Yang-Mills Fields and Gauge Gravity on Generalized Lagrange and Finsler Spaces*, Int. Jour. of Theor. Phys. vol. 34, no. 9, 1995, 1955–1980.

310. VACARU, S. and OSTAF, S., *Nearly Autoparallel Maps of Lagrange and Finsler Spaces*, In vol. of Kluwer Acad. Publ. FTPH no. 76, 1996, 241–254.

311. VAISMAN, I., *Symplectic Geometry and Secondary Characteristic Classes,* Birkhäuser, Basel, 1987.

312. VAGNER, V.V., *Algebraic Theory of Tangent Spaces of Higher Order,* Trudy Seminara Vektornomu i Tensornomu Analizu, 10, 1956, 31–88.

313. VASSILIOU, E., *From Principal Connections to Connections on Principal Sheaves* (in print).

314. VÂLCOVICI, V., *Opere,* vol. 2, Ed. Academiei, Bucureşti, 1978.

315. VINOGRADOV, A.M., *The C-Spectral Sequence, Lagrangian Formalism and Conservation Laws,* I. *The Linear Theory,* II. *The Non-Linear Theory,* Math. Anal. Appl. 100, 1984, 1-40, 41–129.

316. VOINEA, R., *Mecanică teoretică,* (sub redacţia V.Vâlcovici, Şt.Balan, R.Voinea), Ed. Tehnică, Bucureşti, 1959.

317. VOINEA, R. & ATANASIU, M., *Metode analitice noi în teoria mecanismelor,* Ed. Tehnică, Bucureşti, 1964.

318. VRĂNCEANU, GH., *Leçons de géométrie différentielle,* 3 vol., Ed. Acad. Române, Bucureşti, 1957.

319. VRĂNCEANU, GH., *Opera Matematică,* vol. I, II, III, Ed. Academiei, Bucureşti, 1969, 1971, 1973.

320. VRĂNCEANU, GH., MIHĂILEANU, N., *Introducere în teoria relativităţii,* Ed. Tehnică, Bucureşti, 1978.

321. WATANABE, S., *On Special Kawaguchi Spaces,* III, IV, V, VI, Tensor N.S. 11, 1961, 144–153, 254-262, 279–284; 12, 1962, 244–254.

322. YAMAUCHI, S., *On hypersurfaces in Sasakian Manifolds,* Kōdai Math. Sem. Rep. 21, 1969, 64–72.

323. YANO, K. and DAVIES, E.T., *Metrics and Connections in the Tangent Bundle*, Kodai Math. Sem. Rep. 23 (1971), 493–504.

324. YANO, K, and ISHIHARA, S., *Horizontal lifts of tensor Fields to Tangent Bundles*, J. Math. and Mech. 16 (1967), 1015–1029.

325. YANO, K. and ISHIHARA, S., *Tangent and Cotangent Bundles. Differential Geometry,* M. Dekker, Inc., New York, 1973.

326. YANO, K. and KOBAYASHI, S., *Prolongations of Tensor Fields and Connections to Tangent Bundles,* I, J. Math. Soc. Japan 18, 1966, 191–210.

327. YANO, K. and KON, M., *CR–Submanifolds of Kaehlerian and Sasakian Manifolds*, Birkhäuser, 1983.

328. YAWATA, M. *Infinitesimal Transformations on Total Space of a Vector Bundle. Application*, Ph.D. Thesis, "Al.I.Cuza" University of Iaşi, 1993.

329. YAWATA, M., *Infinitesimal Transformations on the Total Space of Vector Bundles,* Ph.D. Thesis, "Al.I.Cuza" University of Iaşi, 1993.

330. YOSHIDA, M., *On the Connections in a Subspace of the Special Kawaguchi Space*, Tensor N.S. 17, 1966, 49–52.

331. ZET, GH., *Applications of Lagrange Spaces to Physics,* In the volume: Physics, Lagrange and Finsler Geometry, ed. P.L. Antonelli and R. Miron, Kluwer Acad. Publ., 1966, 255–262.

332. ZET, GH., *Lagrangian Geometrical Models in Physics*, Math. Comput. Modelling, vol. 20, no. 4/5, 1994, 83–91.

Index

Fundamental Theories of Physics

Series Editor: Alwyn van der Merwe, *University of Denver, USA*

1. M. Sachs: *General Relativity and Matter.* A Spinor Field Theory from Fermis to Light-Years. With a Foreword by C. Kilmister. 1982 ISBN 90-277-1381-2
2. G.H. Duffey: *A Development of Quantum Mechanics.* Based on Symmetry Considerations. 1985 ISBN 90-277-1587-4
3. S. Diner, D. Fargue, G. Lochak and F. Selleri (eds.): *The Wave-Particle Dualism.* A Tribute to Louis de Broglie on his 90th Birthday. 1984 ISBN 90-277-1664-1
4. E. Prugovečki: *Stochastic Quantum Mechanics and Quantum Spacetime.* A Consistent Unification of Relativity and Quantum Theory based on Stochastic Spaces. 1984; 2nd printing 1986 ISBN 90-277-1617-X
5. D. Hestenes and G. Sobczyk: *Clifford Algebra to Geometric Calculus.* A Unified Language for Mathematics and Physics. 1984 ISBN 90-277-1673-0; Pb (1987) 90-277-2561-6
6. P. Exner: *Open Quantum Systems and Feynman Integrals.* 1985 ISBN 90-277-1678-1
7. L. Mayants: *The Enigma of Probability and Physics.* 1984 ISBN 90-277-1674-9
8. E. Tocaci: *Relativistic Mechanics, Time and Inertia.* Translated from Romanian. Edited and with a Foreword by C.W. Kilmister. 1985 ISBN 90-277-1769-9
9. B. Bertotti, F. de Felice and A. Pascolini (eds.): *General Relativity and Gravitation.* Proceedings of the 10th International Conference (Padova, Italy, 1983). 1984 ISBN 90-277-1819-9
10. G. Tarozzi and A. van der Merwe (eds.): *Open Questions in Quantum Physics.* 1985 ISBN 90-277-1853-9
11. J.V. Narlikar and T. Padmanabhan: *Gravity, Gauge Theories and Quantum Cosmology.* 1986 ISBN 90-277-1948-9
12. G.S. Asanov: *Finsler Geometry, Relativity and Gauge Theories.* 1985 ISBN 90-277-1960-8
13. K. Namsrai: *Nonlocal Quantum Field Theory and Stochastic Quantum Mechanics.* 1986 ISBN 90-277-2001-0
14. C. Ray Smith and W.T. Grandy, Jr. (eds.): *Maximum-Entropy and Bayesian Methods in Inverse Problems.* Proceedings of the 1st and 2nd International Workshop (Laramie, Wyoming, USA). 1985 ISBN 90-277-2074-6
15. D. Hestenes: *New Foundations for Classical Mechanics.* 1986 ISBN 90-277-2090-8; Pb (1987) 90-277-2526-8
16. S.J. Prokhovnik: *Light in Einstein's Universe.* The Role of Energy in Cosmology and Relativity. 1985 ISBN 90-277-2093-2
17. Y.S. Kim and M.E. Noz: *Theory and Applications of the Poincaré Group.* 1986 ISBN 90-277-2141-6
18. M. Sachs: *Quantum Mechanics from General Relativity.* An Approximation for a Theory of Inertia. 1986 ISBN 90-277-2247-1
19. W.T. Grandy, Jr.: *Foundations of Statistical Mechanics.* Vol. I: *Equilibrium Theory.* 1987 ISBN 90-277-2489-X
20. H.-H von Borzeszkowski and H.-J. Treder: *The Meaning of Quantum Gravity.* 1988 ISBN 90-277-2518-7
21. C. Ray Smith and G.J. Erickson (eds.): *Maximum-Entropy and Bayesian Spectral Analysis and Estimation Problems.* Proceedings of the 3rd International Workshop (Laramie, Wyoming, USA, 1983). 1987 ISBN 90-277-2579-9

Fundamental Theories of Physics

22. A.O. Barut and A. van der Merwe (eds.): *Selected Scientific Papers of Alfred Landé. [1888-1975]*. 1988 ISBN 90-277-2594-2
23. W.T. Grandy, Jr.: *Foundations of Statistical Mechanics.* Vol. II: *Nonequilibrium Phenomena.* 1988 ISBN 90-277-2649-3
24. E.I. Bitsakis and C.A. Nicolaides (eds.): *The Concept of Probability.* Proceedings of the Delphi Conference (Delphi, Greece, 1987). 1989 ISBN 90-277-2679-5
25. A. van der Merwe, F. Selleri and G. Tarozzi (eds.): *Microphysical Reality and Quantum Formalism, Vol. 1.* Proceedings of the International Conference (Urbino, Italy, 1985). 1988 ISBN 90-277-2683-3
26. A. van der Merwe, F. Selleri and G. Tarozzi (eds.): *Microphysical Reality and Quantum Formalism, Vol. 2.* Proceedings of the International Conference (Urbino, Italy, 1985). 1988 ISBN 90-277-2684-1
27. I.D. Novikov and V.P. Frolov: *Physics of Black Holes.* 1989 ISBN 90-277-2685-X
28. G. Tarozzi and A. van der Merwe (eds.): *The Nature of Quantum Paradoxes.* Italian Studies in the Foundations and Philosophy of Modern Physics. 1988 ISBN 90-277-2703-1
29. B.R. Iyer, N. Mukunda and C.V. Vishveshwara (eds.): *Gravitation, Gauge Theories and the Early Universe.* 1989 ISBN 90-277-2710-4
30. H. Mark and L. Wood (eds.): *Energy in Physics, War and Peace.* A Festschrift celebrating Edward Teller's 80th Birthday. 1988 ISBN 90-277-2775-9
31. G.J. Erickson and C.R. Smith (eds.): *Maximum-Entropy and Bayesian Methods in Science and Engineering.* Vol. I: *Foundations.* 1988 ISBN 90-277-2793-7
32. G.J. Erickson and C.R. Smith (eds.): *Maximum-Entropy and Bayesian Methods in Science and Engineering.* Vol. II: *Applications.* 1988 ISBN 90-277-2794-5
33. M.E. Noz and Y.S. Kim (eds.): *Special Relativity and Quantum Theory.* A Collection of Papers on the Poincaré Group. 1988 ISBN 90-277-2799-6
34. I.Yu. Kobzarev and Yu.I. Manin: *Elementary Particles. Mathematics, Physics and Philosophy.* 1989 ISBN 0-7923-0098-X
35. F. Selleri: *Quantum Paradoxes and Physical Reality.* 1990 ISBN 0-7923-0253-2
36. J. Skilling (ed.): *Maximum-Entropy and Bayesian Methods.* Proceedings of the 8th International Workshop (Cambridge, UK, 1988). 1989 ISBN 0-7923-0224-9
37. M. Kafatos (ed.): *Bell's Theorem, Quantum Theory and Conceptions of the Universe.* 1989 ISBN 0-7923-0496-9
38. Yu.A. Izyumov and V.N. Syromyatnikov: *Phase Transitions and Crystal Symmetry.* 1990 ISBN 0-7923-0542-6
39. P.F. Fougère (ed.): *Maximum-Entropy and Bayesian Methods.* Proceedings of the 9th International Workshop (Dartmouth, Massachusetts, USA, 1989). 1990 ISBN 0-7923-0928-6
40. L. de Broglie: *Heisenberg's Uncertainties and the Probabilistic Interpretation of Wave Mechanics.* With Critical Notes of the Author. 1990 ISBN 0-7923-0929-4
41. W.T. Grandy, Jr.: *Relativistic Quantum Mechanics of Leptons and Fields.* 1991 ISBN 0-7923-1049-7
42. Yu.L. Klimontovich: *Turbulent Motion and the Structure of Chaos.* A New Approach to the Statistical Theory of Open Systems. 1991 ISBN 0-7923-1114-0

Fundamental Theories of Physics

43. W.T. Grandy, Jr. and L.H. Schick (eds.): *Maximum-Entropy and Bayesian Methods.* Proceedings of the 10th International Workshop (Laramie, Wyoming, USA, 1990). 1991 ISBN 0-7923-1140-X
44. P.Pták and S. Pulmannová: *Orthomodular Structures as Quantum Logics.* Intrinsic Properties, State Space and Probabilistic Topics. 1991 ISBN 0-7923-1207-4
45. D. Hestenes and A. Weingartshofer (eds.): *The Electron.* New Theory and Experiment. 1991 ISBN 0-7923-1356-9
46. P.P.J.M. Schram: *Kinetic Theory of Gases and Plasmas.* 1991 ISBN 0-7923-1392-5
47. A. Micali, R. Boudet and J. Helmstetter (eds.): *Clifford Algebras and their Applications in Mathematical Physics.* 1992 ISBN 0-7923-1623-1
48. E. Prugovečki: *Quantum Geometry.* A Framework for Quantum General Relativity. 1992 ISBN 0-7923-1640-1
49. M.H. Mac Gregor: *The Enigmatic Electron.* 1992 ISBN 0-7923-1982-6
50. C.R. Smith, G.J. Erickson and P.O. Neudorfer (eds.): *Maximum Entropy and Bayesian Methods.* Proceedings of the 11th International Workshop (Seattle, 1991). 1993 ISBN 0-7923-2031-X
51. D.J. Hoekzema: *The Quantum Labyrinth.* 1993 ISBN 0-7923-2066-2
52. Z. Oziewicz, B. Jancewicz and A. Borowiec (eds.): *Spinors, Twistors, Clifford Algebras and Quantum Deformations.* Proceedings of the Second Max Born Symposium (Wrocław, Poland, 1992). 1993 ISBN 0-7923-2251-7
53. A. Mohammad-Djafari and G. Demoment (eds.): *Maximum Entropy and Bayesian Methods.* Proceedings of the 12th International Workshop (Paris, France, 1992). 1993 ISBN 0-7923-2280-0
54. M. Riesz: *Clifford Numbers and Spinors* with Riesz' Private Lectures to E. Folke Bolinder and a Historical Review by Pertti Lounesto. E.F. Bolinder and P. Lounesto (eds.). 1993 ISBN 0-7923-2299-1
55. F. Brackx, R. Delanghe and H. Serras (eds.): *Clifford Algebras and their Applications in Mathematical Physics.* Proceedings of the Third Conference (Deinze, 1993) 1993 ISBN 0-7923-2347-5
56. J.R. Fanchi: *Parametrized Relativistic Quantum Theory.* 1993 ISBN 0-7923-2376-9
57. A. Peres: *Quantum Theory: Concepts and Methods.* 1993 ISBN 0-7923-2549-4
58. P.L. Antonelli, R.S. Ingarden and M. Matsumoto: *The Theory of Sprays and Finsler Spaces with Applications in Physics and Biology.* 1993 ISBN 0-7923-2577-X
59. R. Miron and M. Anastasiei: *The Geometry of Lagrange Spaces: Theory and Applications.* 1994 ISBN 0-7923-2591-5
60. G. Adomian: *Solving Frontier Problems of Physics: The Decomposition Method.* 1994 ISBN 0-7923-2644-X
61 B.S. Kerner and V.V. Osipov: *Autosolitons.* A New Approach to Problems of Self-Organization and Turbulence. 1994 ISBN 0-7923-2816-7
62. G.R. Heidbreder (ed.): *Maximum Entropy and Bayesian Methods.* Proceedings of the 13th International Workshop (Santa Barbara, USA, 1993) 1996 ISBN 0-7923-2851-5
63. J. Peřina, Z. Hradil and B. Jurčo: *Quantum Optics and Fundamentals of Physics.* 1994 ISBN 0-7923-3000-5

Fundamental Theories of Physics

64. M. Evans and J.-P. Vigier: *The Enigmatic Photon.* Volume 1: The Field $\boldsymbol{B}^{(3)}$. 1994 ISBN 0-7923-3049-8
65. C.K. Raju: *Time: Towards a Constistent Theory.* 1994 ISBN 0-7923-3103-6
66. A.K.T. Assis: *Weber's Electrodynamics.* 1994 ISBN 0-7923-3137-0
67. Yu. L. Klimontovich: *Statistical Theory of Open Systems.* Volume 1: A Unified Approach to Kinetic Description of Processes in Active Systems. 1995 ISBN 0-7923-3199-0; Pb: ISBN 0-7923-3242-3
68. M. Evans and J.-P. Vigier: *The Enigmatic Photon.* Volume 2: Non-Abelian Electrodynamics. 1995 ISBN 0-7923-3288-1
69. G. Esposito: *Complex General Relativity.* 1995 ISBN 0-7923-3340-3
70. J. Skilling and S. Sibisi (eds.): *Maximum Entropy and Bayesian Methods.* Proceedings of the Fourteenth International Workshop on Maximum Entropy and Bayesian Methods. 1996 ISBN 0-7923-3452-3
71. C. Garola and A. Rossi (eds.): *The Foundations of Quantum Mechanics – Historical Analysis and Open Questions.* 1995 ISBN 0-7923-3480-9
72. A. Peres: *Quantum Theory: Concepts and Methods.* 1995 (see for hardback edition, Vol. 57) ISBN Pb 0-7923-3632-1
73. M. Ferrero and A. van der Merwe (eds.): *Fundamental Problems in Quantum Physics.* 1995 ISBN 0-7923-3670-4
74. F.E. Schroeck, Jr.: *Quantum Mechanics on Phase Space.* 1996 ISBN 0-7923-3794-8
75. L. de la Peña and A.M. Cetto: *The Quantum Dice.* An Introduction to Stochastic Electrodynamics. 1996 ISBN 0-7923-3818-9
76. P.L. Antonelli and R. Miron (eds.): *Lagrange and Finsler Geometry.* Applications to Physics and Biology. 1996 ISBN 0-7923-3873-1
77. M.W. Evans, J.-P. Vigier, S. Roy and S. Jeffers: *The Enigmatic Photon.* Volume 3: Theory and Practice of the $\boldsymbol{B}^{(3)}$ Field. 1996 ISBN 0-7923-4044-2
78. W.G.V. Rosser: *Interpretation of Classical Electromagnetism.* 1996 ISBN 0-7923-4187-2
79. K.M. Hanson and R.N. Silver (eds.): *Maximum Entropy and Bayesian Methods.* 1996 ISBN 0-7923-4311-5
80. S. Jeffers, S. Roy, J.-P. Vigier and G. Hunter (eds.): *The Present Status of the Quantum Theory of Light.* Proceedings of a Symposium in Honour of Jean-Pierre Vigier. 1997 ISBN 0-7923-4337-9
81. *Still to be published*
82. R. Miron: *The Geometry of Higher-Order Lagrange Spaces.* Applications to Mechanics and Physics. 1997 ISBN 0-7923-4393-X
83. T. Hakioğlu and A.S. Shumovsky (eds.): *Quantum Optics and the Spectroscopy of Solids.* Concepts and Advances. 1997 ISBN 0-7923-4414-6

KLUWER ACADEMIC PUBLISHERS – DORDRECHT / BOSTON / LONDON